ANTENNA ARRAYS

ANTENNA ARRAYS

A Computational Approach

Randy L. Haupt
Pennsylvania State University
State College, Pennsylvania

JOHN WILEY & SONS, INC., PUBLICATION

Library of Congress Cataloging-in-Publication Data

Haupt, Randy L.
    Antenna arrays : a computational approach / Randy L. Haupt.
        p. cm.
    Includes bibliographical references and index.
    ISBN 978-0-470-40775-2
    1. Antenna arrays–Mathematical models. I. Title.
    TK7871.67.A77H38 2010
    621.382′4–dc22

                                                    2009041904

10 9 8 7 6 5 4 3 2 1

To my father, Howard Haupt, and the memory of my mother

CONTENTS

PREFACE

This book is intended to be a tutorial on antenna arrays. Each chapter builds upon the previous chapter and progressively addresses more difficult subject material. The many pictures and examples introduce the reader to practical applications.

The book starts with some electromagnetics/antennas/antenna systems information that is relevant to the other eight chapters. The next two chapters deal with the analysis and synthesis of arrays of point sources and their associated array factors. These chapters would be useful for acoustic sensors as well as electromagnetic sensors. Chapter 4 presents a sampling of different kinds of elements that replace the point sources of the previous two chapters. The next chapter shows that the elements do not have to lie along a line or in a plane. Antenna elements may lie conformal to a surface or be distributed in space. Chapter 6 introduces mutual coupling where the elements radiate and receive electromagnetic waves, so they interact. These interactions are extremely complex and difficult to predict. Computer modeling and experiments are needed to predict the performance of arrays where mutual coupling is important. Chapter 7 introduces many different approaches to getting signals to and from the array elements to a computer where the signal detection takes place. Finally, the various numerical techniques behind smart antennas are introduced in Chapter 8.

This book emphasizes the computational methods used in the design and analysis of array antennas. I generated most of the plots presented in this book using one of three commercial software packages. MATLAB (MATLAB Version 2009a, The MathWorks, www.mathworks.com, 2009) is the primary program used to do the plotting and calculations. MATLAB was also used to do many of the computations and is even useful for controlling experimental hardware. I used FEKO (FEKO, Suite 5.4, EM Software and Systems, www. feko.info, 2008) and CST Microwave Studio (CST Microwave Studio, Version 2009.07, Sonnet Software, Inc., www.sonnetsoftware.com, June 16, 2009) to generate the figures listed in the table below. The support staff at FEKO (C. J. Reddy, Ray Sun, and Rohit Sammeta) and Sonnet (Jim Willhite) were very helpful and responsive to my questions while writing this book.

Chapter	CST Microwave Studio Figure	FEKO Figure
4	65,66,67,68,69,70,72,73,74, 75,81,82,83,84	13,14,15,34,35,36,55,56,57,58
5		28,29,30,31,40,41,42,43
6	47,48,49,50,52,53,54,55,56, 57,58,59,60,61,62,63,64	9,10,11,14,15,16,17,18,19,20,21,22,23, 24,25,34,37,38,39,40,42,43,44,45,46
7	34,35,36	31,32
8	68,69,70	58,59

I am grateful for the help I received from many different people. I am especially indebted for the extraordinary help from Northrop Grumman (Dennis Lowes and Dennis Fortner), the National Electronics Museum (Ralph Strong and Michael Simons), and Ball Aerospace (Peter Moosbrugger and Debbie Quintana). The numerous pictures they provided are credited throughout this book. I especially encourage you to visit the National Electronics Museum. It is free, and they have a tremendous collection of antennas as well as other interesting science/engineering items (children-friendly too). Boris Tomasic of the USAF AFRL enlightened me on some very interesting array work and provided me with a lot of pictures and information. Robert Horner and Hale Simonds of USN SPAWAR provided pictures and data on their interesting Vivaldi direction finding array. Junwei Dong and Amir Zaghloul of Virginia Tech gave me many pictures of their horn array and 360° Rotman lens. Régis Guinvarc'h (Supelec-Sondra) provided numerous pictures on several of his innovative research projects. Sue Ellen Haupt, Dan Aten, and Kris Greenert made some editorial comments on the manuscript. Finally, the following people gave me pictures to use in this book: Nick Deplitch (ARA Inc./Seavey Engineering), Roopa Bhide (Raytheon, Corp.), Patricia Smiley (NRAO), Lisa Bell (SKA Program Development Office), Micah Gregory and Doug Werner (Penn State), William Garnier (ALMA), Erik Lier (Lockheed Martin, Corp.), John Maciel (Radant Technologies, Inc.), Dan Schaubert (University of Massachusetts), Jamie M. Knapil and Steve Fast (Remcom, Inc.), and Andrea Massa and Paolo Rocca (University of Trento).

Some of the software that was used in this book is available at ftp://ftp.wiley.com/public/sci_tech_med/antenna_arrays.

Pennsylvania State University   RANDY L. HAUPT

1 Antenna Array Basics

Big antennas can detect faint signals much better than small antennas. A big antenna collects a lot of electromagnetic waves just like a big bucket collects a lot of rain. The largest single aperture antenna in the world is the Arecibo Radio Telescope in Puerto Rico (Figure 1.1). It is 305 m wide and was build inside a giant sinkhole. Mechanically moving this reflector is out of the question.

Another approach to collecting a lot of rain is to use many buckets rather than one large one. The advantage is that the buckets can be easily carried one at a time. Collecting electromagnetic waves works in a similar manner. Many antennas can also be used to collect electromagnetic waves. If the output from these antennas is combined to enhance the total received signal, then the antenna is known as an array. An array can be made extremely large as shown by the Square Kilometer Array radio telescope concept shown in Figure 1.2. This array has an aperture that far exceeds any antenna ever built (hundreds of times larger than Arecibo). It will be capable of detecting extremely faint signals from far away objects.

An antenna array is much more complicated than a system of buckets to collect rain. Collecting N buckets of rain water and emptying them into a large bucket results in a volume of water equal to the sum of the volumes of the N buckets (assuming that none is spilled). Since electromagnetic waves have a phase in addition to an amplitude, they must be combined coherently (all the same phase) or the sum of the signals will be much less than the maximum possible. As a result, not only are the individual antenna elements of an array important, but the combination of the signals through a feed network is also equally important.

An array has many advantages over a single element. Weighting the signals before combining them enables enhanced performance features such as inter-ference rejection and beam steering without physically moving the aperture. It is even possible to create an antenna array that can adapt its performance to suit its environment. The price paid for these attractive features is increased complexity and cost.

This chapter introduces arrays through a short historical development. Next, a quick overview of electromagnetic theory is given. Some basic antenna

Antenna Arrays: A Computational Approach, by Randy L. Haupt
Copyright © 2010 John Wiley & Sons, Inc.

Figure 1.1. Arecibo Radio Telescope (courtesy of the NAIC—Arecibo Observatory, a facility of the NSF).

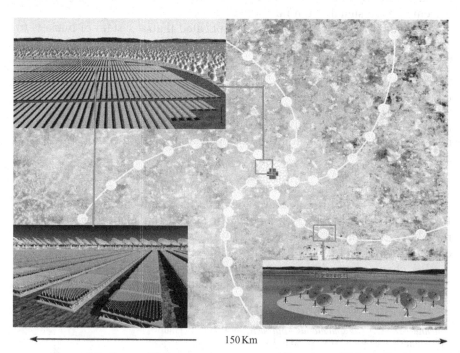

150 Km

Figure 1.2. Square kilometer array concept. (Courtesy of Xilostudios.)

definitions are then presented ends before a discussion of some system considerations for arrays. Many terms and ideas that will be used throughout the book are presented here.

1.1. HISTORY OF ANTENNA ARRAYS

The first antenna array operated in the kilohertz range. Today, arrays can operate at virtually any frequency. Figure 1.3 is a chart of the electromagnetic frequency spectrum most commonly used for antenna arrays. Antenna arrays are extremely popular for use in radars in the microwave region, so that spectrum is shown in more detail.

The development of antenna arrays started over 100 years ago [1]. Brown separated two vertical antennas by half a wavelength and fed them out of phase [2]. He found that there was increased directivity in the plane of the antennas. Forest also noted an increase in gain by two vertical antennas that

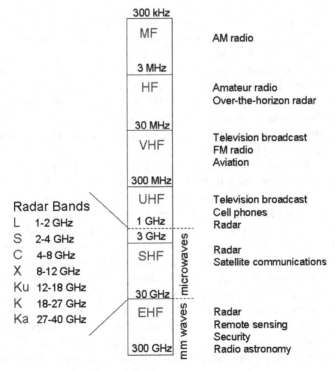

Figure 1.3. Frequency spectrum.

Figure 1.4. Chain Home, AMES Type 1 antenna array. (Courtesy of the National Electronics Museum.)

formed an array [3]. Marconi performed several experiments involving multiple antennas to enhance the gain in certain directions [4]. These initial array experiments proved vital to the development of radar.

World War II motivated countries into building arrays to detect enemy aircraft and ships. The first bistatic radar for air defense was a network of radar stations named "Chain Home (CH)" that received the formal designation "Air Ministry Experimental Station (AMES) Type 1" in 1940 (Figure 1.4) [5]. The original wavelength of 26 m (11.5 MHz) interfered with commercial broadcast, so the wavelength was reduced to 13 m (23.1 MHz). At first, the developers thought that the signal should have a wavelength comparable to the size of the bombers they were trying to detect in order to obtain a resonance effect. Shorter wavelengths would also reduce interference and provide greater accuracy. Unfortunately, the short wavelengths they desired were too difficult to generate with adequate power to be useful. By April 1937, Chain Home was able to detect aircraft at a distance of 160 km. By August 1937, three CH stations were in operation. The transmitter towers were about 107 m tall and spaced about 55 m apart. Cables hung between the towers formed a "curtain" of horizontally half-wavelength transmitting dipoles. The curtain had a main array of eight horizontal dipole transmitting antennas above a secondary "gapfiller" array of four dipoles. The gapfiller array covered the low angles that the main array could not. Wooden towers for the receiving arrays were about 76 m tall and initially had three receiving dipole antennas,

Figure 1.5. SCR-270 antenna array. (Courtesy of the National Electronics Museum.)

vertically spaced on the tower. As the war progressed, better radars were needed. A new radar called the SCR-270 (Figure 1.5) was available in Hawaii and detected the Japanese formation attacking Pearl Harbor. Unlike Chain Home, it could be mechanically rotated in azimuth 360 degrees in order to steer the beam and operated at a much higher frequency. It had 4 rows of 8 horizontally oriented dipoles and operates at 110 MHz [6].

After World War II, the idea of moving the main beam of the array by changing the phase of the signals to the elements in the array (originally tried by F. Braun [7]) was pursued. Friis presented the theory behind the antenna pattern for a two element array of loop antennas and experimental results that validated his theory [8]. Two elements were also used for finding the direction of incidence of an electromagnetic wave [9]. Mutual coupling between elements in an array was recognized to be very important in array design at a very early date [10]. A phased array in which the main beam was steered using adjustable phase shifters was reported in 1937 [11]. The first volume scanning array (azimuth and elevation) was presented by Spradley [12]. The ability to scan without moving is invaluable to military applications that require extremely high speed scans as in an aircraft. As such, the parabolic dish antennas that were once common in the nose of aircraft have been replaced by phased array antennas (Figure 1.6).

Figure 1.6. The old reflector dishes in the nose of aircraft have been replaced by phased array antennas. (Courtesy of the National Electronics Museum.)

Analysis and synthesis methods for phased array antennas were developed by Schelkunoff [13] and Dolph [14]. Their static weighting schemes resulted in the development of low sidelobe arrays that are resistant to interference entering the sidelobes. These later formed that basis of the theory of digital filters. In the 1950s, Howells and Applebaum invented the idea of dynamically changing these weights to reject interence [15]. Their work laid the foundation for adaptive, smart, and reconfigurable antenna arrays that are still being researched today.

Improvements in electronics allowed the increase in the number of elements as well as an increase in the frequency of operation of arrays. The development of transmit–receive (T/R) modules have reduced the cost and size of phased array antennas [16]. Computer technology improved the modeling and design of array antennas as well as the operation of the phased arrays. Starting in the 1960s, new solid-state phase shifters resulted in the first practical large-scale passive electronically scanned array (PESA). A PESA scans a volume of space much more quickly than a mechanically rotating antenna. Typically, a klystron tube or some other high-power source provided the transmit power that was divided amongst the radiating elements. These antennas were ground- and ship-based until the electronics became small and light enough to place on aircraft. The Electronically Agile Radar (EAR) is an example of a large PESA that had 1818 phase shifting modules (Figure 1.7). Active electronically scanned arrays (AESA) became possible with the development of gallium arsenide components in the 1980s. These arrays have many transmit/receive (T/R) modules that control the signals at each element in the array.

Today, very complex phased arrays can be manufactured over a wide range of frequencies and performing very complex functions [17]. As an example, the SBX-1 is the largest X-band antenna array in the world (Figure 1.8) [18]. It is part of the US Ballistic Missile Defense System (BMDS) that tracks and identifies long-range missiles approaching the United States. The radar is mounted on a modified, self-propelled, semi-submersible oil platform that

Figure 1.7. EAR array. (Courtesy of the National Electronics Museum.)

travels at knots and is designed to be stable in high winds and rough seas. Through mechanical and electronic scanning, the radar can cover 360° in azimuth and almost 90° in elevation. There are 45,000 GaAs transmit/receive modules that make up the 284-m^2 active aperture. Figure 1.9 shows the array being placed on the modified oil platform. A radome is placed over the array to protect it from the elements (Figure 1.10).

1.2. ELECTROMAGNETICS FOR ARRAY ANALYSIS

Before delving into the theory of antenna arrays, a review of some basic electromagnetic theory is in order. The frequency of an electromagnetic wave depends on the acceleration of charges in the source. Accelerating charges produce time-varying electromagnetic waves and vice versa. The radiated waves are a function of time and space. Assume that the electromagnetic fields are linear and time harmonic (vary sinusoidally with time). The total electromagnetic field at a point is the superposition of all the time harmonic fields at that point. If the field is periodic in time, the temporal part of the wave has a complex Fourier series expansion of the form

$$E(t) = \sum_{n=-\infty}^{\infty} a_n e^{j2\pi n f_0 t} \tag{1.1}$$

where $a_n = f_0 \int_0^{1/f_0} E(t) e^{-j2\pi n t f_0}$ = Fourier coefficients and f_0 is the fundamental frequency. The fundamental frequency determines where the wave is centered

Figure 1.8. SBX-1 X-band antenna array. (Courtesy of Missile Defense Agency History Office.)

Figure 1.9. SBX-1 array being loaded on board the platform. (Courtesy of Missile Defense Agency History Office.)

Figure 1.10. SBX-1 deployed inside a radome. (Courtesy of Missile Defense Agency History Office.)

on the frequency spectrum in Figure 1.3. If the electromagnetic field is periodic or aperiodic, it has the following temporal Fourier transform pair:

$$E(t) = \int_{-\infty}^{\infty} E(f)e^{-j2\pi ft}df \qquad (1.2)$$

$$E(f) = \int_{-\infty}^{\infty} E(t)e^{j2\pi ft}dt \qquad (1.3)$$

Equations (1.1), (1.2) and (1.3) illustrate how any time-varying electromagnetic field may be represented by a spectrum of its frequency components. $E(t)$ is the superposition of properly weighted fields at the appropriate frequencies. Superimposing and weighting the fields of the individual frequencies comprising the waveform. Traditional electromagnetics analysis examines a single-frequency component, and then it assumes that more complex waves are generated by a weighted superposition of many frequencies.

Equations (1.1), (1.2) and (1.3) do not take the vector nature of the fields into account. A single-frequency electromagnetic field (Fourier component) is represented in rectangular coordinates as

$$\vec{E}(t) = \hat{x}E_x\cos(2\pi ft) + \hat{y}E_y\cos(2\pi ft + \psi_y) + \hat{z}E_z\cos(2\pi ft + \psi_z) \qquad (1.4)$$

where $\hat{x}$, $\hat{y}$, and $\hat{z}$ are the unit vectors in the x, y, and z directions; E_x, E_y, and E_z are the magnitudes of the electric fields in the x, y, and z directions; and ψ_y and ψ_z are the phases of the y and z components relative to the x component. Using Euler's identity, (1.4) may also be written as

$$\vec{E}(t) = \mathrm{Re}\{\mathbf{E}e^{j2\pi ft}\}\tag{1.5}$$

where $\mathbf{E}$ represents the complex steady-state phasor (time independent) of the electric field and is written as

$$\mathbf{E} = \hat{x}E_x + \hat{y}E_y e^{j\psi_y} + \hat{z}E_z e^{j\psi_z}\tag{1.6}$$

and E_x, E_y, and E_z are functions of x, y, and z and are not a function of t.

Maxwell's equations in differential and integral form are shown in Table 1.1. Note that the $e^{j\omega t}$ time factor is omitted, because it is common to all components. Variables in these equations are defined as follows:

E	electric field strength (volts/m)
D	electric flux density (coulombs/m^2)
H	magnetic field strength (amperes/m)
B	magnetic flux density (webers/m^2)
J	electric current density (amperes/m^2)
ρ_{ev}	electric charge density (coulombs/m^3)
J$_m$	magnetic current density (volts/m^2)
ρ_{mv}	magnetic charge density (webers/m^3)
Q_e	total electric charge contained in S (coulombs)
Q_m	total magnetic charge contained in S (coulombs)
S	closed surface (m^2)
C	closed contour line (m)

Electric sources are due to charge. Magnetic sources are fictional but are often useful in representing fields in slots and apertures.

Each of the equations in Table 1.1 is a set of three scalar equations. There are too many unknowns to solve these equations, so additional information is necessary and comes in the form of constitutive parameters that are a function of the material properties. The constitutive relations for a linear, isotropic, homogeneous medium provide the remaining necessary equations to solve for the unknown field quantities.

$$\mathbf{D} = \varepsilon\,\mathbf{E}\tag{1.7}$$

$$\mathbf{B} = \mu\,\mathbf{H}\tag{1.8}$$

$$\mathbf{J} = \sigma\,\mathbf{E}\tag{1.9}$$

TABLE 1.1. Maxwell's Equations in Differential and Integral Form

Law	Differential	Integral
Faraday	$\nabla \times \mathbf{E} = -j\omega\mathbf{B} - \mathbf{J_m}$	$\oint_C \mathbf{E} \cdot d\mathbf{l} = -j\omega\iint_S \mathbf{B} \cdot d\mathbf{s} - \iint_S \mathbf{J_m} \cdot d\mathbf{s}$
Ampere	$\nabla \times \mathbf{H} = j\omega\mathbf{D} + \mathbf{J}$	$\oint_C \mathbf{H} \cdot d\mathbf{l} = j\omega\iint_S \mathbf{D} \cdot d\mathbf{s} + \iint_S \mathbf{J} \cdot d\mathbf{s}$
Gauss electric	$\nabla \cdot \mathbf{D} = \rho_{ev}$	$\oiint_S \mathbf{D} \cdot d\mathbf{s} = Q_e$
Gauss magnetic	$\nabla \cdot \mathbf{B} = \rho_{mv}$	$\oiint_S \mathbf{B} \cdot d\mathbf{s} = Q_m$

where the constitutive parameters describe the material properties and are defined as follows:

μ	permeability (henries/m)
ε	permittivity or dielectric constant (farads/m)
σ	conductivity (siemens/m)

Assuming the constant to be scalars is an over simplification. In today's world, antenna designers must take into account materials with special properties, such as

- Composites
- Semiconductors
- Superconducting materials
- Ferroelectrics
- Ferromagnetic materials
- Ferrites
- Smart materials
- Chiral materials
- Conducting polymers
- Ceramics
- Electromagnetic bandgap (EBG) materials

Antenna design relies upon a complex repitroire of different materials that will provide the desired performance characteristics.

Spatial differential equations have only general solutions until boundary conditions are specified. If these equations still had the time dependence factor, then initial conditions would also have to be specified. The boundary conditions for the field components at the interface between two media are given by

- The tangential electric field:

$$\hat{\mathbf{n}} \times (\mathbf{E}_1 - \mathbf{E}_2) = -\mathbf{J}_m \qquad (1.10)$$

- The normal magnetic flux density:

$$\hat{\mathbf{n}} \cdot (\mathbf{B}_1 - \mathbf{B}_2) = \rho_{ms} \qquad (1.11)$$

- The tangential magnetic field:

$$\hat{\mathbf{n}} \times (\mathbf{H}_1 - \mathbf{H}_2) = \mathbf{J}_s \qquad (1.12)$$

- The normal electric flux density:

$$\hat{\mathbf{n}} \cdot (\mathbf{D}_1 - \mathbf{D}_2) = \rho_{es} \qquad (1.13)$$

where subscripts 1 and 2 refer to the two different media, ρ_{ms} is the magnetic surface charge density (coulombs/m^2), and ρ_{es} is the electric surface charge density (webers/m^2).

Maxwell's equations in conjunction with the constitutive parameters and boundary conditions allow us to find quantitative values of the field quantities.

Power is an important antenna quantity and has units of watts or volts times amps. Multiplying the electric field and the magnetic field produces units of W/m^2 or power density. The complex Poynting vector describes the power flow of the fields via

$$\mathbf{S} = \frac{1}{2} \mathrm{Re}\{\mathbf{E} \times \mathbf{H}^*\} \qquad (1.14)$$

Note that the direction of propagation (direction that $\mathbf{S}$ points) is perpendicular to the plane containing the $\mathbf{E}$ and $\mathbf{H}$ vectors. $\mathbf{S}$ is the power flux density, so $\nabla \cdot \mathbf{S}$ is the volume power density leaving a point. A conservation of energy equation can be derived in the form of

$$\iint \mathbf{E} \times \mathbf{H}^* \cdot \mathbf{ds} = -\iiint \mathbf{E} \cdot \mathbf{J}^* dv - j\omega \iiint \left(\frac{\varepsilon |\mathbf{E}|^2}{2} + \frac{\mu |\mathbf{H}|^2}{2} \right) dv \qquad (1.15)$$

The terms $\frac{1}{2}\varepsilon|\mathbf{E}|^2$ and $\frac{1}{2}\mu|\mathbf{H}|^2$ are the electric and magnetic energy densities, respectively. Finally, $\mathbf{E} \cdot \mathbf{J}^*$ represents the power density dissipated.

1.3. SOLVING FOR ELECTROMAGNETIC FIELDS

The sources that generated the current on the antenna or the voltage across the terminal of the antenna must be known in order to calculate the fields radiated by the antenna. There is an analytical approach to finding fields for some very simple antennas in which the current on the antenna is postulated. In most practical cases, however, the fields must be found using numerical methods. This section presents an approach for analytically finding fields for simple antennas that also forms the basis for some numerical approaches in the frequency domain.

1.3.1. The Wave Equation

A time-varying current on an antenna is the input to a linear system called free space. The output is the radiated electromagnetic field. The simplest conceivable antenna is called an isotropic point source, and it radiates equally in all directions. At a constant distance from the source (the surface of an imaginary sphere), the amplitude and phase of the electromagnetic field radiated by the point source is the same at a given instant in time. Point sources don't really exist. However, certain radiating objects, such as stars, behave as though they were point sources when the observer is far away. If a point source is modeled as a spatial impulse, then an impulse response must exist for free space. Once the impulse response is known, then the output is found by convolving an input with the impulse response. This approach to finding the fields radiated by an antenna is identical to finding the impulse response of a filter.

The quest for the impulse response of free space (also called the free-space Green function) begins with the vector wave equation for the electric field with only electric sources. It is derived by taking the curl of Faraday's law and substituting Ampere's law into the right-hand side.

$$\nabla \times \nabla \times \mathbf{E} = -j\omega\mu\nabla \times \mathbf{H} = -j\omega\mu(j\omega\varepsilon\mathbf{E} + \mathbf{J}) = k^2\mathbf{E} - j\omega\mu\mathbf{J} \qquad (1.16)$$

The left-hand side of this equation may be converted to a more convenient form using the vector identity $\nabla \times \nabla \times \mathbf{E} = \nabla(\nabla \cdot \mathbf{E}) - \nabla^2\mathbf{E}$ and substituting Gauss' law.

$$\nabla^2\mathbf{E} + k^2\mathbf{E} = j\omega\mu\sigma\mathbf{J} + \frac{1}{\varepsilon}\nabla\rho_{ev} \qquad (1.17)$$

This equation is very useful when there are no sources, because $\mathbf{E}$ is easy to find. Unfortunately, the sources are in terms of both $\mathbf{J}$ and ρ_{ev}. Thus, in order to calculate the fields radiated by an antenna or scattering object, both $\mathbf{J}$ and ρ_{ev} must be known.

Our goal is to have one vector quantity on the left-hand side of the equation and one source quantity on the right-hand side. In order to achieve this goal, a wave equation is found for the magnetic vector potential **A**. Then, **E** and **H** are found from **A**. The derivation of the wave equation for the vector magnetic potential starts by defining **A** from Gauss' law, $\nabla \cdot \mathbf{B} = 0$, and the vector identity $\nabla \cdot \nabla \times \mathbf{A} = 0$.

$$\mathbf{B} = \nabla \times \mathbf{A} \tag{1.18}$$

Substituting (1.18) into Faraday's law gives

$$\nabla \times (\mathbf{E} + j\omega \mathbf{A}) = 0 \tag{1.19}$$

Recognizing that (1.19) fits the form of the vector identity $\nabla \times \nabla V = 0$, **E** is defined as

$$-\nabla V = \mathbf{E} + j\omega \mathbf{A} \tag{1.20}$$

or

$$\mathbf{E} = -j\omega \mathbf{A} - \nabla V \tag{1.21}$$

where V is an arbitrary scalar potential. The next step is to substitute (1.18) and (1.21) into Ampere's law to get

$$\frac{1}{\mu}\nabla \times \nabla \times \mathbf{A} = j\omega\varepsilon(-j\omega \mathbf{A} - \nabla V) + \mathbf{J} \tag{1.22}$$

which may be rewritten as

$$\nabla^2 \mathbf{A} + k^2 \mathbf{A} = -\mu \mathbf{J} + \nabla(\nabla \cdot \mathbf{A} + j\omega\mu\varepsilon V) \tag{1.23}$$

by using the vector identity $\nabla \times \nabla \times \mathbf{A} = \nabla(\nabla \cdot \mathbf{A}) - \nabla^2 \mathbf{A}$, defining $k^2 = \omega^2\mu\varepsilon$, and rearranging the terms. Since V and **A** are arbitrary (we took them from some vector identities), we can define our own relationship between them. Looking at (1.23), the choice for relating V and **A** that would greatly simplify the equation is

$$\nabla \cdot \mathbf{A} + j\omega\varepsilon\mu V = 0 \tag{1.24}$$

This relationship between A and V is known as the Lorentz condition. Using the Lorenz condition in (1.23) yields the wave equation.

$$\nabla^2 \mathbf{A} + k^2 \mathbf{A} = -\mu \mathbf{J} \tag{1.25}$$

A similar derivation for magnetic sources yields another wave equation.

$$\nabla^2 \mathbf{F} + k^2 \mathbf{F} = -\varepsilon \mathbf{J}_m \tag{1.26}$$

where $\mathbf{F}$ is the electric vector potential for the fictional magnetic current.

1.3.2. Point Sources

If the source in (1.25) is an impulse function or a point source, then it is represented in rectangular coordinates as

$$J(x', y', z') = J(r') = \delta(x')\delta(y')\delta(z') \tag{1.27}$$

The field characteristics of a point source are most simply defined in terms of θ and ϕ. The z-component of (1.25) outside the origin becomes

$$\frac{1}{r^2}\frac{\partial}{\partial r} r^2 \frac{\partial A_z}{\partial r} + k^2 A_z = 0 \tag{1.28}$$

The θ and ϕ variations are zero, so the wave equation is only a function of r, the distance from the origin to the point of observation. The impulse response of free space, $\mathcal{G}(r)$, is found by substituting $A_z = \mathcal{G}(r)/r$ into (1.28) to get

$$\frac{d^2\mathcal{G}(r)}{dr^2} + k^2\mathcal{G}(r) = 0 \tag{1.29}$$

where $\mathbf{r} = r\hat{\mathbf{r}} = x\hat{\mathbf{x}} + y\hat{\mathbf{y}} + z\hat{\mathbf{z}}$ and $r = |\mathbf{r}|$. Solving this equation for $\mathcal{G}(r)$ results in two solutions. Since the assumed time dependence is $e^{j\omega t}$, the first solution represents waves traveling away from the point source (transmit antenna)

$$\mathcal{G}(r) = \frac{e^{-jkr}}{4\pi r} \tag{1.30}$$

and the second solution represents waves traveling toward the point source (receive antenna)

$$\mathcal{G}(r) = \frac{e^{jkr}}{4\pi r} \tag{1.31}$$

Theoretically, real antennas consist of a collection of point sources. Their far-field patterns are a convolution of the current on the antenna ($\mathbf{J}$) with $\mathcal{G}$. An antenna may be thought to consist of point sources distributed throughout space. When a point source is at (x', y', z') instead of at the origin, it is represented as

$$J(R) = \delta(x - x')\delta(y - y')\delta(z - z')$$

$$= \frac{1}{4\pi r'^2}\delta(r - r')\delta(\theta - \theta')\delta(\phi - \phi') \tag{1.32}$$

If the point source is at the origin, then

$$J(r') = \frac{1}{4\pi r'^2}\delta(r') \tag{1.33}$$

and the free-space Green function is

$$\mathcal{G}(r|r') = \frac{e^{-jkR}}{4\pi R} \tag{1.34}$$

where $\mathbf{r}' = x'\hat{\mathbf{x}} + y'\hat{\mathbf{y}} + z'\hat{\mathbf{z}}$, $r' = |\mathbf{r}'|$, and $R = |\mathbf{r} - \mathbf{r}'|$.

To summarize, the electromagnetic fields radiated by an antenna may be found by the following steps:

1. Postulate the current on the antenna ($\mathbf{J}$). This may be done experimentally, analytically, numerically, or a reasonable guess.
2. Calculate $\mathbf{A}$ by convolving the $\mathbf{J}$ and or $\mathbf{F}$ by convolving the $\mathbf{J_m}$ with $\mathcal{G}$ for each vector component:

$$\mathbf{A} = \mu\iiint_{v'} \mathbf{J}(r')\frac{e^{-jkR}}{4\pi R}\,dv' \tag{1.35}$$

$$\mathbf{F} = \varepsilon\iiint_{v'} \mathbf{J_m}(r')\frac{e^{-jkR}}{4\pi R}\,dv' \tag{1.36}$$

3. Calculate $\mathbf{H}$:

$$\mathbf{H} = \frac{1}{\mu}\nabla \times \mathbf{A} - j\omega\mathbf{F} - j\frac{1}{\omega\mu\varepsilon}\nabla(\nabla \cdot \mathbf{F}) \tag{1.37}$$

4. Calculate $\mathbf{E}$ from Ampere's law:

$$\mathbf{E} = \frac{1}{j\omega\varepsilon}\nabla \times \mathbf{H} - \frac{1}{\varepsilon}\nabla \times \mathbf{F} \tag{1.38}$$

The next two subsections demonstrate this procedure on simple antennas.

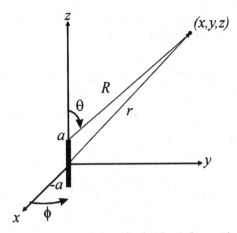

Figure 1.11. Hertzian dipole along the z axis.

1.3.3. Hertzian Dipole

A Hertzian dipole is a straight-wire antenna that is $2a$ long and is very small compared to a wavelength ($2a \ll \lambda$). We follow the steps of the previous section to find the radiated fields. If the antenna lies along the z axis, then it can be modeled as a line of point sources from $z = -a$ to $z = a$ (Figure 1.11). Since the antenna is so small, the current is approximately a constant, $J = \hat{z}I_0\delta(x')\delta(y')$ along the length of the wire. The magnetic vector potential is given by

$$A_z = \mu\int_{-a}^{a}\int_{-\infty}^{\infty}\int_{-\infty}^{\infty}I_0\delta(x')\delta(y')\frac{e^{-jk|r-z'|}}{4\pi|r-z'|}dx'dy'dz' \qquad (1.39)$$

This integral simplifies to

$$A_z = \frac{\mu(2a)I_0e^{-jkr}}{4\pi r} \qquad (1.40)$$

given the following assumptions:

$$R = |r - z'| \simeq r$$

$$2a \ll \lambda$$

$$2a \ll R$$

$$I(z') \text{ is a constant} = I_0$$

This solution has a variable r that is one of the dimensions of a spherical coordinate system yet has a vector component that is in a rectangular coordinate system. In order to put everything in one coordinate system, the z component is converted to spherical coordinates.

$$\mathbf{A} = \frac{2a\mu I_0}{4\pi r} e^{-jkr} \left(\hat{\mathbf{r}} \cos\theta - \hat{\boldsymbol{\theta}} \sin\theta \right) \tag{1.41}$$

The electric and magnetic fields are derived from (1.35) and (1.36):

$$\mathbf{E} = \frac{2a\mu I_0 Z}{4\pi k} \left[2\cos\theta \left(\frac{jk}{r^2} + \frac{1}{r^3} \right) \hat{\mathbf{r}} - \sin\theta \left(-\frac{k^2}{r} - \frac{jk}{r^2} + \frac{1}{r^3} \right) \hat{\boldsymbol{\theta}} \right] e^{-jkr} \tag{1.42}$$

$$\mathbf{H} = \frac{2a\mu I_0 \sin\theta}{4\pi} \left(\frac{jk}{r} + \frac{1}{r^2} \right) e^{-jkr} \hat{\boldsymbol{\phi}} \tag{1.43}$$

where Z is the impedance given by

$$Z = \sqrt{\frac{\mu}{\varepsilon}} \tag{1.44}$$

A short distance from the antenna, the $1/r^2$ and $1/r^3$ terms quickly become negligible compared to the $1/r$ term:

$$\mathbf{E} = j2aZI_0 k \sin\theta \frac{e^{-jkr}}{4\pi r} \hat{\boldsymbol{\theta}} \tag{1.45}$$

$$\mathbf{H} = j2aZI_0 k \sin\theta \frac{e^{-jkr}}{4\pi r} \hat{\boldsymbol{\phi}} \tag{1.46}$$

Equations (1.45) and (1.46) are far-field equations because the electric and magnetic fields are orthogonal to each other and to the direction of propagation. Another property of the far field evident from these equations is that the electric and magnetic fields are related by

$$\mathbf{E} = -Z\hat{\mathbf{r}} \times \mathbf{H} \tag{1.47}$$

$$\mathbf{H} = \frac{1}{Z}\hat{\mathbf{r}} \times \mathbf{E} \tag{1.48}$$

The power flow is shown to be in the radial direction by calculating the complex Poynting vector given by

$$\frac{1}{2}\mathbf{E} \times \mathbf{H}^* = \hat{\mathbf{r}} \frac{|I_0|^2 Z_0 a^2 k^2 \sin^2\theta}{8\pi^2 r^2} \tag{1.49}$$

Thus, the power radiated is a function of $1/r^2$, which is the same as an individual point source. Unlike the isotropic point source, the Hertzian dipole has

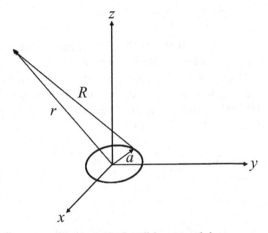

Figure 1.12. Small loop model.

preferred directions of radiation and reception as given by the $\sin\theta$ term. It is also polarized: The electric field is described by a vector.

1.3.4. Small Loop

Point sources may also be placed side-by-side to form a loop as shown in Figure 1.12. Assume the loop is so small that the current is constant on the loop and is given by

$$I(\phi) = \frac{\delta(\theta - 90°)\delta(r - a)}{r}\hat{\phi}, \qquad a \ll \lambda \qquad (1.50)$$

The magnetic vector potential is found from

$$\mathbf{A} = \int_0^{2\pi}\int_0^{\pi}\int_0^{\infty}\hat{\phi}\frac{\mu I(\phi')}{4\pi R}e^{-jkR}r'^2\sin\theta'dr'd\theta'd\phi' \qquad (1.51)$$

where R is given by

$$R = |r - r'| = \sqrt{(r - r')\cdot(r - r')} = \sqrt{r^2 + r'^2 - 2r\cdot r'}$$
$$= \sqrt{r^2 + a^2 - 2ar\{\sin\theta(\cos\phi\cos\phi' - \sin\phi\sin\phi')\}} \qquad (1.52)$$

The observation distance is assumed to be much greater than the loop diameter ($r \gg a$). Factor r out of the radical, then the a^2/r^2 inside the radical is very small and can be ignored. Since a/r is also very small, the binomial expansion for the square root gives an accurate approximation:

$$R \approx r - a\sin\theta(\cos\phi' + \sin\phi\sin\phi') \tag{1.53}$$

The second term contributes little to the amplitude of the magnetic vector potential, because its maximum value is a. For instance, if r is 100 m and a is 1 m, then the amplitude of A decreases by about 1%. Thus, $R \approx r$ in the denominator. However, the same 1% increase in R produces a 180° phase shift at a frequency of 600 MHz, and (1.53) must be used in the phase term. Making the proper substitutions into (1.51) yields

$$\mathbf{A} = \frac{\mu I_0}{4\pi} \int\limits_0^\pi \int\limits_0^\infty \int\limits_0^{2\pi} \frac{\delta(\theta' - 90°)\delta(r' - a)e^{-jk[r - a\sin\theta(\cos\phi\cos\phi' + \sin\phi\sin\phi')]}}{r} r'^2 \sin\theta' dr' d\theta' d\phi' \hat{\boldsymbol{\phi}}$$

$$\tag{1.54}$$

Integrating over θ' and r', substituting the rectangular representation of ϕ', and making the small phase angle approximation $e^{jx} \approx 1 + jx$ reduces the equation to

$$\mathbf{A} = \frac{\mu I_0 a e^{-j\beta r}}{4\pi r} \int\limits_0^{2\pi} (-\hat{x}\sin\phi' + \hat{y}\cos\phi')[1 + jka\sin\theta(\cos\phi\cos\phi' + \sin\phi\sin\phi')] d\phi' \hat{\boldsymbol{\phi}}$$

$$\tag{1.55}$$

After performing the final integration and substituting $\hat{\boldsymbol{\phi}} = -\hat{x}\sin\phi + \hat{y}\cos\phi$, then

$$\mathbf{A} = \frac{j\mu k\pi a^2 I_0 \sin\theta e^{-jkr}}{4\pi r}\hat{\boldsymbol{\phi}} \tag{1.56}$$

The magnetic field is

$$\mathbf{H} = \frac{1}{\mu}\nabla\times\mathbf{A} = -\frac{1}{\mu r}\frac{\partial}{\partial r}(rA_\phi)\hat{\boldsymbol{\theta}} = -\frac{Mk^2\sin\theta}{4\pi r}e^{-jkr}\hat{\boldsymbol{\theta}} \tag{1.57}$$

where $M = \pi a^2 I_0$ is the dipole moment of the small current loop. The electric field is given by

$$\mathbf{E} = -Z\hat{\mathbf{r}}\times\mathbf{H} = -\frac{MZk^2\sin\theta}{4\pi r}e^{-jkr}\hat{\boldsymbol{\theta}} \tag{1.58}$$

These equations have a form similar to those for the Hertzian dipole and are called dual formulations. The analysis of larger loops is more complicated because one cannot assume that a is small compared to λ, so the current is not constant in amplitude and phase around the loop.

1.3.5. Plane Waves

A plane wave is a transverse electromagnetic (TEM) wave having constant amplitude and phase in an infinite plane in space at an instant in time. A TEM wave has the electric and magnetic fields orthogonal to the direction of propagation. The plane wave travels in the direction orthogonal to the plane. Thus, a plane wave is described by a vector or an angle of propagation and magnitude and phase of the field in the plane. The propagation vector points in the direction of propagation and is written as

$$\mathbf{k} = k_x \hat{x} + k_y \hat{y} + k_z \hat{z} \tag{1.59}$$

where the propagation constants in the x, y, and z directions are given by

$$k_x = \frac{2\pi}{\lambda_x} = k \sin\theta \cos\phi$$

$$k_y = \frac{2\pi}{\lambda_y} = k \sin\theta \sin\phi$$

$$k_z = \frac{2\pi}{\lambda_z} = k \cos\theta$$

and the projections of the wavelength onto the x, y, and z directions are given by λ_x, λ_y, and λ_z.

Even though the point source and plane wave are mathematical and conceptual models, we relate them in a very practical way, because we are often only interested in a portion of the angular extent of the field. When the spherical wave of a transmit antenna impinges on the receive antenna, how spherical does it look? As the distance between the antennas increases, the incident wave looks less curved. At some distance R the incident wave can be said to be a plane wave relative to the receive antenna or over a local extent. This approximation is extremely important in antenna measurements. As a rule of thumb (and IEEE definition [19]), a receive antenna is in the far field of a point source when the maximum phase deviation across the aperture is less than $\lambda/16$ or $\pi/8$ radians. Figure 1.13 shows the simple trigonometric derivation for the far-field formula given by

$$R = \frac{2D^2}{\lambda} \tag{1.60}$$

where R is the distance from the point source to the receive antenna and D is the largest dimension of the receive antenna. For high-performance (low-sidelobe) antennas, a stricter error tolerance may be needed, and the far field will be a greater distance from the antenna.

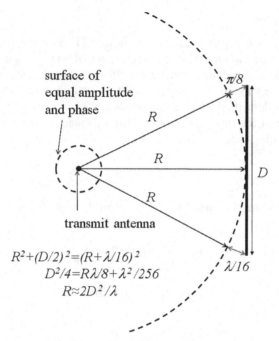

surface of
equal amplitude
and phase

$\pi/8$

R

R

R

D

transmit antenna

$R^2+(D/2)^2=(R+\lambda/16)^2$
$D^2/4=R\lambda/8+\lambda^2/256$
$R\approx 2D^2/\lambda$

$\lambda/16$

Figure 1.13. Derivation of the definition of far field.

1.4. ANTENNA MODELS

Antennas transmit and/or receive signals. From a circuit point of view, the antenna appears as a load on a transmission line. An antenna is matched when the signal from the transmission line is radiated and not reflected back to the transmitter. Determining the impedance of this load and matching it to the feed line is important. An antenna may also be considered a filter. The filter passes electromagnetic waves with desirable frequency, directional, and polarization attributes. These models are widely used in antenna design and are described in the next sections.

1.4.1. An Antenna as a Circuit Element

A radiating system consists of an oscillating source to generate a signal, a transmission line or waveguide, and an antenna to transform that signal to an electromagnetic wave. Not all the power generated by the transmitter goes to the antenna. Transmission lines and connectors between the source and antenna become potential sources for degradation due to mismatches, radiation loss, and heat loss. A guided wave traveling along a transmission line reflects from any discontinuity, or point in the transmission path where the impedance changes. These reflections set up a standing wave in the line which stores energy and reduces the amount of power delivered to the intended load (antenna). The standing wave ratio (SWR) is the ratio of the maximum to minimum value of the voltage standing wave established by the reflections. SWR is a common measure used in matching guided wave components and is calculated by

$$\text{SWR} = \frac{V_{\max}}{V_{\min}} = \frac{1+|\Gamma_L|}{1-|\Gamma_L|} \tag{1.61}$$

where Γ_L is the reflection coefficient at the discontinuity. An SWR of 1 indicates a perfect match. The reflection coefficient is the ratio of the reflected to incident voltages at the discontinuity

$$\Gamma_L = \frac{V_{\text{reflected}}}{V_{\text{incident}}} = \frac{Z_L - Z_0}{Z_L + Z_0} \tag{1.62}$$

where Z_0 is the transmission line impedance and Z_L is the discontinuity impedance. Frequently, Γ_L is also called s_{11} from the s parameters. Impedances are a function of frequency, so SWR is often used to establish the frequency range or bandwidth in which an antenna can be used. In most cases, an SWR < 2 or $s_{11} < -10$ dB define the bandwidth limits. One needs to be careful comparing the bandwidth of two antennas. Sometimes, for receive antennas, the bandwidth is defined over the frequency range when VSWR ≤ 3. It is more important to have a low VSWR for a transmit antenna, because the reflected power can be high enough to damage circuits.

Signal power escapes from the circuit through radiation or heating. Radiation losses occur when the signal leaks from the transmission path by way of connectors or the open sides of microstrip lines. Thermal losses result when resistance in the transmission line converts part of the signal to heat. Resistance comes from the imperfect conductors and dielectrics that make up the transmission line. The reduced power delivered to the antenna terminals is given by

$$P_t = \delta_h \delta_r \left(1 - |\Gamma_L|^2\right) P_{\text{TR}} \tag{1.63}$$

where δ_h is the thermal dissipation efficiency, δ_r is the radiation dissipation efficiency, Γ_L is the reflection coefficient due to reflections within the transmission line, and P_{TR} is the power generated by the transmitter. The intent is to get as much power as possible to radiate in a desired direction and receive as much power from the intended source as possible. Any loss of power or addition of unwanted power is very undesirable.

Example. If a system has $\delta_h = \delta_r = 0.99$ and $Z_0 = 75\ \Omega$ and $Z_L = 77 + j30\ \Omega$, then find P_t.

$$\Gamma_L = \frac{77 + j30 - 75}{77 + j30 + 75} = \frac{2 + j30}{152 + j30}$$

The resulting transmitted power is

$$P_t = .99^2\left(1 - .19^2\right)10W \Rightarrow \%\text{transmitted} = 94.3\%$$

1.4.2. An Antenna as a Spatial Filter

Antennas do not radiate power isotropically (equally in all directions). Instead, an antenna is a spatial filter which concentrates power in certain directions at the expense of decreasing the power radiated in other directions. The power density (W/m^2) radiated by an antenna is given by

$$S_r = \frac{1}{2\eta r^2}\left(E_\theta^2 + E_\phi^2\right) \tag{1.64}$$

Directivity compares the power density in a designated direction to the average power density. Unless otherwise specified, directivity implies that the maximum value of directivity is given by

$$D = \frac{4\pi S_{r\max}}{\int_0^{2\pi}\int_0^\pi S_r \sin\theta d\theta d\phi} \tag{1.65}$$

The gain of the antenna is the ratio of the power radiated in a particular direction to power delivered to the antenna. Gain differs from directivity because gain includes losses.

Directivity is always greater than or equal to gain. The denominator in (1.65) can be replaced by power delivered to the antenna, thus avoiding the double integration. Gain and directivity are related through the radiation efficiency, δ_e, the ratio of the power radiated by the antenna to the power input to the antenna

$$G(\theta, \phi) = \delta_e D(\theta, \phi) \tag{1.66}$$

The realized gain includes the losses due to the mismatch of the antenna input impedance to a specified impedance. Realized gain is frequently used by engineers when integrating the antenna into the system. When gain is written without any angular dependence, G, it implies the maximum gain of the antenna. Since G is a power ratio, it is often expressed in decibels

$$G_{dB} = 10\log_{10}G = 10\log G \tag{1.67}$$

Figure 1.14 shows a three-dimensional plot in cylindrical coordinates of a relative antenna radiation pattern far from the antenna as a function of θ and

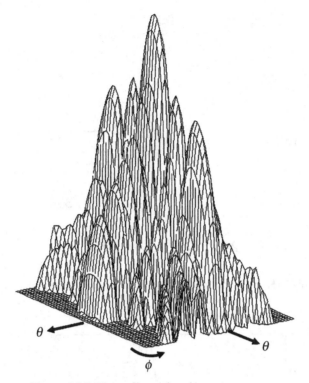

Figure 1.14. Three-dimensional antenna pattern.

ϕ, where θ is measured in the radial direction, ϕ is in the horizontal plane, and the pattern amplitude in the vertical direction. Relative means that no absolute units of power are associated with the pattern, but the power between two different angles are of the correct ratio. A relative antenna pattern means that the maximum value is normalized to 1 or 0 dB. The direction of maximum gain is at the center of a large lobe called the main beam, while smaller lobes are called sidelobes, and the zero-crossings are called nulls. Bigger lobes in some directions indicate greater gain in those directions.

Three-dimensional antenna patterns (Figure 1.14) provide an overall qualitative evaluation of the antenna's spatial response. Accurately determining sidelobe levels, null locations, and beamwidth require the use of two-dimensional cuts, however. An antenna pattern cut is the two-dimensional antenna pattern measured on a great circle around the antenna. Figure 1.15 shows two orthogonal polar magnitude plots of the three-dimensional pattern in Figure 1.14 ($\phi = 0°$ and $\phi = 90°$). These same patterns appear as rectangular plots in Figure 1.16 (dB) and in Figure 1.17 (linear). The polar plot is useful for appreciating the angular layout of the pattern. The rectangular plots are used to precisely locate nulls, determine beamwidth, and establish sidelobe

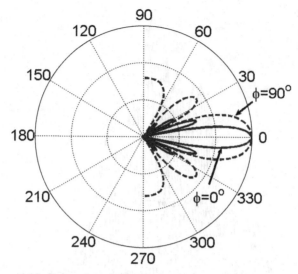

Figure 1.15. Polar plot of the relative antenna pattern in decibels.

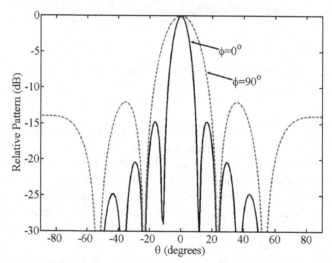

Figure 1.16. Linear antenna pattern plot in decibels.

levels. Note that low sidelobes are difficult to see in the linear plot compared to the dB plot. In this book, ϕ is the azimuth angle and θ is the elevation angle. For linear or nearly linearly polarized antennas, the terms E-plane and H-plane cuts are used. An E-plane cut is the antenna pattern in the plane containing the electric field and the maximum of the main beam, while the H-plane cut is the antenna pattern in the plane containing the magnetic field and the

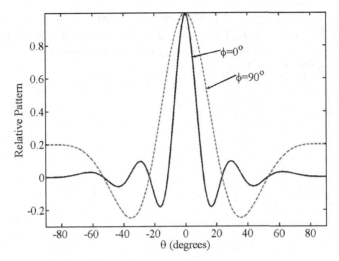

Figure 1.17. Linear rectangular antenna pattern.

maximum of the main beam. Antenna patterns are often normalized to the peak of the main beam.

The beamwidth of an antenna may mean either (a) the angular separation between the half-power (3 dB) points on either side of the peak of the main beam (most common engineering definition) or (b) the angular separation between the first nulls on either side of the main beam (definition often used in optics and physics). If the antenna pattern is not symmetrical, then the beamwidth must be specified in the plane of the antenna pattern cut. Usually the beamwidth is specified in two orthogonal antenna pattern cuts.

Another important antenna gain characteristic is effective (or equivalent) isotropically radiated power (EIRP). EIRP is the gain of the transmitting antenna multiplied by the power delivered to its input.

$$\text{EIRP} = P_t G \tag{1.68}$$

It is the transmitter–antenna combination that determines the transmitted power of a system. EIRP is especially important for satellite antennas where power and antenna size are at a premium.

1.4.3. An Antenna as a Frequency Filter

Antennas transmit and receive certain frequencies better than other frequencies, making the antenna a frequency filter. Antennas that respond to a very small range of frequencies are known as narrowband or resonant antennas, and those that respond over a wide range of frequencies are known as broadband antennas. Usually, a narrowband antenna is quite simple in shape, like a dipole. The simplicity allows the current to resonate over a well-defined region. On the other hand, broadband antennas have a more complex shape, like a helix or spiral. The complex shape gives the antenna the ability to resonate at many different adjacent frequencies.

The bandwidth is usually stated in one of three ways:

- Percent of center frequency

$$BW = \frac{f_{hi} - f_{lo}}{f_{center}} \times 100 \qquad (1.69)$$

- Ratio of high and low frequencies

$$BW = \frac{f_{hi}}{f_{lo}} \qquad (1.70)$$

- Range of frequencies

$$BW = f_{hi} - f_{lo} \qquad (1.71)$$

Broadband implies that the antenna has a 10% or higher bandwidth, or it operates over at least an octave ($f_{hi}/f_{lo} = 2$). The term ultra-wide band (UWB) refers to antennas that have very broad bandwidths [20]. The Defense Advanced Research Projects Agency (DARPA) defines UWB as BW $\geq 25\%$ and the Federal Communications Commission (FCC) defines UWB as BW $\geq 20\%$.

Defining the values of f_{hi} and f_{lo} are not easy. Some ways this is done include:

- A function of antenna gain. f_{center} is the frequency of the highest antenna gain, f_{hi} is the highest frequency at which the gain has not fallen below −3 dB, and f_{lo} is the lowest frequency at which the gain has not fallen below −3 dB.
- A function of SWR. f_{center} is the frequency at which the antenna is best matched, f_{hi} is the highest frequency at which the SWR is still less than 2, and f_{lo} is the lowest frequency at which the SWR is still less than 2. An equivalent definition is the reflection coefficient (s_{11}) is less than 1/3 or −10 dB. Sometimes receive antennas may be specified using a $VSWR > 2$.
- A function of some important antenna performance feature. f_{hi} and f_{lo} define the bandwidth over which the performance indicator lies within acceptable bounds.

The bandwidth can refer to either the instantaneous bandwidth or operational bandwidth. Instantaneous bandwith is the bandwidth of the signal at the antenna. The operational bandwidth is the bandwidth of the antenna and is greater than the instantaneous bandwidth.

Example. Is an antenna that has a bandwidth over the AM broadcast frequencies a broadband antenna? f_{hi} = 1600 kHz and f_{lo} = 540 kHz $\Rightarrow$ f_{center} = 1070 kHz BW = $f_{hi} - f_{lo}$ = 1060 kHz, BW = $f_{hi} - f_{lo}/f_{center} \times 100$ = 1060/1070 $\times$ 100 = 99.065%, and BW = f_{hi}/f_{lo} = 1600/540 = 2.963. This antenna would be broadband.

1.4.4. An Antenna as a Collector

As mentioned previously, an antenna collects electromagnetic waves in a similar manner that a bucket collects rain. A time-varying electromagnetic field incident on an antenna causes charges in the receiving antenna to oscillate. If the charges oscillate at the same rate as the incident field, some of the electromagnetic wave re-radiates as a wave at the same frequency as the incident wave. The remainder of the wave converts into heat or is delivered to a load such as a radio receiver. The amount of current induced by an incident wave may be represented by a current density distributed over an area called the collecting aperture (A_c). The areas over which the collected energy is coupled to a receiver, scattered, and dissipated are represented respectively by the effect aperture (A_e), the scattering aperture (A_s), and the loss aperture (A_L) [21].

$$A_c = A_e + A_s + A_L \tag{1.72}$$

All the aperture terms have units of area, but they are not necessarily related to the projected area of the antenna. The effective aperture represents that part of the incident power density delivered to the receiving system, while the scattering and loss apertures represent those parts of the incident power density that are scattered and dissipated as heat.

The power delivered to the output of a receiving antenna is the same as the incident power density multiplied by the effective aperture.

$$P_r = S_i A_e \tag{1.73}$$

Equation (1.73) is very similar to EIRP, as we would expect from a reciprocal device. The effective aperture is related to gain by

$$G = \frac{4\pi A_e}{\lambda^2} \tag{1.74}$$

Effective aperture is a term reserved for receive antennas, whereas gain describes both transmitting and receiving antennas.

Example. Find the gain of a 50-m-diameter radio telescope parabolic reflector antenna at 1 GHz. Assume that $A_c = A_e$ = area of the reflector aperture.

$$A_e = \pi 25^2$$

Then

$$G = \frac{4\pi^2 25^2}{\left(3\times10^8 / 1\times10^9\right)^2} = 54.4 \text{ dB}$$

1.4.5. An Antenna as a Polarization Filter

Polarization of an electromagnetic wave describes how the magnitude and orientation of the electric field vector changes as a function of time at a given point in space. The polarization of an antenna is defined as the polarization of the wave transmitted by the antenna. The orientation of the time-varying electric field is important because it determines the orientation of the current induced in an object. Remember that the current flows in the same direction as the electric field. Thus, a time-varying electric field with z-directed polarization will produce a time-varying current in a wire parallel to the field, no current in a wire perpendicular to the z direction, and some time-varying current in a wire oriented between parallel and perpendicular. The orientation of a transmitting antenna, receiving antenna, and any scatterer in between affects the amount of power received.

If we assume that the electric field vector is a plane wave traveling in the z direction, the electric field lies in the x–y plane. The time harmonic representation of a single frequency electric field is

$$\vec{E}(t) = E_{x0}\cos(\omega t - kz)\hat{\mathbf{x}} + E_{y0}\cos(\omega t - kz + \Psi_y)\hat{\mathbf{y}} \tag{1.75}$$

We may examine the vector at a point in space ($z = 0$):

$$\mathbf{E} = E_x\hat{\mathbf{x}} + E_y\hat{\mathbf{y}} \tag{1.76}$$

Equating (1.75) to (1.76) results in these definitions:

$$E_x = E_{x0}\cos(\omega t) \tag{1.77}$$

$$E_y = E_{y0}\cos(\omega t + \Psi_y) \tag{1.78}$$

Solving for cos(ωt) produces

$$\cos(\omega t) = \frac{E_x}{E_{x0}} \tag{1.79}$$

Using trigonometry, (1.79) can be written as

$$\sin(\omega t) = \sqrt{1 - \left(\frac{E_x}{E_{x0}}\right)^2} \qquad (1.80)$$

With a little manipulation, the following equation describes the orthogonal components of the propagating plane wave:

$$aE_x^2 - bE_xE_y + cE_y^2 = 1 \qquad (1.81)$$

where

$$a = \frac{1}{E_{x0}^2 \cos^2 \Psi_y}, \qquad b = \frac{2\sin\Psi_y}{E_{x0}E_{y0}\cos^2 \Psi_y}, \qquad c = \frac{1}{E_{y0}^2 \cos^2 \Psi_y} \qquad (1.82)$$

This equation for an ellipse tells us that at any point in space, the tip of the electric field vector traces an ellipse over a period of time. Conversely, if a wave is frozen in time, the tip of the E vector along the propagation path traces out the same ellipse. For this reason, we say that the wave is elliptically polarized.

The electric field vector rotates either clockwise or counterclockwise. If you place your right thumb in the direction of wave propagation, and your fingers curve in the direction of the E field trajectory, the wave is said to be right-hand polarized (RHP). On the other hand (literally), if the trajectory is such that the thumb of the left hand can be pointed in the direction of wave propagation, and the fingers curve in the direction of the E field trajectory, the wave is left-hand polarized (LHP). The relative phase determines the handedness of the wave. For $0° < \Psi_y < 180°$ the wave is LHP, and for $180° < \Psi_y < 360°$ the wave is RHP. Figure 1.18 shows the electric field rotation for left-hand and right-hand elliptical polarization.

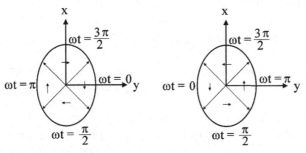

right-hand polarization   left-hand polarization

Figure 1.18. Rotation of the electric field for right-hand and left-hand polarization.

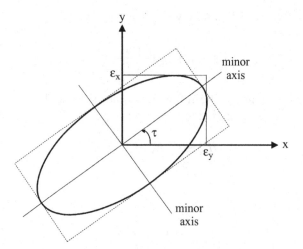

Figure 1.19. Polarization ellipse.

An ellipse (Figure 1.19) is characterized by (a) its axial ratio (AR), defined by the ratio of the major axis to the minor axis of the ellipse, and (b) the orientation, represented by the angle the major axis makes with the x axis of the coordinate system (τ). The AR has values ranging from 1 for a circle to ∞ for a line. Sometimes the inverse of the AR is given, because it has values between zero and one which are more computer-friendly. The axial ratio is positive for right-hand polarization and negative for left-hand polarization.

Two extremes of elliptical polarization are when AR = ∞ and AR = 0. When AR = ∞ the minor axis of the ellipse is zero, so the trajectory describes a straight line.

Linear Polarization (AR = ∞)

$E_{x0} = 0$ (linearly polarized in y direction)
$E_{y0} = 0$ (linearly polarized in x direction)
$E_{x0} = E_{y0}$ and $\Psi_y = 0$ (linearly polarized with $\tau = 45°$)

Since an x-polarized wave has $E_y = 0$, and a y-polarized wave has $E_x = 0$, any linearly polarized wave is the sum of an x-polarized wave and a y-polarized wave.

The other special case occurs when the length of the major axis equals the minor axis (AR = 1). Since both the longest and the shortest chords through the center are the same length, the trajectory is a circle. Circular polarization occurs when $E_{x0} = E_{y0}$ and they are 90° out of phase.

Circular Polarization (AR = 1)

$E_{x0} = E_{y0}$, $\Psi_y = +90°$ (left-circularly polarized)
$E_{x0} = E_{y0}$, $\Psi_y = -90°$ (right-circularly polarized)

If the receive antenna is not polarization-matched to the incoming electromagnetic wave, then it will not receive the maximum possible power. The receive polarization of an antenna is defined as the polarization of an incident wave that results in maximum power at the antenna terminals. It is related to the (transmit) polarization of the antenna in the same plane of polarization by having the same

1. Axial ratio
2. Sense of polarization
3. Spatial orientation

The power received by an antenna is multiplied by a polarization efficiency or polarization mismatch factor to account for the polarization mismatch between an incident wave and an antenna's receive polarization. This polarization efficiency is calculated by taking the inner product of the incident wave polarization vector and the complex conjugate of the receive antenna polarization vector.

$$p = \hat{e}_i \cdot \hat{e}_r^* \qquad (1.83)$$

where

$$\hat{e}_i = \text{polarization vector of incident wave} = \frac{\mathbf{E}_{incident}}{|\mathbf{E}_{incident}|} \qquad (1.84)$$

$$\hat{e}_r = \text{polarization vector of receive antenna} = \frac{\mathbf{E}_{antenna}}{|\mathbf{E}_{antenna}|} \qquad (1.85)$$

The received power is given by

$$P_r = p A_e S \qquad (1.86)$$

If the receive antenna has the same polarization as the transmit antenna, then there is a perfect match.

Example. Given the following values of E_x, E_y, and Ψ_y, what is the polarization of the field?

E_x	E_y	Ψ_y	Answer
1	0	45°	x linear
0.707	0.707	0°	linear 45° from x axis
0.707	0.707	90°	LHP circular
0.867	0.5	180°	linear 60° from x axis
0.867	0.5	90°	elliptical

1.5. ANTENNA ARRAY APPLICATIONS

Antenna arrays find applications over a wide range of frequencies. Some common types of systems that depend on arrays are described in this section.

1.5.1. Communications System

A communications system sends information from one point to another. For the receiver to detect the signal, the signal must be strong enough to be distinguished from noise. Radio receivers are typically rated by the minimum detectable ratio of received power to noise power, also known as the signal-to-noise ratio (SNR).

Average power density at a distance R from an isotropic radiator is the total radiated power divided by the surface area of a sphere, $P_t/4\pi R^2$. Increasing R to $2R$ reduces the average power density on the new imaginary sphere by one forth. The transmitter power density incident on an object, therefore, depends on the transmitted power, the antenna gain (which depends upon the antenna efficiency and the directivity function of azimuth and elevation), and the range from the radiator to the target:

$$S_i = \frac{P_t G_t}{4\pi R^2} \tag{1.87}$$

In reality, electromagnetic waves encounter such problems as atmospheric absorption, particulate scattering, and obstacle scattering. To account for these additional losses, a loss factor ($L < 1.0$) is included in the calculation of power density.

$$S_i = \frac{P_t G_t L}{4\pi R^2} \tag{1.88}$$

The power density incident on the receiving antenna is multiplied by the effective aperture to get the power delivered to the output terminals of the antenna. The resulting equation is known as the Friis transmission formula (Figure 1.20) [22].

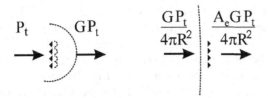

Figure 1.20. Friis transmission formula.

$$P_r = \frac{P_t G_t L A_e}{4\pi R^2} \tag{1.89}$$

Example. A cellular phone transmits 1 W of power at 840 MHz. Assume the phone is always between 100 m and 3 km of a base station. What is the minimum sensitivity of the receiver at the base station? The antennas are monopoles with gains of 1.5.

$$P_r = \frac{P_t G_t G_r L \lambda^2}{(4\pi R)^2} = \frac{1 \times 1.5 \times 1.5 \times 1 \times 0.357^2}{(4\pi \times 3000)^2}$$

1.5.2. Radar System

A radar system determines the characteristics of a target by radiating electromagnetic waves toward a target and analyzing the waves re-radiated toward the radar receiver. Radar can determine up to five different target parameters: angular location (azimuth and elevation), range, speed, size (in RCS terms), and identification.

The angular location of a target is found from the orientation of the antenna beam. When a target is detected, the position of the antenna pattern main lobe corresponds to the target location within a beamwidth of accuracy. In order to accurately determine location, radar antennas must have narrow main beamwidths, meaning antennas with high gain or directivity, and the beams must be movable to search the space around the radar. Antenna beams are scanned by either physically moving the antenna or electronically scanning the beam.

Monopulse is a more sophisticated method of locating a target. A monopulse antenna simultaneously employs two beams: a sum beam and a difference beam. The sum beam has a high gain in the direction of the target to determine the presence of the target. The difference beam has a sharp, deep null in the direction of the target to accurately determine its angular location. If the target is kept inside the deep null, the angular location of the target can be accurately determined. Since the difference pattern beam null is deep and narrow, it is easy to precisely locate a target.

Other target parameters are determined by characteristics of the received signal. A radar signal is an information signal; and consequently, the information extracted depends on the signal bandwidth and the type of information transmitted in the first place. Different types of radar modulation provide different information. One common type of modulation is pulse modulation where the carrier is switched on and off at a particular rate (called the PRF or pulse repetition frequency) for a short period of time (or pulse width). Another method of modulating a radar signal is to sweep the frequency linearly over a bandwidth (this is a sawtooth FM signal). Frequency and pulse modulation are combined in pulse compression radars.

The simplest method of determining target distance comes from accurately timing a radar pulse from the time it leaves the radar until it returns. The target distance is given by [23]

$$R = \frac{c \Delta t}{2} \qquad (1.90)$$

where c is the speed of light and Δt the time delay between pulse transmission and reception.

The range resolution depends on the pulse width.

$$\Delta R = \frac{c \tau}{2} \qquad (1.91)$$

where ΔR is range resolution and τ is pulse width. The maximum unambiguous range is the range beyond which a target appears closer because multiple pulses were transmitted before a return pulse is received.

$$R_{\text{unamb}} = \frac{c}{2 \text{PRF}} \qquad (1.92)$$

where PRF is the pulse repetition frequency.

When an object stands in the free-space propagation path of an electromagnetic wave, the wave induces current on that object. Some of the current induced on the object reradiates or scatters, but not equally in all directions. Like the effective aperture of an antenna, the radar cross section, (σ), has units of area (typically square meters) and is only partially related to the physical size of the scatterer. RCS is a function of the size, shape, and material composition of the target, as well as the frequency and polarization of the incident wave.

The power density incident on a scattering object is given by (1.88). The power scattered in any direction is determined by multiplying the incident power density by the area represented by the radar cross section.

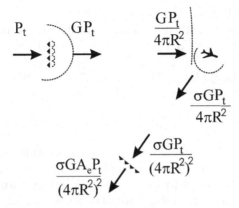

Figure 1.21. Derivation of the bistatic radar range equation.

$$P_s = \frac{P_t G_t L_t \sigma}{4\pi R^2} \tag{1.93}$$

If the scattered power travels a distance R_r to receive with a gain of G_r, then the final equation for the received power (Figure 1.21) is

$$P_s = \frac{P_t G_t L_t \sigma A_e L_r}{16\pi^2 R_t^2 R_r^2} \tag{1.94}$$

This equation is known as the bistatic radar range equation because the transmitter and receiver are at two different locations [24].

Like an antenna pattern, the RCS pattern has a main lobe, sidelobes, and nulls. Also like antenna patterns, two-dimensional plots are frequently used to evaluate various properties of RCS. Since RCS has units of m^2, when expressed in logarithmic form it is usually compared to a 1-m^2 target. Thus, the units are dBsm or dB relative to a square meter.

When the radar uses one antenna to transmit and receive, the bistatic radar range equation reduces to the monostatic radar range equation or more simply the radar range equation. The RCS in this case represents only power scattered directly back to the radar (backscattering). For clarity, the path loss (L) has been ignored.

$$P_r = \frac{P_t G A_e \sigma}{\left(4\pi R^2\right)^2} = \frac{P_t G^2 \lambda^2 \sigma}{\left(4\pi\right)^3 R^4} \tag{1.95}$$

Example. An over-the-horizon radar transmits a pulse with 1-MW average power. This waveform bounces from the ionosphere to the ocean and back to a receive station that is 300 miles away from the transmit station. If the distance from the transmitter to the ocean is 2000 km and the distance from the ocean to the receive antenna is 2400 km, then how much power arrives at the receiver. Both the transmit and receive antennas have gains of 30 dB. The bistatic RCS of the ocean at these angles is 10 m^2. The radar operates at 10 MHz. $G = 10^{30/10} = 1000$, $\lambda = 3 \times 10^8/(10 \times 10^6) = 30.0$ m, $A_e = 30^2(1000)/(4\pi) = 71,620.0$ m^2, $P_r = 1 \times 10^6(1000)(71,620)(10)/[(4\pi \cdot 2,000,000)^2(4\pi \cdot 2,400,000)^2] = 1.2466 \times 10^{-15}$ W

1.5.3. Radiometer

Communications and radar systems use both transmitting and receiving subsystems. A radiometer, on the other hand, uses only the receiver subsystem [25]. The radiometer listens to electromagnetic waves naturally emitted by objects. All objects with a temperature above absolute zero have vibrating charges. Because accelerating charges radiate electromagnetic waves, the random thermal motion of charges in any object results in the radiation of electromagnetic waves. Temperature indicates the amount of random molecular motion. At higher temperatures, more molecular collisions take place, and molecules move faster because more energy is stored in the material; therefore, more waves will be radiated at higher frequencies. Thus, temperature and electromagnetic radiation are closely related.

A blackbody is a perfect radiator and absorber of electromagnetic energy. Planck's radiation law states that a blackbody radiates uniformly in all directions with a spectral brightness given by

$$B_f = \frac{2hf^3}{c^2}\left(\frac{1}{e^{hf/k_B T} - 1}\right) \tag{1.96}$$

where B_f is spectral brightness, h is Planck's constant, f is temporal frequency (Hz), c is the speed of light in a vacuum (3×10^8 m/s), $k_B =$ Boltzman's constant (1.23×10^{-23} JK^{-1}), and T is absolute temperature (K). This power is radiated over a broad range of frequencies; but for objects with temperatures near the ambient reference temperature (300 K), most of the power is concentrated in the thermal infrared region of the electromagnetic spectrum. At microwave frequencies, although these signals are only about one-millionth as strong as the thermal infrared signal, good microwave antenna systems can detect the blackbody radiation. The brightness is found by integrating the blackbody spectral brightness over a frequency bandwidth (f) for a blackbody at temperature T. This equation is known as the Stefan–Boltzmann law:

$$B = \int_0^\infty B_f \, df = \frac{\sigma_s}{\pi} T^4 \tag{1.97}$$

where $\sigma_s = 5.673 \times 10^{-8}$ Wm^{-2}K^{-4} is the Stefan–Boltzmann constant. No natural objects emit perfect blackbody radiation; however, all objects such as terrain, sea, or the atmosphere emit a fraction of the ideal thermal radiation. The emissivity (e) is the ratio of the brightness of an object to the brightness of a blackbody at the same temperature.

$$e(\theta, \phi) = \frac{B(\theta, \phi)}{B_{bb}} \qquad (1.98)$$

where $B(\theta, \phi)$ represents brightness of material at temperature T and B_{bb} represents brightness of a blackbody at temperature T. Emissivity ranges between zero for a perfect reflector to unity for a blackbody. Emissivity varies with the material composition and the shape of the radiating object as well as with wavelength. At some frequencies, a particular body looks a lot more like a blackbody than at other frequencies.

Brightness temperature, T_B, is another way to represent the thermal radiation emitted from a gray body. For a blackbody, the temperature equals the absolute temperature of the object. Note that the emissivity and brightness temperature vary with orientation.

$$T_B(\theta, \phi) = e(\theta, \phi)T \qquad (1.99)$$

The output of an antenna receiving only thermal radiation is frequently represented by an antenna temperature, T_A, which is proportional to the total power resulting from the thermal radiation incident on the antenna. The antenna temperature is given by

$$T_A = \frac{1}{4\pi} \iint_{4\pi} T_B(\theta, \phi) G(\theta, \phi) \, d\Omega \qquad (1.100)$$

where $G(\theta, \phi)$ is the antenna gain pattern and $T_B(\theta, \phi)$ is the brightness temperature distribution incident on the antenna, and $d\Omega$ is the differential solid angle. The antenna temperature is therefore the spatially filtered sum of the radiation emitted by the bodies surrounding the antenna.

A receiving antenna generates power due to the increased thermal activity. If the antenna is modeled as a noise-generating resistor at temperature, T_A, the available noise power from the antenna is given by

$$P_r = k_B T_A \Delta f \qquad (1.101)$$

where $k_B = 1.23 \times 10^{-23}$J K^{-1}J (Boltzmann's constant) and Δf is the bandwidth of the receiver. A radiometer uses an antenna and receiving system to measure emission from objects. The brightness temperature distribution incident on a spaceborne microwave radiometer directed toward the earth is due both to radiation from the earth's surface and its atmosphere. At microwave frequencies below 10 GHz, atmospheric absorption and emission is small and may be neglected. At higher frequencies, the atmospheric contributions are significant and must be included.

Since emissivity is a characteristic of target size, shape, and composition, the brightness temperature for any aspect maps the emissivity of the observed target to a power level. The radiometer uses a highly directional antenna to scan in azimuth and elevation, and the data are recorded to produce a pixel map of the emissivity of the surface being scanned.

Example. Calculate the power received by an isotropic point source if the emissivity of the observed object is isotropic at 300 K.

First, find T_A: $T_A = \frac{1}{4\pi}\int_0^{2\pi}\int_0^{\pi}(1)(300)(1)\sin\theta d\theta d\phi$ and then substitute into (1.101) to get $P_r = 300 \times 1.23 \times 10^{-23} \times \Delta f$.

1.5.4. Electromagnetic Heating

Electromagnetic heating systems radiate electromagnetic waves for the sole purpose of heating an object. When an electromagnetic wave strikes an object, it induces both a displacement current and a conduction current. Conduction current results from the free movement of electrons in an object, while displacement current results from the constrained motion of electric dipoles, a polarized pair of charges. If the material has high conductivity, conduction current predominates, and the surface current density is expressed by Ohm's law,

$$\mathbf{J}_\sigma = \sigma\mathbf{E} \tag{1.102}$$

If the material has a large real-valued dielectric constant, most of the induced current will be a displacement current density equal to the time rate of change of the electric flux density (**D**).

$$\mathbf{J}_\varepsilon = \frac{\partial\mathbf{D}}{\partial t} \tag{1.103}$$

The total current density is the sum of the displacement current density and the conduction current density.

$$\mathbf{J}_T = \mathbf{J}_\varepsilon + \mathbf{J}_\sigma = \frac{\partial\mathbf{D}}{\partial t} + \sigma\mathbf{E} \tag{1.104}$$

Ordinarily, charged particles and dipoles are randomly distributed and oriented in a media, so the thermal activity is totally random. An electric field, however, induces organized motion of charges (current). A time-varying field causes free electrons, ions, and dipoles to move in a target. They collide and transfer some of their energy to other particles. Since molecular dipoles have larger mass than electrons, heating is more effective in dielectrics, where disno field

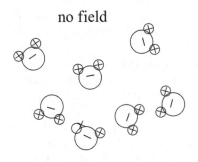

applied electric field

E

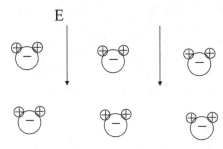

Figure 1.22. Water molecules aligning with the electric field.

placement current is large. The field induces a torque on the dipoles that makes each molecule attempt to rotate in order to align its dipole moment with the electric field. For instance, water molecules, which are dipoles, become polarized by an applied electric field (Figure 1.22). Due to the inertia of the molecule, it takes time for this torque to polarize the media. Energy is transferred to surrounding molecules and the dipoles rotate, thereby increasing the temperature. Conversely, when the electric field is removed, the increased random molecular motion destroys the alignment of the dipole moments and reduces the polarization exponentially with time.

The response time of a dielectric is a measure of the rate at which the polarization decays if the electric field is suddenly removed. The amount of displacement current density is time-dependent. Some of the dipole alignment energy becomes random motion (heat) every time a dipole is knocked out of alignment and then realigned. The response time indicates whether the dipole moments can keep in step with a time-varying electric field. At low frequencies the electric fields change direction slower than the response time of the dipoles, so the dipoles orient quickly, and the media only absorbs energy for a relatively short period of time. If the electric field changes direction faster than the response time of the dipoles, the dipoles do not rotate, no energy is absorbed, and the dielectric does not heat. When the electric field changes at about the same rate that the dipoles can respond, they rotate, but the resulting polarization lags behind the changes in the direction of the electric field. This lag indicates that the dielectric absorbs energy from the field and its temperature increases [26].

A microwave heating system consists of a microwave source and antenna. The source generates power at a frequency selected to correspond to the response time of the dielectric being heated. Heating of dielectrics has two familiar applications: microwave ovens and cancer therapy (induced hyperthermia). These applications work because both food and tumors contain mostly water (a molecular dipole). The heater uses an appropriate frequency (high MHz to low GHz region) to excite the water dipoles at a rate near the response time of water, and the target absorbs the transmitted energy.

Example. If a microwave oven is placed in a room at a temperature less than 0°C, can the oven melt an ice cube?

Answer: As explained above, the microwave oven excites dipoles in the water. Ice is a crystal. Consequently, the ice will not melt. If there is a small amount of water on the ice, then this water will heat and the ice will melt through microwave heating of the water and heat conduction.

1.5.5. Direction Finding

Finding the direction of a signal can be done in two ways. The first is to point the antenna main beam at the signal, so the direction of the signal occurs at the maximum received power. This approach requires a large antenna for accurate direction finding. On the other hand, nulls are precisely defined and have large variations in gain over a short angular sector. A small loop has a distinct null that has been used for direction finding since the early 1900s. Figure 1.23 shows an example of an early DF loop antenna that operated at HF.

1.6. ORGANIZATION AND OVERVIEW

This book is organized as a progression from relatively simple antenna arrays consisting of point sources to very complex digital beamforming arrays that can perform extremely complex signal processing. Most research on arrays was limited to point sources due to the computational limits of computers. The next two chapters summarize many of the developments surrounding the analysis and synthesis of these simple arrays. Real antenna arrays have real antenna elements, however. These elements are introduced in Chapter 4. Chapter 5 extends the narrow view of an array lying in a plane to an array consisting of antenna elements that can lie anywhere on a surface or in space. Thus, an array becomes even more versatile than a single aperture antenna. Placing array elements close together results in the elements interacting with each other. Each element in the array receives signals for all the other ele-

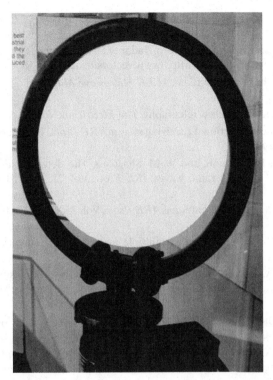

Figure 1.23. Early DF loop antenna. (Courtesy of the National Electronics Museum.)

ments in the array. This mutual coupling can significantly change the array performance and must be accounted for in the design. This complicated mutual coupling concept is described in Chapter 6. Coherently combining the signals in an array or beamforming is addressed in Chapter 7. Finally, the array has the potential to change its ability to receive and transmit signals based upon the environment and feedback. These adaptive arrays can reject interference, form multiple beams, and change performance characteristics. An emphasis is placed upon computational aspects of antenna arrays.

REFERENCES

1. J. A. Fleming, *The Principles of Electric Wave Telegraphy and Telephony*, 3rd ed., New York: Longmans, Green, and Co., 1916.
2. S. G. Brown, British Patent No. 14,449, 1899.
3. L. De Forest, Wireless-signaling apparatus, U.S. Patent 749,131, January 5, 1904.
4. G. Marconi, On methods whereby the radiation of electic waves may be mainly confined to certain directions, and whereby the receptivity of a receiver may be

restricted to electric waves emanating from certain directions, *Proc. R. Soc. Lond. Ser. A*, Vol. 77, 1906, p. 413.

5. G. Goebel, The British invention of radar, v2.0.3 / *chapter 1 of 12 / 01 may 09 / greg goebel / public domain*, http://www.vectorsite.net/ttwiz_01.html#m2.

6. S. N. Stitzer, The SCR-270 radar, *IEEE Microwave Mag.*, Vol. 8, No. 3, June 2007, pp.88–98.

7. F. Braun, Directed wireless telegraphy, *The Electrician*, Vol. 57, May 25, p. 222.

8. H. T. Friis, A new directional receiving system, *IRE Proc.*, Vol. 13, No. 6, December 1925, pp. 685–707.

9. H. T. Friis, C. B. Feldman, and W. M. Sharpless, The determination of the direction of arrival of short radio waves, *IRE Proc.*, Vol. 22, No. 1, January 1934, pp. 47–77.

10. G. H. Brown, Directional antennas, *IRE Proc.*, Vol. 25, No. 1, Part 1, January 1937, pp. 78–145.

11. H. T. Friis and C. B. Feldman, A multiple unit steerable antenna for short-wave reception, *IRE Proc.*, Vol. 25, No. 7, July 1937, pp. 841–917.

12. J. L. Spradley, A volumetric electrically scanned two-dimensional microwave antenna array, *IRE Int. Convention Record*, Vol. 6, Part 1, March 1958, pp. 204–212.

13. S. A. Schelkunoff, A mathematical theory of linear arrays, *Bell Syst. Tech. J.*, Vol. 22, 1943, pp. 80–107.

14. C. L. Dolph, A current distribution for broadside arrays which optimizes the relationship between bream width and side-lobe level, *Proc. IRE*, Vol. 34, June 1946, pp. 335–348.

15. S. P. Applebaum, Adaptive arrays, Syracuse University Research Corporation Report SPL TR 66-1, August 1966.

16. R. J. Mallioux, A history of phased array antennas, in *History of Wireless*, T. K. Sarkar et al., ed., New York: John Wiley & Sons, 2006, pp. 567–604.

17. E. Brookner, Phased-array radars: Pase, astounding breakthroughs and future trends, *Microwave J.*, Vol. 51, No. 1, January 2008, pp. 30–50.

18. A brief history of the sea-based X-band radar-1, Missile Defense Agency History Office, May 1, 2008.

19. IEEE standard definitions of terms for antennas, *IEEE AP Trans.*, AP-31, No. 6, November 1993.

20. H. Schantz, *Ultrawideband Antennas*, Norwood, MA: Artech House, 2005.

21. J. D. Kraus, *Antennas*, 2nd ed., McGraw-Hill, New York, 1988.

22. H. T. Friis, A note on a simple transmission formula, *IRE Proc.*, Vol. 33, No. 2, May 1946, pp. 254–256.

23. G. W. Stimson, *Introduction to Airborne Radar*, Hughes Aircraft Co., El Segundo, CA, 1983.

24. M. I. Skolnik, *Introduction to Radar Systems*, New York: McGraw-Hill, 2002.

25. F. T. Ulaby, R. K. Moore, and A. K. Fung, *Microwave Remote Sensing: Active and Passive*, Vol. II, Reading, MA: Addison-Wesley, 1982.

26. J. Walker, The secret of a microwave oven's rapid cooking action is disclosed, *Sci. Am.*, February 1987, pp. 134–138.

2 Array Factor Analysis

This chapter presents the fundamental approaches to the analysis of linear and planar arrays of point sources. Element patterns, polarization, and mutual coupling are delayed until future chapters. Keeping the elements along a straight line or in a plane are the most common array configurations. Other types of nonplanar arrays will be discussed in future chapters.

2.1. THE ARRAY FACTOR

A single isotropic point source transmits a field as derived in Chapter 1. If that point source transmits to an array of point sources, then the output of the array is proportional to the weighted sum of the received signal from each element in the array.

$$E\left(x_f, y_f, z_f\right) \propto \sum_{n=1}^{N} w_n \frac{e^{jkR_n}}{R_n} \tag{2.1}$$

where R_n = distance from element n to the point at $(x_f,\ y_f,\ z_f)$. As shown in Chapter 1, the phase of the received signal at the element is positive, because the signal is traveling toward the element. A transmit array has a minus sign in the phase, because the radiation is going away from the antenna. Figure 2.1 is a diagram of a point source transmitting to an array of point sources. When the array is very far from the point source, then all the R_n in the denominator of (2.1) are approximately the same. Consequently, the field is proportional to the sum of the weighted phase factors.

$$E\left(x_f, y_f, z_f\right) \simeq \sum_{n=1}^{N} w_n e^{jkR_n} \tag{2.2}$$

Most arrays are either linear or planar. A linear array has all of its elements lying along a straight line. To make calculations easy, assume that the array lies along the x, y, or z axes. The phase reference, or point of zero phase, is

Antenna Arrays: A Computational Approach, by Randy L. Haupt
Copyright © 2010 John Wiley & Sons, Inc.

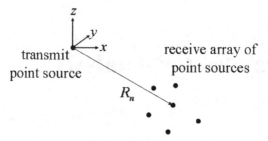

Figure 2.1. Near field of the array.

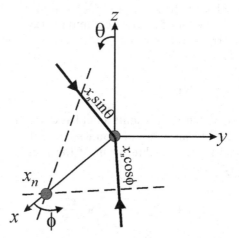

Figure 2.2. Phase difference between two elements on the x axis. The dashed line is the plane wave, and the arrows indicate the direction of propagation.

chosen to be either the first element or the physical center of the array. The origin of the coordinate system is placed at the phase center. An incident plane wave arrives at all of the elements at the same time when the incident field is normal or broadside to the array. When the plane wave is off-normal, then the plane wave arrives at each element at a different time. Thus, the phase difference between the signals received by the elements is accounted for by an appropriate phase delay before summing the signals to get the array output. An example of the phase difference between two elements along the x axis is shown in Figure 2.2. If the incident wave vector is in the x–y plane, then the phase is a function of ϕ. If the incident wave vector is in the x–z plane, then the phase is a function of θ. The array factor or antenna pattern due to isotropic point sources is a weighted sum of the signals received by the elements.

$$\mathrm{AF} = \sum_{n=1}^{N} w_n e^{j\psi_n} \qquad (2.3)$$

where N is the number of elements, $w_n = a_n e^{j\delta_n}$ is the complex weight for element n, $k = 2\pi/\lambda$, (x_n, y_n, z_n) is the location of element n, (θ, ϕ) is the direction in space, and

$$\psi_n = \begin{cases} kx_n u = kx_n \cos\phi \text{ or } kx_n \sin\theta & \text{along } x \text{ axis} \\ ky_n u = ky_n \sin\phi \text{ or } ky_n \sin\theta & \text{along } y \text{ axis} \\ kz_n u = kz_n \cos\theta & \text{along } z \text{ axis} \end{cases}$$

Equations (2.2) and (2.3) are the same when $\psi_n = kR_n$. The definition of the variable, u, depends upon which plane contains the array and the incident field vector. In digital signal processing terms, this equation is a spatial finite impulse response (FIR) filter. The diagram of a linear array in Figure 2.3 shows the signals sampled at discrete points in space/time and then weighted and summed.

A planar array has all of its elements in the same plane. By convention, the elements of a planar array usually lie in the x–y plane with the z axis pointing away from broadside. Thus, θ is measured from broadside and is often called the elevation angle, while ϕ is measured from the x axis and is often called the azimuth angle. Figure 2.4 is a diagram of a planar array with elements lying in arbitrary positions in the x–y plane. Of course, the array can lie in the x–y, x–z, or y–z planes. The definition of ψ_n depends upon the plane in which it lies.

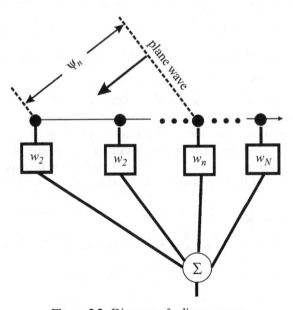

Figure 2.3. Diagram of a linear array.

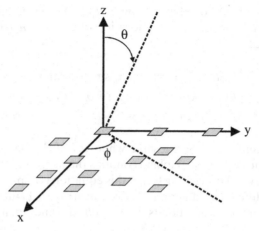

Figure 2.4. Arbitrary distribution of elements in the x–y plane.

$$\psi_n = \begin{cases} k(x_n u + y_n v) & x\text{–}y \text{ plane} \\ k(x_n u + z_n w) & x\text{–}z \text{ plane} \\ k(y_n v + z_n w) & y\text{–}z \text{ plane} \end{cases} \qquad (2.4)$$

where

$$(x_n, y_n) = \text{location of element } n$$

$$u = \sin\theta\cos\phi$$

$$v = \sin\theta\sin\phi$$

$$w = \cos\theta$$

$$u^2 + v^2 + w^2 = \sin^2\theta\left(\sin^2\phi + \cos^2\phi\right) + \cos^2\theta \le 1$$

The array factor is a function of amplitude weights, phase weights, element placement, and frequency. This chapter examines the effects on the array factor due to varying these variables. The next chapter provides recipes for finding the values of these variables that produce a desired array factor.

2.1.1. Phase Steering

The maximum value of the array factor is at the peak of the main beam which occurs when $\psi_n = 0$ for all n.

$$\text{AF}_{\max} = \sum_{n=1}^{N} w_n \qquad (2.5)$$

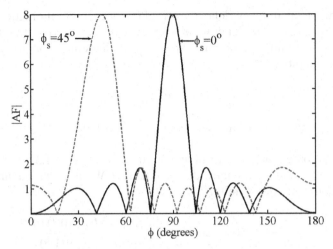

Figure 2.5. Beam of an 8-element array steered to 45°.

This maximum can be moved without moving the antenna by adding a constant phase shift, δ_n, to ψ_n. For a linear array along the x axis, the phase at element n is

$$\psi_n = kx_n u + \delta_n \qquad (2.6)$$

If δ_n is selected such that $\psi_n = 0°$ in the desired steering direction, u_s, then the peak of the main beam will appear at $u = u_s$.

$$\delta_n = -kx_n u_s \qquad (2.7)$$

Phase shifters placed at each element implement the steering phase.

Example. Consider an 8-element uniform array along the x axis with an element spacing of 0.5λ. Steering its beam to an angle of $\phi = 45°$ (Figure 2.5) requires a phase at element n of $\delta_n = -.707\pi (n - 1)$ radians.

A phase shift only delays a signal by up to one period or 2π radians. If the signal received at the first element is within the same period as the signal received at the last element, then

$$k(x_N - x_1)u_s \le 2\pi \qquad (2.8)$$

Thus, one pulse width can illuminate the entire aperture at once when it is incident from the maximum steering angle. Technically, to reproduce the signal with 100% accuracy, time delay units ($\delta_n > 2\pi$) would be required whenever (2.8) is violated. A phase shifter only delays a signal up to one period, T, while a time delay unit can delay the signal many periods. In practice, the rule of thumb is to use time delay units whenever the first element of the array receives the beginning of the pulse and the last element of the array receives the end of the pulse [1].

$$(x_N - x_1)u_s \leq c\tau_{3\,dB} \tag{2.9}$$

where τ_{3dB} is the pulse width and c is the speed of light.

Example. A 60-element uniform linear array with $d = 1.5\,cm$ at $10\,GHz$ receives a pulse having a bandwidth of $100\,MHz$. What is the maximum scan angle without time delay units?

$$\tau_{3\,dB} = \frac{1}{100 \times 10^6} = 10\,ns \quad so \quad u_s \leq \frac{c\tau_{3\,dB}}{(N-1)d} = \frac{(3 \times 10^{10})10^{-9}}{59(1.5)} = 0.34$$

which means that the main beam is limited to scanning $\pm 19.9°$ off broadside.

The phase shift required at element n in a planar array to steer a beam to (ϕ_s, θ_s) is

$$\delta_n = \begin{cases} k(x_n u_s + y_n v_s) & x\text{-}y \text{ plane} \\ k(x_n u_s + z_n w_s) & x\text{-}z \text{ plane} \\ k(y_n v_s + z_n w_s) & y\text{-}z \text{ plane} \end{cases} \tag{2.10}$$

where $u_s^2 + v_s^2 + w_s^2 \leq 1$. The phase in (2.10) reduces to that of a linear array if the beam is steered in one of the principal planes (either $\phi = 0°$ or $\phi = 90°$).

Figure 2.6 shows the effects on the array factor in three dimensions as it is steered from 90° to 0°, assuming that the 8-element linear array lies along the x axis and has a constant $\lambda/2$ element spacing. The array factor starts with a doughnut shape at broadside and ends with a dumbbell shape at end-fire.

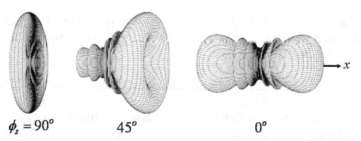

$\phi_s = 90°$ 45°   0°

Figure 2.6. Beam of an 8-element linear array steered from 90° to 0° in three dimensions.

2.1.2. End-Fire Array

An end-fire array is a linear array having the peak of its main beam pointing in the same direction as the axis of the array. The simplest end-fire array is a uniform array with its peak steered to $u = 1$ (see 0 degrees in Figure 2.6), which means that $\psi = 0°$ or

$$\delta_s = -kd \qquad (2.11)$$

A polar plot of the array factor for an end-fire array with $N = 5$ and $d = \lambda/2$ is shown as the solid line in Figure 2.7. This array factor has a very wide beamwidth and two main beams. Shrinking the element spacing to 0.2λ produces the dashed line in Figure 2.8. The closer spacing eliminates the main beam at 180°.

A narrower main beam, hence better resolution and a higher directivity, is possible by altering the spacing and phase in an optimal fashion. The Hansen–Woodyard end-fire array [2] achieves greater directivity by forcing the maximum of the main beam to be in invisible space or $u > 1$. An optimum directivity occurs when an extra π phase shift occurs across the extent of the array. This optimum phase shift is given by

$$\delta_s = \begin{cases} -kd + \dfrac{\pi}{N} & \text{for a maximum at } u = -1 \\[2ex] -kd - \dfrac{\pi}{N} & \text{for a maximum at } u = 1 \end{cases} \qquad (2.12)$$

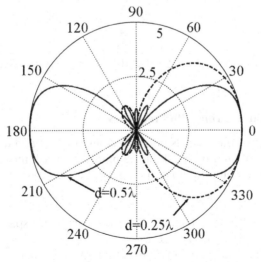

Figure 2.7. Polar plot of the array factors of an end-fire array for $d = 0.5\lambda$ and $d = 0.2\lambda$.

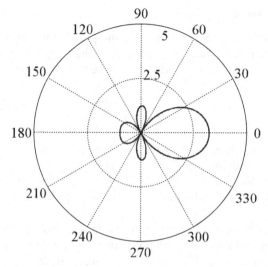

Figure 2.8. Polar plot of the array factors of a Hansen–Woodyard end-fire array for optimum spacing ($d = 0.2\lambda$).

when the array element spacing is

$$d = \left(\frac{N-1}{N}\right)\frac{\lambda}{4} \tag{2.13}$$

which for large arrays becomes

$$d \simeq \frac{\lambda}{4} \tag{2.14}$$

Figure 2.8 is the Hansen–Woodyard end-fire array factor with $d = 0.2\lambda$ for a 5-element array.

2.1.3. Main Beam Steering with Frequency

If the main beam of the array is pointing at an angle of u_s due to phase steering the main beam, then the peak of the main beam will move with a change in frequency. As a result, the main beam can be steered without moving the array.

Example. At 10 GHz, the elements of a linear array are spaced 0.5λ and the array operates from 8.2 to 12.4 GHz.

The phase delay between elements is then given by $\delta_s = 0.79$ radians. This phase shift steers the beam to an angle of

$$u_s = \frac{\delta_s}{kd} = 0.25 = 14.5° \qquad \text{from broadside}$$

At 8.2 GHz ($\lambda = 3.66$ cm), the beam points at $u_s = 0.305 = 17.8°$.
At 12.4 GHz ($\lambda = 2.42$ cm), the beam points at $u_s = 0.202 = 11.7°$.

2.1.4. Focusing

Focusing an array concentrates the field radiated/received by the array at/from a specific point in the near field rather than the far field. Antenna focusing has been applied to antenna measurements [3] and medical treatments [4]. The focusing occurs over a small area rather than at a single point due to the finite aperture size. Ricardi [5] and Hansen [6] show that the minimum spot size of a uniform array antenna is about 0.35λ in diameter. Since the focal point is in the near field, the array factor must take into account the exact distance from each element to the focal point at (x_f, y_f). The relative array factor in the near field is given by

$$AF = \sum_{n=1}^{N} w_n \frac{e^{jkR_n}}{R_n} \qquad (2.15)$$

where $R_n = \sqrt{(x_n - x_f)^2 + (y_n - y_f)^2}$. In order to focus the array, the array factor should coherently sum to a maximum at the focal point. In other words, the phase of the field from each element is made identical at the focal point by setting the weights to

$$w_n = e^{-jkR_n} \qquad (2.16)$$

Focusing an array results in the highest possible field level at the focal point without any amplitude weighting.

Example. Find the phase weights of an 8-element array centered at $x = 0$ with $d = 0.5\lambda$ that has a beam focused at $x = -0.5\lambda$ and $y = 2\lambda$. Compare results with an unfocused array.

The phase weights (in radians) needed to focus the array are

$$[-2.25 \quad -0.85 \quad -0.0978 \quad -0.0978 \quad -0.855 \quad -2.25 \quad 2.15 \quad -0.0653]$$

Figure 2.9 shows the near-field array factor for the array focused at $(x_f, y_f) = (-0.5\lambda, 2\lambda)$. The power is much higher (10.8 dB) than if the far-field beam is steered in the direction of the focal point (1.9 dB) as shown in Figure 2.10.

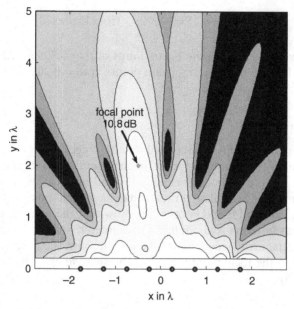

Figure 2.9. Near-field array factor when array is focused in the near field at $(x_f, y_f) =$ $(-0.5\lambda, 2\lambda)$.

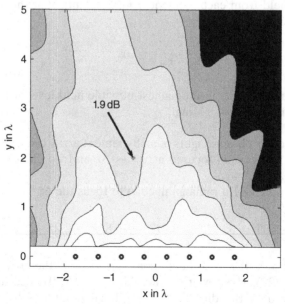

Figure 2.10. Near-field array factor is focused in the far field at $(x_f, y_f) = (-50\lambda, 200\lambda)$.

2.2. UNIFORM ARRAYS

Uniform arrays have periodic element spacing, and the weights at each element are the same, except for beam-steering phase shifts.

2.2.1. Uniform Sum Patterns

A uniform linear sum array has $w_n = 1$ and equally spaced elements with the peak of the main beam pointing at broadside. Assuming the phase center is at the first element of the array, then the array factor is given by

$$\text{AF} = 1 + e^{j\psi} + e^{j2\psi} + \cdots + e^{j(n-1)\psi} = \sum_{n=1}^{N} e^{j(n-1)\psi} \tag{2.17}$$

where

$$\psi = k\,du = \begin{cases} kd\cos\phi \text{ or } kd\sin\theta & \text{along } x \text{ axis} \\ kd\sin\phi \text{ or } kd\sin\theta & \text{along } y \text{ axis} \\ kd\cos\theta & \text{along } z \text{ axis} \end{cases}$$

d = spacing between elements

The distance a plane wave needs to travel between adjacent elements along the x axis is shown in Figure 2.11. Multiplying the distance by k yields the phase, ψ. A similar derivation is possible when the array lies along the y or z axes.

The phase reference does not have to be the first element in the array. Moving the phase reference results in multiplying the array factor by a

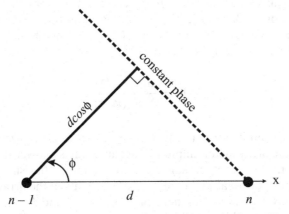

Figure 2.11. Phase difference between adjacent elements of a uniform linear array along the x axis.

constant phase but will not change the magnitude of the array factor. Another convenient place for the phase reference is the physical center of the array. When the origin of the coordinate system moves to the center of the array, the array factor becomes

$$AF = \sum_{n=1}^{N} e^{j\left[n-\frac{N+1}{2}\right]\psi} \tag{2.18}$$

Equations (2.17) and (2.18) differ only by the phase factor

$$e^{j\left[1-\frac{N+1}{2}\right]\psi} \tag{2.19}$$

which is the phase difference between element 1 and the center of the array.

Multiplying both sides of (2.17) by $e^{j\psi}$ and subtracting the resulting product from (2.17) results in a simpler expression for the array factor

$$AF - AFe^{j\psi} = 1 - e^{j\psi} \tag{2.20}$$

Solving (2.20) for AF produces

$$AF = \frac{1 - e^{jN\psi}}{1 - e^{j\psi}} = \frac{\sin(N\psi/2)}{\sin(\psi/2)} e^{j\frac{N-1}{2}\psi} \tag{2.21}$$

The maximum of (2.21) occurs when $\psi = 0°$.

$$AF = \sum_{n=1}^{N} e^{j(n-1)0} = N \tag{2.22}$$

Often, only the relative array factor is important, and dividing (2.21) by N normalizes the array factor. If the phase center is at the physical center of the array, then the phase term in (2.21) disappears. Thus, the array factor normalized to the peak of a uniform array factor (N) when the phase center is at the physical center of the array is

$$AF_N = \frac{\sin(N\psi/2)}{N\sin(\psi/2)} \tag{2.23}$$

Note that the denominator in (2.21) is independent of array size and forms an envelope for the array factor amplitude, because it has a lower frequency than the numerator. On the other hand, the frequency of the sine wave in the numerator is proportional to the number of elements in the array. A comparison of the array factors and numerators for $N = 10$ and $N = 20$ is shown in Figure 2.12. The dotted line in the figure is the envelope or denominator and is the same for both arrays.

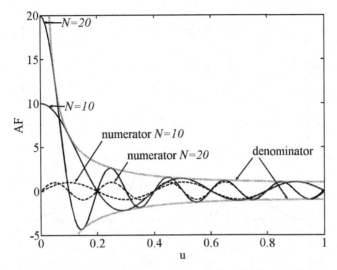

Figure 2.12. Uniform array factors for $N = 10$ and $N = 20$ along with the numerators and denominator of the expression for the uniform array.

The zeros of (2.23) or nulls in the array factor occur when $AF_N(\psi_{null}) = 0$, and the denominator is not equal to zero or

$$\frac{N\psi_{null}}{2} = \pm m\pi \Rightarrow \psi_{null} = \pm\frac{2m\pi}{N}, \qquad m = 1, 2, \ldots \qquad (2.24)$$

Sidelobes are local maxima that occur between two nulls. Their approximate location occurs when the numerator of (2.23) is a maximum or

$$\sin\left(\frac{N\psi}{2}\right) = \pm 1 \Rightarrow \psi = \pm\frac{2n+1}{N}\pi, \qquad n = 1, 2, \ldots \qquad (2.25)$$

For a large array, the numerator of (2.23) quickly varies while the denominator slowly varies. The first sidelobe occurs at a small value of ψ; and the small argument approximation for sin, or $x \simeq x$ for small values of x, is used in the denominator. Consequently, the sidelobe envelope close to the mainbeam is given by the denominator of (2.23). Substituting the location of the first sidelobe into (2.23) provides an estimate of the sidelobe level of the first sidelobe of a uniform linear array.

$$sll \simeq \frac{1}{N\psi/2} = \frac{1}{\dfrac{N}{2}\dfrac{3\pi}{N}} = 0.212 = -13.5 \text{ dB} \qquad (2.26)$$

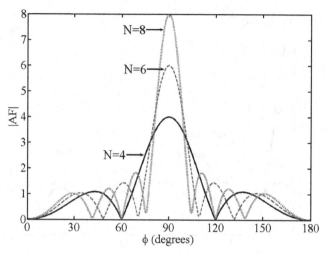

Figure 2.13. Array factors for $N = 4$-, 6-, and 8-element arrays having elements spaced $d = \lambda/2$ along the x axis.

Smaller uniform linear arrays have sidelobe levels close to 13 dB below the peak of the main beam.

Example. Plot the unnormalized array factors for $N = 4$-, 6-, and 8-element arrays having elements spaced $d = \lambda/2$ along the x axis.

Figure 2.13 shows the three array factors. Increasing the number of elements increases the main beam peak and the number of sidelobes and decreases the width of the main beam.

As mentioned in Chapter 1, the antenna beamwidth determines the resolution and gain of the array. A useful approximation for the null-to-null beamwidth for a uniform array comes from (2.24) with $m = 1$

$$\theta_{\text{nullBW}} = \frac{2\lambda}{Nd} \tag{2.27}$$

In most electromagnetics systems, the 3-dB beamwidth is more useful than the null-to-null beamwidth. It is found by taking the difference between the two angles found by solving the following transcendental equation:

$$\text{AF}_N = \frac{\sin(N\psi/2)}{N\sin(\psi/2)} = \frac{1}{\sqrt{2}} \tag{2.28}$$

When N is large, then the argument in the denominator is small, so the array factor can be approximated by a sinc function, and the the 3-dB beamwidth is found by

$$\frac{\sin(N\psi/2)}{N\psi/2} = \text{sinc}(N\psi/2) = \frac{1}{\sqrt{2}} \qquad (2.29)$$

The 3-dB point on a sinc function is given by

$$N\psi/2 = 1.3916 \Rightarrow u_{3\,dB} = \frac{0.443}{\left(\dfrac{d}{\lambda}\right)N} \qquad (2.30)$$

For half-wavelength spacing the 3-dB beamwidth is approximately twice the angle found in (2.30).

$$\theta_{3\,dB} \approx 101.5°/N \qquad (2.31)$$

Example. Find all the maxima and nulls for a 6-element array having elements spaced $d = \lambda/2$ along the x axis. What is the null-to-null and 3-dB beamwidths?

$$\psi_{\text{null}} = \pm\frac{2m\pi}{6} \Rightarrow \phi_{\text{null}} = \cos^{-1}(\pm m/3) \qquad \text{for } m = 1, 2, 3$$

$$\phi_{\text{null}} = 70.5°, 48.2°, 0°, 109.5°, 131.8°, 180°$$

$$\psi_{\text{sl}} \approx \pm\frac{2n+1}{6}\pi \Rightarrow \phi_{\text{sl}} = \cos^{-1}\left(\pm\frac{2n+1}{6}\right) \qquad \text{for } m = 1, 2$$

$$\phi_{\text{sl}} \approx 60°, 33.6°, 120°, 146.4°$$

$$\text{null-to-null beamwidth} = 109.5° - 70.5° = 39°$$

The 3-dB beamwidth is found by finding the angles for which

$$\frac{\sin(3\pi\cos\phi)}{\sin(\pi\cos\phi/2)} = \frac{6}{\sqrt{2}} \Rightarrow \phi = 0.3 \text{ radians or } 17.2°$$

This value is quite close to that obtained using (2.31), $\theta_{3dB} \approx 16.9°$.

A planar array can be constructed by stacking linear arrays. Start with an N_{x1} element linear array along the x axis with element spacing d_x and weights w_{1n}. Another linear array having N_{x2} elements is placed d_y above the first array and displaced Δ_2 in the x direction. Linear arrays are added d_y above the last one until the planar array is complete (see Figure 2.14). The array factor for this array is given by

$$\text{AF} = \sum_{m=1}^{N_y}\sum_{n=1}^{N_{xm}} w_{mn}e^{jk[x_n u + y_m v]} = \sum_{m=1}^{N_y}\sum_{n=1}^{N_{xm}} w_{mn}e^{jk\{[(n-1)d_x + \Delta_m]u + (m-1)d_y v\}} \qquad (2.32)$$

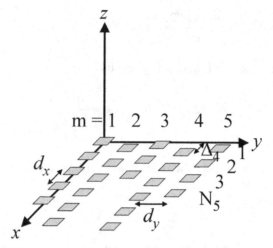

Figure 2.14. Planar array can be built from a set of equally spaced linear arrays.

where N_y is the number of linear arrays, N_{xm} is the number of elements in linear array m, and Δ_m represents x displacement of array m.

Two common element layouts are rectangular spacing

$$\Delta_m = 0 \tag{2.33}$$

and triangular or hexagonal spacing [7]

$$\Delta_m = \begin{cases} 0 & m \text{ is even/odd} \\ d_x/2 & m \text{ is odd/even} \end{cases}$$

$$d_y = d_x\sqrt{3}/2 \tag{2.34}$$

These spacings for a rectangular array are shown with two microstrip arrays in Figure 2.15. The array on the left has 16 elements in a rectangular lattice within a square boundary and operates at X band. The array on the right has 19 elements in a triangular lattice within a hexagonal boundary and operates at L band. Although the equilateral triangle spacing is most common, isosceles and general triangular spacing is also possible [8].

If all the weights are equal to one, then the array factor for rectangular spacing is

$$AF = \frac{\sin\left(\dfrac{N_x \psi_x}{2}\right)}{N_x \sin\left(\dfrac{\psi_x}{2}\right)} \frac{\sin\left(\dfrac{N_y \psi_y}{2}\right)}{N_y \sin\left(\dfrac{\psi_y}{2}\right)} \tag{2.35}$$

Figure 2.15. Common element lattices for a planar array. (Courtesy of Ball Aerospace & Technologies Corp.)

where

$$\psi_x = kd_x(u - u_s)$$

$$\psi_y = kd_y(v - v_s)$$

and for triangular spacing is

$$\mathrm{AF} = \frac{\sin(N_{xe}\psi_x)}{N_{xe}\sin(\psi_x)}\frac{\sin(N_{ye}\psi_y)}{N_{ye}\sin(\psi_y)} + e^{-j(\psi_x + \psi_y)}\frac{\sin(N_{xo}\psi_x)}{N_{xo}\sin(\psi_x)}\frac{\sin(N_{yo}\psi_y)}{N_{yo}\sin(\psi_y)} \qquad (2.36)$$

The triangular spacing array factor consists of two uniform arrays with spacings of $2d_x$ and $2d_y$. The first array has its bottom left element at the origin, and the second array has its bottom left element at (d_x, d_y).

Usually, the beamwidth of a planar array is defined for two orthogonal planes. For example, the beamwidth is usually defined in θ for $\phi = 0°$ and $\phi = 90°$. As with a linear array, nulls in the array factor of a planar array are found by setting the array factor equal to zero. Unlike linear arrays, the nulls are not single points.

2.2.2. Uniform Difference Patterns

A difference pattern has a null at broadside instead of a peak. This null can precisely locate the direction of a signal, because the null has a very narrow angular width compared to the width of the main beam of a corresponding sum pattern. When the array output is zero, the signal is in the null. Slight movement of the null dramatically increases the gain of the array factor and the reception of a signal. A 20-element uniform sum array can be converted into a 20-element difference array by giving half the elements a 180° phase shift, or half the elements are one and half are minus one. The array factor for this uniform difference array is shown in Figure 2.16.

The array factor for a uniform difference array is written as

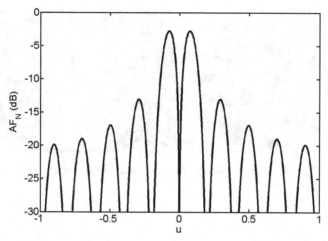

Figure 2.16. Twenty-element uniform difference array factor normalized to N.

$$AF = 1 + e^{j\psi} + \cdots + e^{j(N/2-1)\psi} - e^{j(N/2)\psi} - \cdots - e^{j(N-1)\psi} \tag{2.37}$$

Multipling both sides of (2.37) by $e^{j\psi}$ and subtracting the resulting product from (2.37) results in a simpler expression for the array factor

$$AF - AFe^{j\psi} = 1 - 2e^{jN\psi/2} + e^{jN\psi} \tag{2.38}$$

Solving (2.38) for AF produces

$$AF = \frac{1 + e^{jN\psi} - 2e^{jN\psi/2}}{1 - e^{j\psi}} = \frac{1 - \cos(N\psi/2)}{j\sin(\psi/2)} e^{j\frac{N-1}{2}\psi} \tag{2.39}$$

Moving the phase center to the middle of the array reduces (2.39) to

$$AF = \frac{1 - \cos\left(\dfrac{N\psi}{2}\right)}{j\sin\left(\dfrac{\psi}{2}\right)} \tag{2.40}$$

Nulls occur when

$$\frac{N\psi_{\text{null}}}{2} = \pm 2m\pi \Rightarrow \psi_{\text{null}} = \pm\frac{4m\pi}{N}, \qquad m = 0, 1, 2, \ldots \tag{2.41}$$

Sidelobes in the array factor approximately occur when the numerator is a maximum.

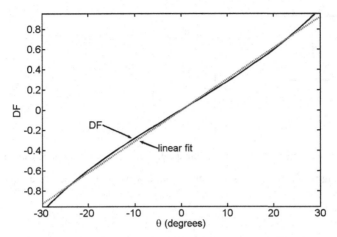

Figure 2.17. Plot of the magnitude of the ratio of the sum and difference patterns for a 2-element uniform array from $\theta = -30°$ to 30°.

$$\cos\left(\frac{N\psi}{2}\right) = -1 \Rightarrow \psi = \pm\frac{2(2n+1)}{N}\pi, \qquad n = 1, 2, \ldots \qquad (2.42)$$

This equation does not yield a good approximation to the height of the difference main lobes; hence n starts at one instead of zero.

Example. Plot the magnitude of the ratio of the sum and difference patterns (DF) for a 2-element uniform array from $\theta = -30°$ to 30°.

Figure 2.17 has a plot of the ratio, DF, (solid line) and a linear fit of DF (dotted line). This plot shows that the angle of incidence can be found from the ratio of the difference pattern to the sum pattern.

2.3. FOURIER ANALYSIS OF LINEAR ARRAYS

Antenna arrays sample the signals incident on them through elements at discrete locations. In order to avoid aliasing, the array must sample the incident signal at the Nyquist rate. The Nyquist rate stipulates that two samples are taken during the period of the highest frequency. Since an electromagnetic signal has time (T) and spatial (λ) periods, the samples are taken at

$$\frac{T}{2} \qquad \text{in time}$$

$$\frac{\lambda}{2} \qquad \text{in space}$$

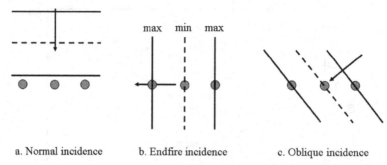

a. Normal incidence   b. Endfire incidence   c. Oblique incidence

Figure 2.18. Array sampling of plane wave from three directions.

Of course, these are related by

$$\frac{\lambda/2}{T/2} = c \tag{2.43}$$

where c is the speed of light. Consequently, the elements of an array must be separated at most by half a wavelength to avoid under sampling an incident plane wave for any angle of incidence. A plane wave impinges on all the elements of an array simultaneously at normal incidence (Figure 2.18a). At this angle ($u = 0$) the element spacing is irrelevant. Figure 2.18b demonstrates the need for $\lambda/2$ sampling for an incoming wave from the $u = 1$ direction. Undersampling is not a problem at normal incidence but can be at angles off-normal. The angle at which aliasing begins (Figure 2.18c) is given by u_A. This wave is sampled at the Nyquist rate if

$$\lambda = 2du_A \tag{2.44}$$

Thus,

$$d < \frac{\lambda}{2}, \quad \text{then } u_A > 1 \quad \text{(oversampled)}$$

$$d = \frac{\lambda}{2}, \quad \text{then } u_A = 0 \quad \text{(perfect sampling)}$$

$$d > \frac{\lambda}{2}, \quad \text{then } u_A < 0 \quad \text{(undersampled)}$$

Aliasing manifests itself as pattern replication. As the spacing gets wider, grating lobes or clones of the main beam appear in the pattern. These extra main beams appear at regular intervals given by

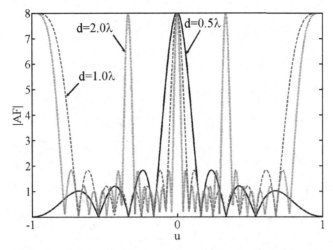

Figure 2.19. Array factors of 8-element arrays for three different spacings.

$$u_g = u_s \pm \frac{m\lambda}{d} \tag{2.45}$$

When the array with grating lobes receives a signal, the direction of the signal cannot be determined, because the angle of arrival is ambiguous: Did the signal enter the main beam or the grating lobe? The broadside array factors of an 8-element array with element spacings of $d = \lambda/2$, $d = \lambda$, and $d = 2\lambda$ are shown in Figure 2.19. Increasing the frequency causes d/λ to increase as well. An array designed for optimum sampling at the center frequency undersamples a plane wave at frequencies above the center frequency and oversamples at frequencies below the center frequency.

Example. An array operating at a center frequency of $f_0 = 10\,GHz$ ($\lambda_0 = 3\,cm$) has a 5% bandwidth. What is the maximum element spacing (in centimeters) that adequately samples signals up to ±30° from broadside?

The highest frequency will limit the element spacing, so $f = 10.25\,GHz$ and

$$\lambda_{hi} = 2.93 \text{ cm}, \qquad d = \frac{\lambda_{hi}}{2\cos 30°} = 1.69 \text{ cm or } 0.56\lambda_0$$

An equally spaced linear array is the superposition of many 2-element arrays. If the amplitude weights are symmetric about the center of the array, then symmetric pairs of exponent terms in the array factor can be combined into a cosine term. An even element array has $N = 2M$ elements and an odd-element array has $N = 2M+1$ elements. The symmetric elements in an odd array with the phase center at the physical center of the array combine as follows:

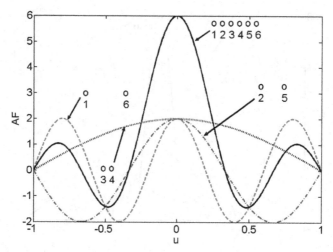

Figure 2.20. A 6-element array is composed of three Fourier components added together. The 6-element array is shown at the top. Pairs of elements point to their corresponding array factors.

$$a_m e^{jm\psi} + a_m e^{-jm\psi} = 2a_m \cos(m\psi) \qquad (2.46)$$

The array pattern, then, is the superposition of all the 2-element array patterns that make up the array. Figure 2.20 has plots of the three Fourier components of a 6-element array superimposed on the array factor for a 6-element uniform array. At $u = 0$, the three components have peaks that add in phase to form the main beam. If an array has an odd number of elements, then there is a single element at the center (zero spatial frequency) with an amplitude weight a_0. The array factors for the even- and odd-element arrays can be written as

$$\mathrm{AF}_{2M} = 2 \sum_{m=1}^{M} a_m \cos\left[\left(\frac{2m-1}{2}\right)\psi\right] \qquad \text{even} \qquad (2.47)$$

$$\mathrm{AF}_{2M+1} = a_0 + 2 \sum_{m=1}^{M} a_m \cos(m\psi) \qquad \text{odd} \qquad (2.48)$$

These equations have the familiar form of a Fourier series, especially when written as

$$\mathrm{AF}_{2M} = 2 \sum_{m=1}^{M} a_m \cos\left[(2m-1)\pi\left(\frac{d}{\lambda}\right)u\right] \qquad \text{even} \qquad (2.49)$$

$$\mathrm{AF}_{2M+1} = a_0 + 2 \sum_{m=1}^{M} a_m \cos\left[2m\pi\left(\frac{d}{\lambda}\right)u\right] \qquad \text{odd} \qquad (2.50)$$

A difference pattern with low sidelobes requires a different set of weights than a corresponding sum pattern. The amplitude taper for sidelobes in a sum pattern will not result in the same low sidelobes in a difference pattern and vice versa. Difference arrays have an even number of elements, because the center element of an array with an odd number of elements serves as a constant in the Fourier series. This constant would not be canceled by the other elements, so the difference null would not be as deep. The array factor for a symmetric linear difference array is given by

$$\text{AF} = \frac{j}{N} \sum_{n=1}^{N} b_n \sin[kd(n-.5)u] \qquad (2.51)$$

where b_n is the difference amplitude weight for element n.

The fast Fourier transform (FFT) is an efficient way to calculate the array factor. Usually, the antenna weights are placed in an array then padded with zeros (zeros added onto the end of the array weights). For instance, if there are N elements in the array and P zeros used for padding, then the vector looks like

$$\left[w_1 \cdots w_N \underbrace{0 \cdots 0}_{P} \right] \qquad (2.52)$$

The number of zeros in the pad determines how smooth the array factor will look. Basically, the interpolation in u space is in increments of $\Delta u = 2/(N + P)$ from $u = -1$ to $u = (N + P)\Delta u - 1$. Note that $u = 1$ is not included. This fact is important in relating the FFT results to points in space.

Example. Find the array factor of an $N = 8$ element uniform linear array using a FFT.

The MATLAB command is given by

```
AFdb=10*log10(abs(fftshift(fft(ones(1,N),P+N))))
```

An FFT algorithm usually returns the dc component in the first cell of the vector. A command like fftshift rearranges the vector to put the dc component in the center. As the number of zeros in the pad increases, the array factor becomes smoother. Figure 2.21 shows the FFT array factor for $P = 0, 8, 24$, and 56.

2.4. FOURIER ANALYSIS OF PLANAR ARRAYS

The elements of a planar array are usually laid out in a periodic lattice in the x–y plane. For example, when elements are in a rectangular lattice, (2.32) is written as a two-dimensional discrete Fourier series

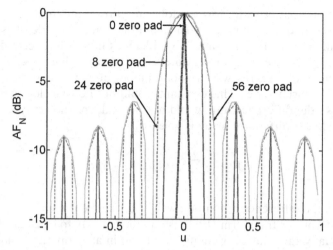

Figure 2.21. FFT of an 8-element uniform linear array with $P = 0, 8, 24$, and 56.

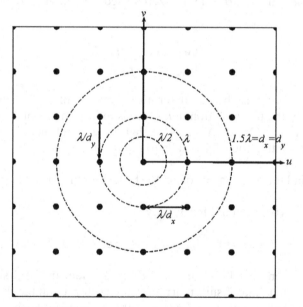

Figure 2.22. Grating lobe plot for a rectangular array.

$$AF = \sum_{m=1}^{N_y} \sum_{n=1}^{N_x} w_{mn} e^{j2\pi\left\{(n-1)\frac{d_x}{\lambda}(u-u_s)+(m-1)\frac{d_y}{\lambda}(v-v_s)\right\}} \tag{2.53}$$

Figure 2.22 is a plot of the maxima of (2.53) as a function of u and v and is known as a grating lobe plot. The three dashed lines correspond to the angular space $0 \le \theta \le 90°$ and $0 \le \phi \le 306°$ (real space) when $d_x = d_y = \lambda/2, \lambda$, and 1.5λ. If the element spacing is not the same or the number of elements are different in the x and y directions, then the circles in the u–v plane are ellipses. For the rectangular element spacing, grating lobes appear at regular intervals given by

$$
\begin{aligned}
u_m &= u_s + m\lambda/d_x && \text{for } m = 0, \pm 1, \pm 2, \ldots \\
v_n &= v_s + n\lambda/d_y && \text{for } n = 0, \pm 1, \pm 2, \ldots
\end{aligned}
\tag{2.54}
$$

All grating lobes in (2.54) that satisfy

$$
(u_m - u_s)^2 + (v_n - v_s)^2 \le 1
\tag{2.55}
$$

appear in the array factor in real space. As can be seen from (2.55), steering the main beam moves the center of the circle in the u–v plane to the point (u_s, v_s). A planar array (with no steering) in the x–y plane with uniform spacing sees the grating lobes enter real space at $\theta = 90°$ and $\varphi = 0°, 90°, 180°, 270°$ when $d = 1.0\lambda$ in the x and y directions. When $d = 1.41\lambda$, then additional grating lobes enter real space at the $45°, 135°, 225°$, and $315°$ azimuth angles as shown in Figure 2.22.

Triangular spacing modifies the sampling strategy and produces grating lobes at the locations (u_m, v_n) defined by

$$
\left(u_s + \frac{m\lambda}{2d_x}, v_s + \frac{n\lambda}{2d_y} \right) \quad \text{and} \quad \left(u_s + \frac{(2m-1)\lambda}{2d_x}, v_s + \frac{(2n-1)\lambda}{2d_y} \right)
$$
$$
\text{for } m = 0, \pm 1, \pm 2, \ldots
$$
$$
\text{for } n = 0, \pm 1, \pm 2, \ldots
\tag{2.56}
$$

The spacing d_x is the length of the bottom side of the triangle, while d_y is the height of the triangle. Only grating lobes that satisfy (2.55) appear in the array factor. The grating lobe plot for equilateral triangular spacing is shown in Figure 2.23. The first set of six grating lobes appear in real space when $d_x = 2/\sqrt{3}\lambda$ which is larger than the minimum spacing for grating lobes for rectangular element spacing. Grating lobes enter real space at $\theta = 90°$ when $\varphi = 30°, 90°, 150°, 210°, 270°, 330°$. The element spacing in those directions is $2d_y$. Since $d_y = d_x \sqrt{3}/2$ for an equilateral triangle, then in those six directions

$$
d_y = d_x \sqrt{3}/2 = 2/\sqrt{3}\lambda \sqrt{3}/2 = \lambda
\tag{2.57}
$$

and grating lobes appear. Thus, we would expect grating lobe peaks to enter real space when the element spacing between elements in any row (x direction) to be $d_x = 2/\sqrt{3}\lambda = 1.15\lambda$. Now grating lobes at $\varphi = 0°, 60°, 120°, 180°, 240°$, and $300°$ due to element spacings in the x direction should occur when $d = 2.0\lambda$ as verified by Figure 2.23.

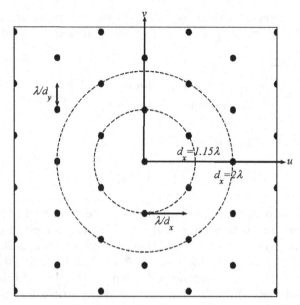

Figure 2.23. Grating lobe plot for triangular spacing.

Triangular spacing not only delays the appearance of grating lobes but allows larger elements in the array. The maximum area that an element can occupy in rectangular spacing is $d_x d_y$ while for triangular spacing it is $2d_x^2/\sqrt{3}$. As a result, an array with triangular spacing has 86.6% fewer elements than the same array with a square lattice. At least from a throretical point of view, triangular spacing is superior to rectangular spacing. Practical considerations, such as the feed network, may dictate the need for rectangular spacing.

Grating lobes appear in directions where the periodic sampling is one wavelength or more. By controlling the sampling in certain directions through varying Δ_m, d_x, and d_y in (2.32), the grating lobes can be moved to a direction that can accommodate limited scanning without bringing the grating lobes into real space [9]. Another approach is to randomize the Δ_m, so there is no periodic spacing in any direction.

The element lattice that samples the received signal occurs inside a defined perimeter or shape. The three most common planar array shapes are rectangular (including square), elliptical (including circular), and hexagonal. Figure 2.24 has examples of hexagonal, circular, elliptical, and rectangular arrays. Other shapes, including fractal, have also been used. Oftentimes, the shape must conform to the surface area available. For instance, the AN/APG-77 radar antenna shown in Figure 2.25 fits inside the nose cone of a USAF F-22A. It has 1500 elements arranged in an irregular shape. Grating lobes are a function of the periodic spacing of the elements and not the array shape. Directivity of a array factor is a function of array area but not shape. The shape does

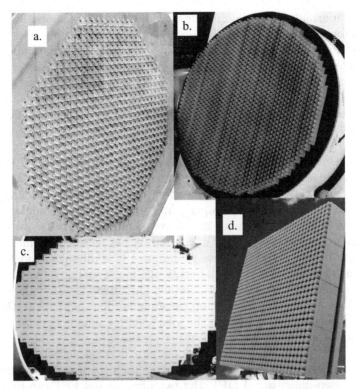

Figure 2.24. Some planar array shapes: a. hexagonal b. circular c. elliptical d. rectangular. (Courtesy of Northrop Grumman and available at the National Electronics Museum.)

Figure 2.25. AN/APG-77 radar array. (Courtesy of Northrop Grumman and available at the National Electronics Museum.)

play an important role in determining beamwidth and sidelobe level of the array factor, however.

The element locations inside any shaped boundary can be found by first selecting an element lattice (rectangular or triangular) that is larger than the desired array. Next, overlay the array shape on the element lattice. Elements in the lattice that are outside of the shape are discarded. Certain element configurations can be generated without this process, such as rectangular arrays and fractal arrays.

Example. Compare the uniform array factors for a square, circle, and hexagon-shaped planar array with rectangular element spacings of $d_x = d_y = 0.5\lambda$.

The element layouts and corresponding uniform array factors appear in Figure 2.26 to Figure 2.31. The square array has the highest sidelobes. They only occur in two directions of ϕ, however. Outside of those two cuts, the sidelobes are extremely low. The circular array has nearly constant sidelobes as a function of ϕ, especially close to the main beam. Its peak sidelobes are lower than those of the rectangular array. The hexagonal array has its highest sidelobes along four cuts in the ϕ direction. Directivity is nearly the same for all the shapes, since they have about the same number of elements. Table 2.1 lists the directivity and maximum sidelobe level for square, circular, and hexagonal arrays in Figures 2.26–2.31.

When the elements lie on a rectangular grid, then the array factor takes the form of a two-dimensional DFT. The two-dimensional FFT can be used to

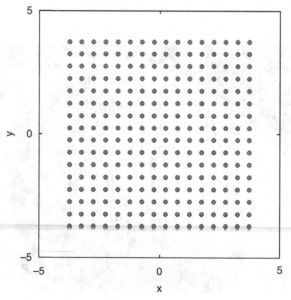

Figure 2.26. Square array with rectangular spacing.

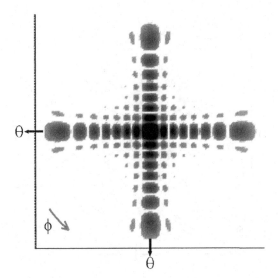

Figure 2.27. Array factor for the square array.

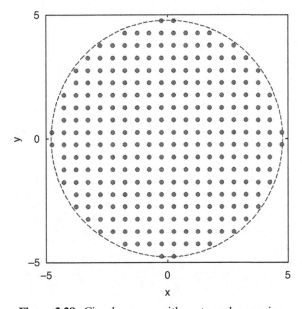

Figure 2.28. Circular array with rectangular spacing.

calculate an array factor [10]. Zero padding is necessary in both rows and columns of the weight matrix in order to get sufficient sampling. In this case, both u and v are a function of θ and ϕ, so there is not a nice relationship between the Fourier transform variable and an angle as in the case of a linear array.

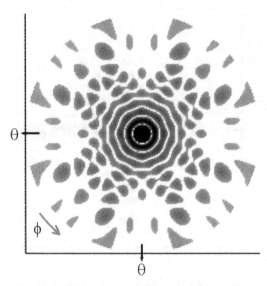

Figure 2.29. Array factor for the circular array.

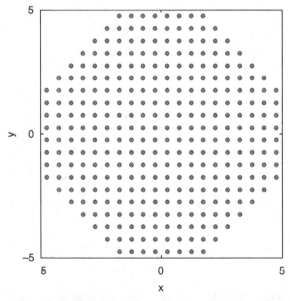

Figure 2.30. Hexagon array with rectangular spacing.

Example. Calculate the array factor of the triangular array in Figure 2.32. The array has 36 elements on a rectangular grid with $d_x = 0.5\lambda$ and $d_y = 1.0\lambda$.

A DFT multiplies a row vector of N element weights with a $N \times N_{ang}$ phase matrix. This results in about $N \times N_{ang}$ operations. To invoke an FFT, the weight

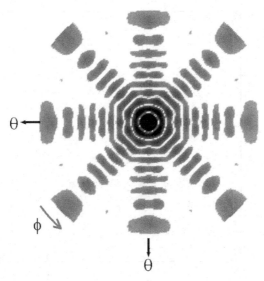

Figure 2.31. Array factor for the hexagon array.

TABLE 2.1. Characteristics Due to Planar Array Shapes when Elements Are in a Square Grid

Shape	Directivity	N	Max Sidelobe Level (dB)
Square	27.72 dB	$14 \times 14 = 196$	−13.1
Circle	27.75 dB	197	−17
Hexagon	27.77 dB	196	−15.4

matrix is filled with zeros in order to make the weight matrix rectangular. Next, the weight matrix is padded with zeros in order to sufficiently interpolate the array factor so that it looks smooth when graphed. Figure 2.33 is the array factor calculated using the FFT. The array factor points outside the circle are ignored, because they are associated with physically impossible angles. They account for $(1-\pi)/4 \times 100\%$ or 21.5% of all the points calculated.

2.5. ARRAY BANDWIDTH

The array bandwidth is determined by a number of factors:

1. Bandwidth of the elements in the array
2. Element spacing
3. Maximum steering angle
4. Array size

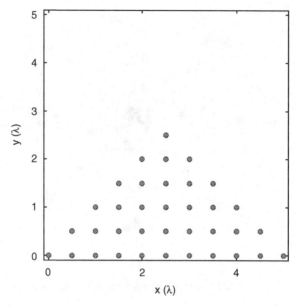

Figure 2.32. Weight matrix for triangular array for $N = 36$, $d_x = 0.5\lambda$, and $d_y = 1.0\lambda$.

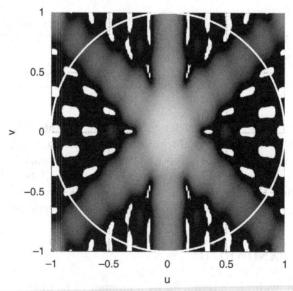

Figure 2.33. Array factor calculated using an FFT. The points outside the circle should be ignored.

The bandwidth of the elements will be discussed in Chapter 4. The other three factors can be examined using isotropic point sources. Increasing the frequency increases the sampling in terms of wavelength between elements. As a result, the grating lobes appear sooner for higher frequencies than for lower frequencies as indicated by (2.54) and (2.56). Element spacing and grating lobe formation sets an upper frequency limit to the bandwidth.

As presented earlier, when a beam is scanned, a change in frequency causes a change in main beam pointing direction. The main beam shift as a function of frequency is most pronounced at the highest scan angle. At frequency f, the main beam squints (moves) from $u_s = \sin\theta_s$ at $f_0 = $ center frequency to angle $u_{\text{squint}} = \sin\theta_{\text{squint}}$ according to

$$kdu_{\text{squint}} = k_0 du_s$$

$$u_{\text{squint}} = u_s \frac{k}{k_0} = u_s \frac{f_0}{f} \tag{2.58}$$

Note that u_{squint} gets smaller for frequencies higher than f_0 and smaller for frequencies less than f_0. If f_{hi} and f_{lo} define the upper and lower limits of the bandwidth, then for a constant phase shift between element (steering phase)

$$\delta_s = \frac{2\pi}{\lambda_{\text{lo}}} du_{\text{lo}} = \frac{2\pi}{\lambda_{\text{hi}}} du_{\text{hi}} \tag{2.59}$$

The angular difference between the upper and lower frequencies is given by

$$u_{\text{lo}} - u_{\text{hi}} = u_s \left(\frac{\lambda_{\text{hi}} - \lambda_{\text{lo}}}{\lambda_0} \right) = u_s f_0 \left(\frac{1}{f_{\text{hi}}} - \frac{1}{f_{\text{lo}}} \right) \tag{2.60}$$

Figure 2.34 is a plot of the steering angle as a function of frequency for $\theta_s = 10°, 20°, 30°, 40°, 50°, 50°$ at f_{lo}. The change in beam pointing direction over the same frequency range increases as θ_s increases.

Increasing the size of an array decreases its bandwidth. If the bandwidth is bound by a 3-dB reduction in the main beam, then the bandwidth is calculated by

$$BW = \frac{\theta_{3\,\text{dB}}}{u_s} < \frac{\lambda}{N du_s} \tag{2.61}$$

Figure 2.35 is a plot of the array bandwidth as a function of the center frequency for array sizes with $N = 20, 40, 60,$ and 100 for a uniform array with $d = 0.5\lambda$. A good approximation for the bandwidth is given by [1]

$$BW(\%) = 2\theta_{3\,\text{dB}}(\text{degrees}) \tag{2.62}$$

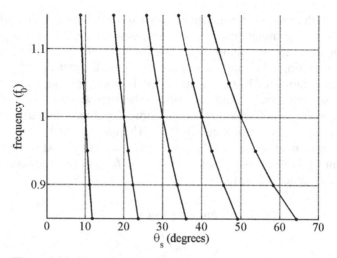

Figure 2.34. Plot of steering angle as a function of frequency.

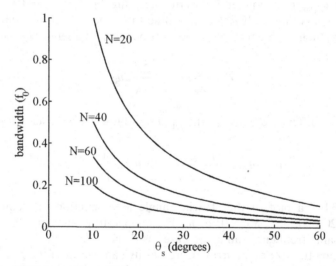

Figure 2.35. Array bandwidth as a function of steering angle and array size.

2.6. DIRECTIVITY

To find the directivity of an array, substitute the array factor into the equation for directivity in Chapter 1.

$$D = \frac{4\pi |AF_{max}|^2}{\int\limits_0^{2\pi}\int\limits_0^{\pi} |AF|^2 \sin\theta\, d\theta\, d\phi} \tag{2.63}$$

A linear array with arbitrary element spacing along the z axis is symmetric with respect to ϕ, so the integral with respect to ϕ in the denominator reduces to 2π, and the directivity becomes

$$D = \frac{2\left|\sum\limits_{n=1}^{N} w_n\right|^2}{\int\limits_{0}^{\pi}\left[\sum\limits_{n=1}^{N} w_n e^{jkz_n(\cos\theta-\cos\theta_s)}\right]\left[\sum\limits_{m=1}^{N} w_m^* e^{-jkz_m(\cos\theta-\cos\theta_s)}\right]\sin\theta\,d\theta}$$

$$= \frac{\left|\sum\limits_{n=1}^{N} w_n\right|^2}{\sum\limits_{n=1}^{N}\sum\limits_{m=1}^{N} w_n w_m^* e^{jk(z_m-z_n)\cos\theta_s}\int\limits_{0}^{\pi/2} e^{jk(z_n-z_m)\cos\theta}\sin\theta\,d\theta} \qquad (2.64)$$

The integral in (2.64) is easy to numerically compute. If the elements have a constant spacing, d, then an analytical form exists [20]:

$$D = \frac{\left|\sum\limits_{n=1}^{N} w_n\right|^2}{\sum\limits_{n=1}^{N}\sum\limits_{m=1}^{N} w_n w_m^* e^{jkd(m-n)\cos\theta_s}\operatorname{sinc}[(n-m)kd]} \qquad (2.65)$$

where $\operatorname{sinc}(x) = \sin(x)/x$. If the elements are spaced 0.5λ apart, then the directivity formula simplifies to

$$D = \frac{\left|\sum\limits_{n=1}^{N} w_n\right|^2}{\sum\limits_{n=1}^{N} |w_n|^2} \qquad (2.66)$$

Uniform arrays with constant spacing have a directivity of

$$D = \frac{N^2}{\sum\limits_{n=1}^{N}\sum\limits_{m=1}^{N} e^{jkd(m-n)\cos\theta_s}\operatorname{sinc}[(n-m)kd]}$$

$$= \frac{N^2}{N+2\sum\limits_{n=1}^{N-1}(N-n)\operatorname{sinc}(nkd)\cos(nkd\cos\theta_s)} \qquad (2.67)$$

When $d = 0.5\lambda$, the directivity becomes

$$D \simeq N \qquad (2.68)$$

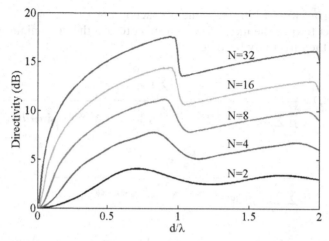

Figure 2.36. Directivity of a uniform linear array as a function of element spacing.

Graphs of the directivity as a function of element spacing for several values of N are shown in Figure 2.36. The directivity increases until a grating lobe appears (d is a multiple of λ), and then it sharply decreases. This decrease is due to an increase in the denominator while the maximum value of AF in the numerator remains the same. The decrease in directivity due to the grating lobe becomes more dramatic as the number of elements increases, because the main beam and grating lobes have narrower beamwidths: A small change in θ produces a large change in AF.

Weighting the elements in the array reduces the directivity and efficiency of the array. A weighted aperture collects less total electromagnetic waves than a uniformly weighted aperture, so it is less efficient. Taper efficiency (also called illumination or aperture efficiency) is the ratio of the directivity of the tapered array to that of a uniform linear array or

$$\eta_T = \frac{D_{\text{taper}}}{D_{\text{uniform}}} = \frac{\left| \sum\limits_{n=1}^{N} w_n \right|^2 \sum\limits_{n=1}^{N} \sum\limits_{m=1}^{N} \text{sinc}[(n-m)kd]}{N^2 \sum\limits_{n=1}^{N} \sum\limits_{m=1}^{N} w_n w_m^* \text{sinc}[(n-m)kd]} \tag{2.69}$$

When $d = 0.5\lambda$, this equation simplifies to

$$\eta_T = \frac{\left| \sum\limits_{n=1}^{N} w_n \right|^2}{N \sum\limits_{n=1}^{N} |w_n|^2} \tag{2.70}$$

Taper efficiency is an important figure of merit when evaluating low sidelobe tapers.

The directivity for a planar array is found by substituting (2.4) into (2.63) to get

$$D = \frac{4\pi \left| \sum_{n=1}^{N} w_n \right|^2}{\int_0^{2\pi} \int_0^{\pi} \sum_{n=1}^{N} w_n e^{jk[x_n(u-u_s)+y_n(v-v_s)]} \sum_{m=1}^{N} w_m^* e^{-jk[x_m(u-u_s)+y_m(v-v_s)]} \sin\theta\, d\theta\, d\phi} \quad (2.71)$$

It is unusual to have a planar array radiate out the front and back, so the limit of integration for θ only goes from 0° to 90°, because the pattern is zero from θ equals 90° to 180°. Note that this assumption was not made for the linear array. Some terms in the denominator can be pulled outside the integrals to get

$$D = \frac{4\pi \left| \sum_{n=1}^{N} w_n \right|^2}{\sum_{n=1}^{N} \sum_{m=1}^{N} w_m^* w_n e^{jk[(x_n-x_m)u_s+(y_n-y_m)v_s]} \int_0^{2\pi} \int_0^{\pi/2} e^{jk[(x_n-x_m)u+(y_n-y_m)v]} \sin\theta\, d\theta\, d\phi} \quad (2.72)$$

The integrals in the denominator can be evaluated analytically for certain element spacings, such as rectangular. Directivity formulas tend to be quite complicated and severely restricted to the element layout [20]. Performing the numerical integration in (2.72) is relatively easy and has the advantage of being geometry-independent.

Many approximate formulas exist for quickly estimating the directivity of an array. First, the directivity can be calculated from the projected area of the array (A_p) as long as the element spacing is not much larger than 0.5λ:

$$D = \frac{4\pi A_p}{\lambda^2} \eta_t \quad (2.73)$$

If the array has an irregular shape, then assume that each element occupies an area equal to $d_x d_y$ for rectangular spacing. Thus, an N element array has an approximate directivity given by

$$D = \eta_t N d_x d_y / \lambda^2 \quad (2.74)$$

Separable apertures are rectangular planar arrays whose array factors can be written as the product of two linear arrays and have an approximate directivity of

$$D = \pi D_x D_y \quad (2.75)$$

where D_x is the directivity of the linear array in the x direction and D_y is the directivity of the linear array in the y direction. The directivity can also be estimated if the 3-dB beamwidths are known in orthogonal directions.

$$D = \frac{4\pi}{\theta_{3\,dB\phi=0}^{radians}\,\theta_{3\,dB\phi=\pi/2}^{radians}} = \frac{32{,}400}{\theta_{3\,dB\phi=0}^{\circ}\,\theta_{3\,dB\phi=90}^{\circ}} \tag{2.76}$$

where $\theta_{3\,dB\phi=0}^{\circ}$ is the 3-dB beamwidth in degrees at $\phi = 0°$, $\theta_{3\,dB\phi=90}^{\circ}$ is the 3-dB beamwidth in degrees at $\phi = 90°$, $\theta_{3\,dB\phi=0}^{radians}$ is the 3-dB beamwidth in radians at $\phi = 0$, and $\theta_{3\,dB\phi=\pi/2}^{radians}$ is the 3-dB beamwidth in radians at $\phi = \pi/2$. These formulas result from approximating the 3-dB beamwidth by $\lambda/(Nd)$. When the array scans its beam, then the directivity decreases due to the decrease in the projected area of the array.

$$D(\theta_s) = D\cos\theta_s \tag{2.77}$$

Example. Find the directivity of a rectangular planar array with $N_x = 6$, $N_y = 10$, $d_x = 0.5$, and $d_y = 0.7$ using (2.72) and (2.73) for both a uniform aperture and an aperture that is uniform in the y direction and has the amplitude weights [0.541 0.777 1 1 0.777 0.541] in the x direction.

The taper efficiency for this array is 0.944.

Equation for D	Uniform	x-Amplitude Taper
(2.72)	23.8 dB	23.6
(2.73)	24.2 dB	24.0

Since grating lobes can enter space from more than one ϕ direction, the directivity of a planar array has more abrupt changes as the element spacing increases than a linear array. Figure 2.37 has plots of directivity versus element spacing in λ for a square array of 10 by 10 elements with square and hexagonal element lattices. These curves relate to the grating lobe plots in Figures 2.22 and 2.23. The square spacing has its first minimum at $d = 1.0\lambda$ and the second at $d = 1.4\lambda$. The hexagonal spacing has its first minimum at $d = 1.15\lambda$ and the second at $d = 2.0\lambda$.

2.7. AMPLITUDE TAPERS

The element weights control the directivity and sidelobe level of the array factor. Low-sidelobe amplitude tapers have high amplitude weights in the center of the array. The weights generally decrease from the center to the edges. Some examples are shown in Table 2.2. The results in Table 2.2 were obtained through numerical simulations (also see reference 11). The sidelobe levels and taper efficiencies assume that Nd is large. As the taper efficiency decreases, the 3-dB beamwidth increases and sidelobe levels decrease.

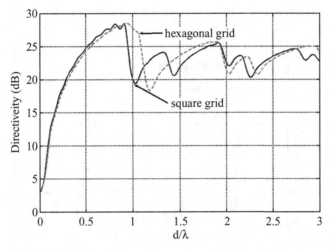

Figure 2.37. Directivity of planar arrays as a function of element spacing.

TABLE 2.2. Properties of Some Simple Low-Sidelobe Amplitude Tapers for a Linear Array with Large Nd [11]

w_n	First Sidelobe Level (dB)	η_t	$\theta_{\text{null BW}}$	θ_{3dB}
1.0	−13.3	1.0	$\dfrac{114.5°}{Nd/\lambda}$	$\dfrac{50.8°}{Nd/\lambda}$
$\left\lvert\dfrac{2n-1}{N}-1\right\rvert$	−26.5	0.75	$\dfrac{229.2°}{Nd/\lambda}$	$\dfrac{73.1°}{Nd/\lambda}$
$\left\lvert\left(\dfrac{2n-1}{N}\right)^2-1\right\rvert$	−15.8	0.83	$\dfrac{163.8°}{Nd/\lambda}$	$\dfrac{66.2°}{Nd/\lambda}$
$\cos\left[\left(\dfrac{2n-1}{N}-1\right)\dfrac{\pi}{2}\right]$	−23.0	0.81	$\dfrac{172.0°}{Nd/\lambda}$	$\dfrac{68.2°}{Nd/\lambda}$
$\cos^2\left[\left(\dfrac{2n-1}{N}-1\right)\dfrac{\pi}{2}\right]$	−31.5	0.67	$\dfrac{229.3°}{Nd/\lambda}$	$\dfrac{82.6°}{Nd/\lambda}$
$\cos^3\left[\left(\dfrac{2n-1}{N}-1\right)\dfrac{\pi}{2}\right]$	−39.3	0.58	$\dfrac{286.3°}{Nd/\lambda}$	$\dfrac{95.0°}{Nd/\lambda}$
$\cos^4\left[\left(\dfrac{2n-1}{N}-1\right)\dfrac{\pi}{2}\right]$	−46.7	0.51	$\dfrac{344.0°}{Nd/\lambda}$	$\dfrac{106.2°}{Nd/\lambda}$
$0.5+0.5\cos\left[\left(\dfrac{2n-1}{N}-1\right)\dfrac{\pi}{2}\right]$	−17.6	0.97	$\dfrac{132.3°}{Nd/\lambda}$	$\dfrac{56.0°}{Nd/\lambda}$
$0.33+0.67\cos\left[\left(\dfrac{2n-1}{N}-1\right)\dfrac{\pi}{2}\right]$	−19.8	0.93	$\dfrac{142.4°}{Nd/\lambda}$	$\dfrac{58.8°}{Nd/\lambda}$

Example. Plot the array factors for a 20-element $d = 0.5\lambda$ array with triangular, cosine, and cosine-squared amplitude tapers.

Figure 2.38 shows the three low sidelobe array factors superimposed. The beamwidths and first sidelobe levels are shown in Table 2.3.

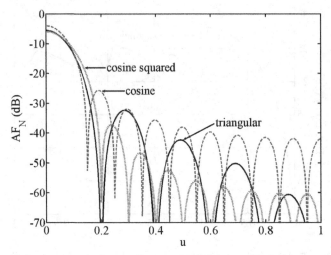

Figure 2.38. Array factors for a 20-element $d = 0.5\lambda$ array with triangular, cosine; and cosine-squared amplitude tapers.

TABLE 2.3. Array Factor Characteristics for Three Low-Sidelobe Amplitude Tapers

Taper	First Sidelobe Level (dB)	θ_{3dB}	$\theta_{null\ BW}$
Cosine	–21.7	7°	17.3°
Cosine-squared	–31.3	8.4°	23°
Triangular	–26.4	7.4°	23°

2.8. z TRANSFORM OF THE ARRAY FACTOR

The z transform converts the linear array factor into a polynomial using the substitution [12]

$$z = e^{j\psi} \tag{2.78}$$

Substitute z into (2.17) to get

$$AF = \sum_{n=1}^{N} w_n z^{(n-1)} \tag{2.79}$$

It is convenient to write the array factor in (2.79) as

$$AF = w_N \left(z^{N-1} + \frac{w_{N-1}}{w_N} z^{N-2} + \frac{w_{N-2}}{w_N} z^{N-3} + \cdots + \frac{w_1}{w_N} \right) \tag{2.80}$$

Factoring the above polynomial yields

$$\text{AF} = w_N(z - z_1)(z - z_2)\cdots(z - z_{N-1}) \tag{2.81}$$

Each root $(z = z_n)$ corresponds to a null in the array pattern. The roots have a magnitude of one and phase of ψ_n. A 4-element uniform array with $d = 0.5\lambda$ has the following polynomial form:

$$z^3 + z^2 + z + 1 = (z+1)(z+j)(z-j) \tag{2.82}$$

Its roots are graphed on the unit circle in Figure 2.39, and the array factor is shown in Figure 2.40.

The magnitude of the array factor between zeros relates to the angular separation of the zeros. Closely spaced zeros have small lobes between them, while widely spaced zeros have large lobes between them. The uniform array example has zeros 1 and 2 and zeros 2 and 3 closely spaced, while zeros 1 and 3 are widely spaced. The array factor between the closely spaced zeros are sidelobes, and the array factor between the widely spaced zeros is the main beam. Thus, the sidelobe levels of the uniform array may be lowered by moving the zeros on the unit circle as shown in Figure 2.41. For instance, moving the zeros of the 4-element uniform array closer to the negative real axis lowers the pattern's sidelobe levels and widens the main beam. Null locations at $\psi = \pm120°$ instead of $\psi = \pm90°$ result in

$$\text{AF} = (z+1)\left(z - e^{j0.67\pi}\right)\left(z - e^{-j0.67\pi}\right) = z^3 + 2z^2 + 2z + 1 \tag{2.83}$$

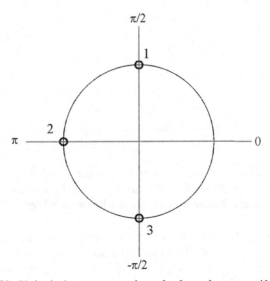

Figure 2.39. Unit circle representation of a four element uniform array.

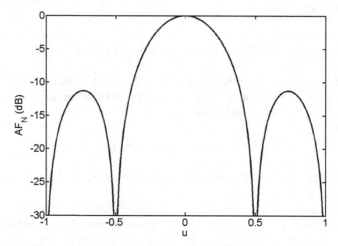

Figure 2.40. Array factor of a four element uniform array with $d = 0.5\lambda$.

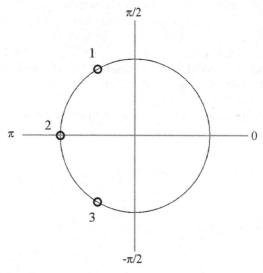

Figure 2.41. The zeros are moved closer together on the unit circle.

The amplitude weights for the array elements are 1, 2, 2, and 1. These weights produce the normalized low-sidelobe array factor shown in Figure 2.42. Note that the sidelobes are lower and the main beam is wider than those of the uniform array.

The examples so far only considered zeros lying on the unit circle. What happens to the array factor when the zeros move off the unit circle. The four element array has roots at $z = -1$ and $\pm j$ as indicated by the zeros labeled with a 1 in Figure 2.43. Changing the roots to $z = -1.2$ and $\pm 1.2j$ moves them

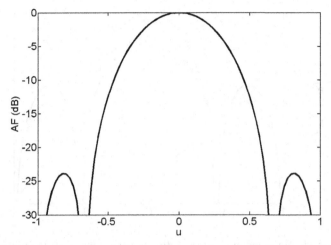

Figure 2.42. Moving the zeros closer together on the unit circle lowers the sidelobes and expands the main beam.

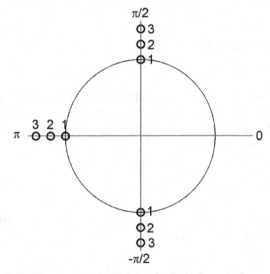

Figure 2.43. Zeros moved off the unit circle for a 4-element array.

a radial distance of 0.2 outside of the unit circle as shown by the zeros labeled with a 2 in Figure 2.43. The corresponding array weights are real since the roots are complex conjugate pairs and not symmetric: $w = [0.5787 \ 0.6944 \ 0.8333 \ 1.0]$. Moving the roots off the unit circle by an additional 0.2 results in roots at $z = -1.4$ and $\pm 1.4j$ which are labeled by a 3. The corresponding array weights are real since the roots are complex conjugate pairs and not

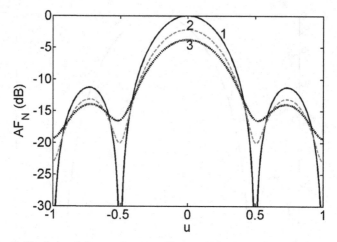

Figure 2.44. Array factors corresponding to zeros moved off the unit circle.

symmetric: $w = [0.3644\ 0.5102\ 0.7143\ 1.0]$. The effects of moving the zeros is apparent on the array factor shown in Figure 2.44 and include

- Decreased directivity and efficiency
- Increased relative sidelobe levels
- Filled-in nulls

Moving zeros inside the unit circle produces similar effects to the array factor. For instance, moving the roots to $z = -0.8$ and $\pm 0.8j$ results in the exact same amplitude weights as when $z = -1.2$ and $\pm 1.2j$, and the array factor looks like the plot labeled 2 in Figure 2.44.

2.9. CIRCULAR ARRAYS

A circular or ring array with elements lying on a circle of radius r_c in the x–y plane has an array factor given by

$$
\begin{aligned}
\mathrm{AF}_{\mathrm{cir}} &= \sum_{n=1}^{N} w_n e^{jk(r_c \cos\phi_n \sin\theta \cos\phi + r_c \sin\phi_n \sin\theta \sin\phi)} \\
&= \sum_{n=1}^{N} w_n e^{jkr_c \sin\theta(\cos\phi_n \cos\phi + \sin\phi_n \sin\phi)} \\
&= \sum_{n=1}^{N} w_n e^{jkr_c \sin\theta \cos(\phi-\phi_n)}
\end{aligned}
\tag{2.84}
$$

where ϕ_n is the angular location of element n. The beam is steered by applying a phase at element n given by

$$e^{-jkr_c \sin\theta_s \cos(\phi_s-\phi_n)}$$ (2.85)

Normally, the elements are equally spaced around the circle, so they are separated by an angle

$$\Delta\phi = \frac{2\pi}{N}$$ (2.86)

As a result, the radius of the circle and the arc distance between adjacent elements is given by

$$d = 2r_c \sin\left(\frac{\pi}{N}\right)$$ (2.87)

If r_c is known, then d can be found, or if d is known then r_c can be found. If the array weights are uniform, then the array factor can be written as [13]

$$AF = N \sum_{n=-\infty}^{\infty} J_{nN}(kr)e^{jnN(\pi/2-\xi)}$$ (2.88)

where $\xi = \tan^{-1}\left(\frac{v-v_s}{u-u_s}\right)$ and J_n is a Bessel function of order n. The principal component is the $n = 0$ term, and all other terms are referred to as residuals. For large arrays, the principal component is the dominant term, and the residuals can be ignored. Thus, for a large radius, the array factor is approximately

$$AF = NJ_0(kr_c)$$ (2.89)

If several circular arrays with different radii share a common center, then the resulting planar array is known as a concentric ring array [14,15]. Figure 2.45 is a diagram of a concentric ring array with ring n having N_n elements and a radius of r_n. The physical distance between elements on ring n is constant and given by d_n. Ring arrays are either designed to have a main beam at $\theta = 90°$ and scan only in azimuth or have a main beam at $\theta = 0°$ and scan in azimuth and elevation.

The array factor for the concentric ring array with a single element at the center (Figure 2.45) is given by

$$AF = 1 + \sum_{n=1}^{N_r} w_n \sum_{m=1}^{N_n} e^{jkr_n[\cos\phi_m u+\sin\phi_m v]}$$ (2.90)

where N_n is the number of elements in ring n, N_r is the number of rings, w_n represents elements weights for ring n, r_n is the radius of ring n, (x_n, y_n) is the

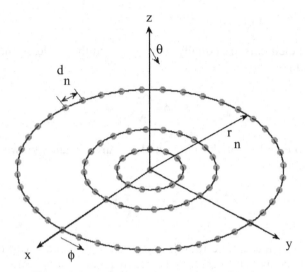

Figure 2.45. Diagram of a concentric ring array.

location of element n, $x_n = r_n \cos \phi_m$, $y_n = r_n \sin \phi_m$, and $\phi_m = 2\pi(m-1)/N_n$. In this formula, all the elements in the same ring have the same weight.

The array factor for a circular array is often written in terms of Bessel functions as shown in (2.88). If each ring is represented by (2.88), then (2.90) can be rewritten as [13]

$$AF = 1 + \sum_{n=1}^{N_r} w_n N_n \sum_{m=-\infty}^{\infty} J_{mN_n}(kr_n \sin\theta) e^{jmN_n(\pi/2-\phi)} \qquad (2.91)$$

Assuming that all the radii are large, then the array factor is independent of ϕ and only the principal component terms for each ring remain.

$$AF = 1 + \sum_{n=1}^{N_r} w_n N_n J_0(kr_n \sin\theta) \qquad (2.92)$$

Figure 2.46 is a diagram of a 279-element concentric ring array. There are nine rings with $r_n = n\lambda/2$ and $d_n = \lambda/2$. The number of equally spaced elements in ring n is given by

$$N_n = 2\pi r_n/d_n = 2\pi n \qquad (2.93)$$

Since the number of elements must be an integer, the value in (2.93) must be rounded up or down. To keep $d \geq \lambda/2$, the digits to the right of the decimal point are dropped. Table 2.4 lists the ring spacing and number of elements in each ring for a uniform concentric ring array with nine rings as shown in Figure 2.46.

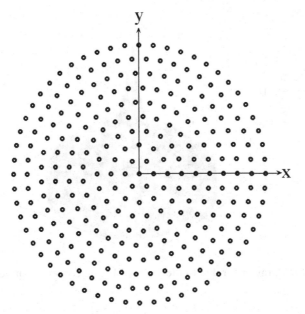

Figure 2.46. Concentric ring array with nine rings spaced $\lambda/2$ apart and having $d_n \simeq \lambda/2$.

TABLE 2.4. Ring Radius and Number of Elements per Ring for a 9-Ring Uniform Concentric Ring Array

n:	1	2	3	4	5	6	7	8	9
r_n (λ):	0.5	1.0	1.5	2.0	2.5	3.0	3.5	4.0	4.5
N_n:	6	12	18	25	31	37	43	50	56

A uniform array has equal element spacing and weighting. For a uniform concentric ring array, the ring spacing, r_n, is a constant times the ring number and the spacing between elements within a ring, d_n, is approximately constant for all rings. A nine-ring concentric ring array has the array factor shown in Figure 2.47. The Bessel function behavior in (2.92) is quite evident in the array factor. It has a directivity of 29.4 dB, a peak sidelobe level of −17.4 dB, and is symmetric in ϕ. As long as the array factor is predominantly a function of θ, the maximum can be found from a slice of the array factor for a single value of ϕ.

2.10. DIRECTION FINDING ARRAYS

Direction finding with linear arrays is limited to either the θ or ϕ directions. In order to direction find in both azimuth and elevation directions, a planar array is needed. Circular arrays are also commonly used for direction finding.

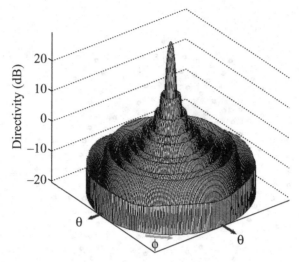

Figure 2.47. Array factor due to the array with nine concentric uniform rings.

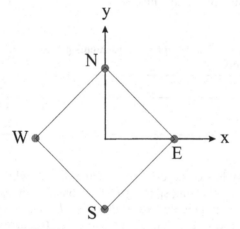

Figure 2.48. Diagram of a 4-element Adcock antenna.

2.10.1. Adcock Array

The original Adcock array has four uniformly weighted elements situated at the four corners of a square with sides less than $\lambda/2$ (Figure 2.48) [16]. It was developed to find the direction of arrival of a signal in both azimuth and elevation. The north (N) and south (S) antennas on the y axis are out of phase, and the east (E) and west (W) antennas along the x axis are out of phase as well. The N–S array has an array factor given by

$$\mathrm{AF_{NS}} = 2j\sin\left(k\frac{d}{2}\sin\theta\cos\phi\right) \tag{2.94}$$

Likewise, the array factor for the E–W array is given by

$$\mathrm{AF_{EW}} = 2j\sin\left(k\frac{d}{2}\sin\theta\sin\phi\right) \tag{2.95}$$

Sir Watson-Watt developed the principle for finding the elevation and azimuth of a source incident on an Adcock array [17]. An estimate of the tangent of the azimuth angle is the ratio of the output from the E–W array to the output of the N–S array.

$$\tan\phi \approx \frac{\mathrm{AF_{EW}}}{\mathrm{AF_{NS}}} = \frac{\sin\left(k\dfrac{d}{2}\sin\theta\sin\phi\right)}{\sin\left(k\dfrac{d}{2}\sin\theta\cos\phi\right)} \tag{2.96}$$

An estimate of the elevation angle is given by

$$\cos\theta \approx \frac{1}{kd}\sqrt[4]{\mathrm{AF_{EW}^2 + AF_{NS}^2}} \tag{2.97}$$

More accurate estimates of the arrival angles are possible by adding element pairs to the 4-element Adcock antenna on opposite sides of a circle with a center at the origin of the x–y axes.

2.10.2. Orthogonal Linear Arrays

A planar array is needed to locate sources in both azimuth and elevation. The Adcock array is the simplest version. Adding more elements to an array increases the cost of the components and computational complexity of the signal processing algorithms. As a result, direction of arrival (DOA) arrays have just enough elements to locate a desired number of sources. Orthogonal linear arrays are often used in place of fully populated arrays for direction finding [18].

The array factor for the linear array along the y axis can be written as a polynomial.

$$\mathrm{AF}_y = 1 + w_{y1}z_y + w_{y2}z_y^2 + \cdots \tag{2.98}$$

where $z_y = e^{j\psi_y}$. A similar polynomial for AF_x is written in terms of $z_x = e^{j\psi_x}$. The polynomials for orthogonal arrays along the x and y axes can be factored to find the zeros:

$$\left(z_x - z_{x1}\right)\left(z_x - z_{x2}\right)\cdots\left(z_x - z_{x(N_x-1)}\right) = 0 \qquad (2.99)$$

$$\left(z_y - z_{y1}\right)\left(z_y - z_{y2}\right)\cdots\left(z_y - z_{y(N_y-1)}\right) = 0 \qquad (2.100)$$

If several signals are incident upon the array, then the weights are adjusted until the array output is minimized. Factoring the array factor polynomial yields the polynomial zeros, hence the location of the nulls and the incident signals. The azimuth angle of a null is found by [19]

$$\frac{\psi_{x1}}{\psi_{y1}} = \frac{kd_x \sin\theta_1 \cos\phi_1}{kd_y \sin\theta_1 \sin\phi_1} = \frac{d_x}{d_y}\tan\phi_1 \qquad (2.101)$$

or

$$\phi_1 = \tan^{-1}\left(\frac{d_y\psi_{x1}}{d_x\psi_{y1}}\right) \qquad (2.102)$$

The elevation angle is found from

$$\psi_{x1}^2 + \psi_{y1}^2 = \left(kd_x \sin\theta_1 \cos\phi_1\right)^2 + \left(kd_y \sin\theta_1 \sin\phi_1\right)^2$$
$$= \left(k\sin\theta_1\right)^2\left(d_x^2 \cos^2\phi_1 + d_y^2 \sin^2\phi\right) \qquad (2.103)$$

Solving for θ_1 yields

$$\theta_1 = \sin^{-1}\left(\frac{1}{k}\sqrt{\frac{\psi_{x1}^2 + \psi_{y1}^2}{d_x^2 \cos^2\phi_1 + d_y^2 \sin^2\phi}}\right) \qquad (2.104)$$

Once the zeros of the array factor polynomials are known, then the source locations in (θ_1, ϕ_1) can be found.

2.11. SUBARRAYS

Large phased array antennas are often divided into many smaller subarrays. The array panel in Figure 2.49 consists of 5 rows and 8 columns of 2 by 2 (quad) elements in a subarray. Control electronics are mounted on back of the subarrays. An artist's concept of the fully deployed antenna appears in Figure 2.50. The antenna operates from 1.215 to 1.3 GHz. Element spacing is 12.7 cm or 0.55λ at 1.3 GHz. Another example of a subarrayed antenna is shown in Figure 2.51. This array operates from 2.2 to 2.3 GHz. It has 36 subarray with 4 by 8 elements per subarray. The elements lie in a square lattice with $d = 6.6$ cm.

Subarrays are modular and allow amplitude and phase weighting to occur at the subarray outputs or ports as well as at the individual elements.

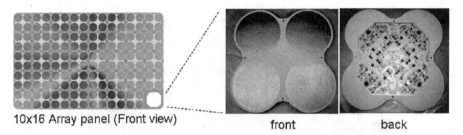

10x16 Array panel (Front view)   front   back

Figure 2.49. Array panel of 10 by 16 elements or 5 by 8 quad modules. (Courtesy of Ball Aerospace & Technologies Corp.)

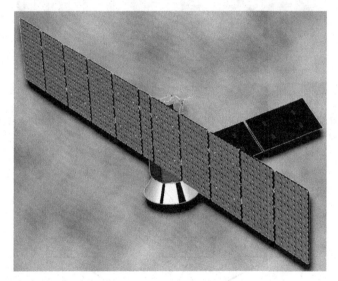

Figure 2.50. Artist concept of the L band array with 11 panels. (Courtesy of Ball Aerospace & Technologies Corp.)

Considerable savings is possible if amplitude and phase weights are only located at the subarray ports and all the element weights are uniform. Such an array appears in Figure 2.52. Unfortunately, this savings comes at an unacceptable performance cost of introducing grating lobes into the array factor.

The effective weight of an element in an array is the product of the element weight times its corresponding subarray weight, or

$$w_n = a_{mn} \times b_m \qquad (2.105)$$

where a_{mn} is the weight of element n which is in subarray m, b_m is the weight at subarray port m, N_s is the number of subarrays, and N_e is the number of

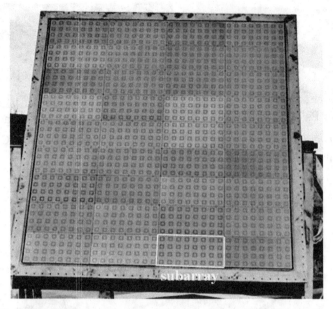

Figure 2.51. Planar array with 36 subarrays. (Courtesy of Ball Aerospace & Technologies Corp.)

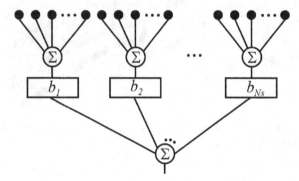

Figure 2.52. Linear array with uniform element weights and weights at the subarray ports.

elements in a subarray. Subarray weighting alone assumes that $a_{mn} = 1$ and for N_e constant assumes that array weights are represented as

$$w = [\underbrace{b_1 b_1 \cdots b_1}_{N_e} \; \underbrace{b_2 b_2 \cdots b_2}_{N_e} \cdots \underbrace{b_{N_s} b_{N_s} \cdots b_{N_s}}_{N_e}] \tag{2.106}$$

where w is a vector containing all the w_n. Substituting the element weights into the equation for the array factor results in the following simplification:

$$AF = \frac{\sin\left(\dfrac{N_e kd\sin\theta}{2}\right)}{\sin\left(\dfrac{kd\sin\theta}{2}\right)} \sum_{m=1}^{N_s} b_m e^{jk[m-(N_s+1)/2]dN_e\sin\theta}$$

$$= AF_e \times AF_s \qquad\qquad (2.107)$$

where AF_e is the array factor due to a single uniform subarray and AF_s is the array factor due to subarray weighting alone. Equation (2.107) is the product of a uniform array factor due to the elements in one subarray and the array factor due to the weighted sum of isotropic point sources at the phase centers of all the subarrays. Usually, $dN_e > \lambda$, so grating lobes appear in AF_s at [21]

$$\theta_g = \sin^{-1}\left(\frac{g\lambda}{dN_e}\right) \qquad \text{for } g = 1, 2, \dots \quad \text{and} \quad g \le \frac{d}{\lambda}N_e \qquad (2.108)$$

Since AF_e has nulls at the same angles as the grating lobes appear, the grating lobes have nulls in their centers. As a result, the grating lobes do not get as large as those associated with large element spacing. An approximate expression for the grating lobe heights relative to the peak of the main beam is [21]

$$|AF(\theta_g)|^2 \approx \frac{B_b^2}{N_T^2 \sin^2\left(\dfrac{g\pi}{N_e}\right)} \qquad\qquad (2.109)$$

where $B_b \ge 1$ is the beam broadening factor (the ratio of the 3-dB beamwidth of the weighted array to the 3-dB beamwidth of the corresponding uniform array). In general, decreasing the sidelobe level increases B_b, which in turn increases the grating lobe heights.

A 64-element array divided into 8 subarrays with a low-sidelobe taper applied to its subarray ports has the effective element weights shown in Figure 2.53. AF_e is the broad beam uniform array factor in Figure 2.54. AF_s has low sidelobes due to the subarray weights and grating lobes due to $dN_e > \lambda$. Multiplying these patterns results in the array factor shown in Figure 2.55. Except at u = 0, the peaks of AF_s occur at the nulls of AF_e. The nulls in AF_e place nulls in the grating lobe peaks. These nulls reduce the grating lobes but do not eliminate them.

Example. A 128-element array with $d = 0.5\lambda$ can be divided into several different subarray configurations. Table 2.5 shows four possible divisions with their associated maximum grating lobe locations and approximate heights when a low-sidelobe amplitude taper is applied at the subarray ports. This amplitude taper has a beam broadening factor of $B_b = 1.288$. Increasing the number of elements in a subarray causes the grating lobe to get larger and move closer to the main beam as indicated in Table 2.5.

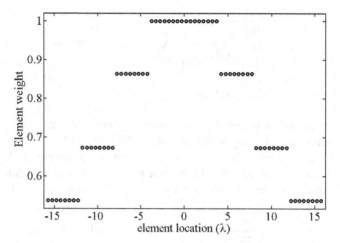

Figure 2.53. Effective element weights for a 64-element array divided into 8 subarrays with a low-sidelobe taper applied to its subarray ports.

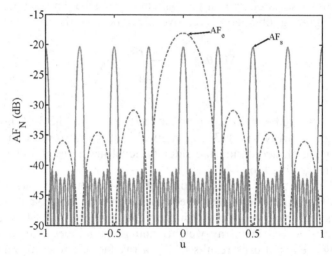

Figure 2.54. Graphs of AF_e (dashed line) and AF_s (solid line).

The beam broadening factor is a function of the subarray weighting. For most practical cases, a low-sidelobe amplitude tape will have a higher B_b than a high-sidelobe amplitude taper. B_b has a lower bound of one. Thus, there is some tradeoff between the lowest design sidelobe level and the highest grating lobe level. A higher taper efficiency leads to a lower B_b, which in turn results in a lower peak grating lobe. As a result, optimizing the amplitude taper should lead to a peak sidelobe level that is the same height as the peak grating lobe. One can only expect a small improvement to the maximum sidelobe level

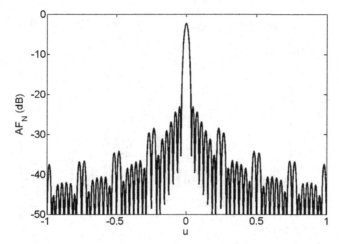

Figure 2.55. Array factor due to the subarray amplitude taper in Figure 2.53.

TABLE 2.5. Some Possible Subarray Divisions of a 128-Element Array and the Location and Approximate Height of the First Grating Lobe

N_e	N_s	θ_1	Approximate Grating Lobe Level
4	32	30°	−36.9 dB
8	16	14.5°	−31.6 dB
16	8	7.2°	−25.8 dB
32	4	3.6°	−19.8 dB

through optimization, since increasing B_b to get lower sidelobes also causes the peak grating lobe to increase.

Steering the main beam using phase shifters at the subarray ports results in very large grating lobes. The array factors shown in Figure 2.56 compare steering the pattern in Figure 2.55 to $u = 0.12$ using phase shifters at the subarray ports only (solid line) and phase shifters at the individual elements (dashed line). Steering AF_s moves the peak of its pattern out of the null of AF_e (which is not steered). These nulls in AF_e are what cause the grating lobes to have a split down the middle in Figure 2.55. Without the nulls in the center of the grating lobes, the grating lobes dramatically increase. Practical beam steering is done at the element level.

2.12. ERRORS

The signals at each element in the array have errors due to manufacturing tolerances, element failures, aging, and quantization. Errors are either modeled

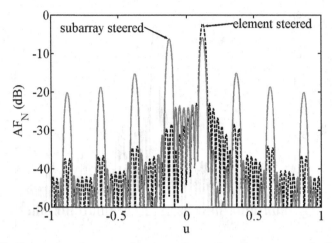

Figure 2.56. Comparison between phase steering at the subarray ports (solid line) and phase steering at the element level (dashed line) when the patterns are steered to $u = 0.12$.

as statistically independent from element to element or statistically independent between groups of elements. The main concern with errors is the rise in the sidelobe levels. Other problems include beam pointing errors, loss in gain, and need for recalibration.

2.12.1. Random Errors

Random errors are statistically independent from element to element. These errors occur within individual array elements and associated components and have no effect on surrounding elements. For the most part, four types of random errors are possible:

1. Random amplitude error, δ_n^a
2. Random phase error, δ_n^p
3. Random position error, δ_n^s
4. Random element failure, $P_n^v = \begin{cases} 1 & \text{element functioning properly} \\ 0 & \text{element failure} \end{cases}$

Random amplitude and phase errors appear in the signal weighting while the position errors appear in the relative element phases. An array factor with errors can be written as

$$\mathrm{AF_{err}} = \sum_{n=1}^{N} \left(a_n + \delta_n^a \right) e^{j\left(p_n + \delta_n^p \right)} e^{jk\left(s_n + \delta_n^s \right)u} \qquad (2.110)$$

If the position and phase errors are relatively small, then the small phase approximation can be used ($e^{jx} \simeq 1 + jx$ for $x \ll 1$) for the phase terms in (2.110):

$$\mathrm{AF}_{\mathrm{err}} \simeq \sum_{n=1}^{N} \left(a_n + \delta_n^a\right)\left(1 + j\delta_n^p\right) e^{jp_n} \left(1 + j\delta_n^s\right) e^{jksn_u} \tag{2.111}$$

Multiplying the quantities in (2.111) and collecting terms results in

$$\mathrm{AF}_{\mathrm{err}} = \sum_{n=1}^{N} a_n e^{jp_n} e^{jkus_n} + \sum_{n=1}^{N} \left\{ \left[\delta_n^a - \delta_n^a \delta_n^p k u \delta_n^s - a_n \delta_n^p k u \delta_n^s \right] \right.$$
$$\left. + j\left[\delta_n^a \delta_n^p + \delta_n^a k u \delta_n^s + a_n \delta_n^p + a_n k u \delta_n^s \right] \right\} e^{jp_n} e^{jkus_n} \tag{2.112}$$

Note that the first summation is the array factor with no errors. The second summation is an error array factor that is added to the array factor with no errors. Since the errors are assumed to be small, all terms that consist of the product of at least two errors can be dropped, leaving

$$\mathrm{AF}_{\mathrm{err}} \simeq \mathrm{AF}_0 + \sum_{n=1}^{N} \left\{ \delta_n^a + j\left(a_n \delta_n^p + a_n k u \delta_n^s \right) \right\} e^{jp_n} e^{jkus_n} \tag{2.113}$$

where AF_0 is the error free array factor.

The error pattern effects are easy to see at a null in the error-free array factor, because the error-free pattern is zero and only the error term in (2.113) remains. In order to analyze the contribution of each type of error, assume that only one error at a time exists. The power pattern due to the amplitude weights alone is proportional to the magnitude of the error array factor squared.

$$\left| \mathrm{AF}_{\mathrm{amperr}} \right|^2 = \sum_{n=1}^{N} \delta_n^a e^{jp_n} e^{jkus_n} \sum_{m=1}^{N} \delta_m^a e^{-jp_m} e^{-jkus_m} \tag{2.114}$$

Taking the average value of the power pattern (line over quantity) results in

$$\overline{\left| \mathrm{AF}_{\mathrm{amperr}} \right|^2} = \sum_{n=1}^{N} \sum_{m=1}^{N} \overline{\delta_n^a \delta_m^a} e^{j(p_n - p_m)} e^{jku(s_n - s_m)} = \sum_{n=1}^{N} \overline{\delta_n^{a^2}} = N\overline{\delta_n^{a^2}} \tag{2.115}$$

Similar average power patterns occur for the other two error types:

$$\overline{\left| \mathrm{AF}_{\mathrm{pherr}} \right|^2} = \sum_{n=1}^{N} \sum_{m=1}^{N} a_n a_m \overline{\delta_m^p \delta_n^p} e^{j(p_n - p_m)} e^{jku(s_n - s_m)} = \left(\sum_{n=1}^{N} a_n^2 \right) \overline{\delta_n^{p^2}} \tag{2.116}$$

$$\overline{\left| \mathrm{AF}_{\mathrm{poserr}} \right|^2} = (ku)^2 \sum_{n=1}^{N} \sum_{m=1}^{N} a_n a_m \overline{\delta_m^s \delta_n^s} e^{j(p_n - p_m)} e^{jku(s_n - s_m)} = (ku)^2 \left(\sum_{n=1}^{N} a_n^2 \right) \overline{\delta_n^{s^2}} \tag{2.117}$$

Since δ_n^a is just as likely to add as to subtract from the main beam, it does not change the peak value of AF but has a relatively constant effect in the sidelobe region. Thus, the relative sidelobe level caused by the amplitude errors is the ratio of the sidelobe power to the power in the peak of the main beam:

$$\text{sll}_{\text{rms}} = \frac{N\overline{\delta_n^{a^2}}}{\left(\sum\limits_{n=1}^{N} a_n\right)^2} \tag{2.118}$$

On the other hand, the phase errors cause small subtractions from the main beam as well as produce a relatively constant pattern in the sidelobe region. As a result, the average sidelobe level due to random phase errors is given by

$$\text{sll}_{\text{rms}} = \frac{\left(\sum\limits_{n=1}^{N} a_n^2\right)\overline{\delta_n^{p^2}}}{\left(\sum\limits_{n=1}^{N} a_n\right)^2\left(1-\overline{\delta_n^{p^2}}\right)} \tag{2.119}$$

The relative sidelobe level due to position errors alone is

$$\text{sll}_{\text{rms}} = \frac{(ku)^2\left(\sum\limits_{n=1}^{N} a_n^2\right)\overline{\delta_n^{s^2}}}{\left(\sum\limits_{n=1}^{N} a_n\right)^2} \tag{2.120}$$

Position errors have little effect in the main beam region (u is very small) but have an increasing effect toward $u = \pm 1$.

Example. A 50-element linear array ($d = 0.5\lambda$) has a low-sidelobe taper. Plot $\text{AF}_{\text{amperr}}$, AF_{pherr}, and $\text{AF}_{\text{poserr}}$ when $\delta_n^a = 0.1$, $\delta_n^p = 0.1$, and $\delta_n^s = 0.1$.

Figure 2.57 shows the plots. Random amplitude and phase errors have approximately the same magnitude over all the angles, while the position errors increase with $|u|$.

A different type of random error occurs when elements stop functioning. Element failures result from amplifier, receiver, phase shifter, and so on, malfunctions. The peak of the main beam is now at $N\eta_t P^e$, where P^e is the probability of element failure. Assuming that the element failures are uniformly distributed across the aperture, the formula for rms sidelobe levels due to failed elements is

$$\text{sll}_{\text{rms}} = \frac{1-P_e}{N\eta_t P_e} \tag{2.121}$$

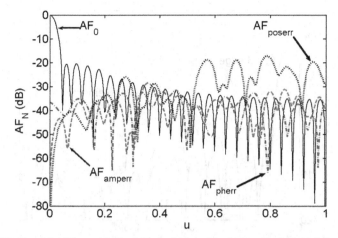

Figure 2.57. Plot of AF_{amperr} (dashed line), AF_{pherr} (dash–dot line), and AF_{poserr} (dotted line) when $\delta_n^a = 0.1$, $\delta_n^p = 0.1$, and $\delta_n^s = 0.1$. The error-free pattern is the solid line.

Comparing this formula with (2.118) reveals that the probability that an element has failed, $1 - P_e$, is the same as an rms amplitude error, $\overline{\delta_n^{a^2}}$. Position errors are relatively small compared to the other three errors, so a formula to calculate the rms sidelobe level of the array factor for amplitude and phase errors with element failures is [22]

$$sll_{rms} = \frac{(1 - P_e) + \overline{\delta_n^{a^2}} + P_e \overline{\delta_n^{p^2}}}{P_e \left(1 - \overline{\delta_n^{p^2}}\right) \eta_t N} \tag{2.122}$$

Example. A 30-element linear array ($d = 0.5\lambda$) has a 30-dB, $\bar{n} = 7$ low-sidelobe taper. Show the effects of a single element failure at (1) the edge and (2) the center of the array.

Figure 2.58 shows the array factors superimposed on each other. Since the edge element has a low amplitude, it has little effect on the pattern when it fails. When the center element fails, the sidelobes significantly increase. The original pattern has $\eta_T = 0.8535$, while the edge failed taper has $\eta_T = 0.8332$ and the center element failed taper has $\eta_T = 0.8243$.

So far, only random, uncorrelated errors have been mentioned. If the same error occurs for groups of elements, then that error is correlated within that group of elements even though the error is random. An example would be a random amplitude error at the subarray port. That random error is passed on to each element in the subarray, so it is the same for all the elements of that subarray; hence there is a correlated error between elements of that subarray.

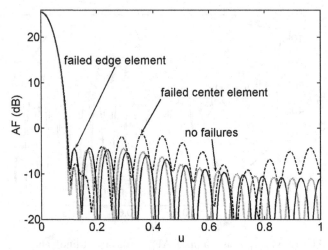

Figure 2.58. Array factor of a 30-element linear array ($d = 0.5\lambda$) with a 30 dB, $\bar{n} = 7$ low-sidelobe taper (dotted line) superimposed on the same array factor with element 1 failed (solid line) and element 15 failed (dashed line).

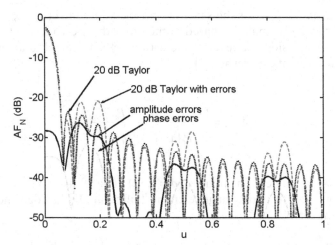

Figure 2.59. Array factor of a 36-element linear array ($d = 0.5\lambda$) with a low-sidelobe taper (dotted line) and divided into 6 subarrays of 6-elements each. The dashed line in the pattern with random errors at the subarray ports only ($\delta_n^a = 0.1$ and $\delta_n^p = 0.1$). Plots of AF_{amperr} (solid line) and AF_{pherr} (dash–dot line) are superimposed.

Example. A 30-element linear array ($d = 0.5\lambda$) has a 20-dB, $\bar{n} = 2$ taper applied at the elements. The array is divided into 6 subarrays, each having 6 elements. Plot the error patterns when $\delta_n^a = 0.1$ and $\delta_n^a = 0.1$ at the subarray port.

Figure 2.59 has graphs of the array factor with no errors, with both the phase and amplitude errors, and AF_{amperr} and AF_{pherr}. The errors have greatest impact at the grating lobe locations.

2.12.2. Quantization Errors

Phase shifters and attenuators have digital controls with a finite number of possible values. For instance, the least significant bit associated with amplitude and phase have the following values:

$$\Delta_a = 2^{-N_{ba}} \tag{2.123}$$

$$\Delta_p = 2\pi \times 2^{-N_{bp}} \tag{2.124}$$

where N_{ba} is the number of amplitude bits and N_{bp} is the number of phase bits. A quantized amplitude weight has the value of

$$a_n^q = \Delta_a \text{floor} \left\{ \frac{a_n}{\Delta_a} \right\} + \frac{\Delta_a}{2} \tag{2.125}$$

and the quantized phase has a value of

$$p_n^q = \text{floor} \left\{ \frac{\delta_n - ks_n u_s - 0.5\Delta_p}{\Delta_p} \right\} \tag{2.126}$$

where *floor* rounds down to the closest integer. If only the phase steering is quantized, then δ_n is not included in this equation.

If the difference between the desired and quantized amplitude weight is assumed to be a uniformly distributed random number with the bounds being the maximum amplitude error of $\pm\Delta_a/2$, then the rms amplitude error is $\delta_n^a = \Delta_a/\sqrt{12}$. This value can then be substituted into (2.122) to find the rms sidelobe level.

If no two adjacent elements receive the same quantized phase shift, then the difference between the desired and quantized phase shifts are treated as uniform random variables between $\pm\Delta_p/2$. As with the amplitude error, the random phase error formula in this case is $\delta_n^p = \Delta_p/\sqrt{12}$. These same quantized phase errors result in beam-pointing errors too. The difference in phase between the desired and quantized steering phase shifts is

$$q_n = kd(n-1)(u_s - u_q) \tag{2.127}$$

Solving for the angular difference at element n yields

$$u_s - u_q = \frac{q_n}{kd(n-1)} \tag{2.128}$$

Taking the average value of the right-hand side of (2.128) for all the elements in the array and the rms quantization error leaves

$$\overline{u_s - u_q} \approx \pm \frac{\lambda}{Nd2^{N_{bp}}} \tag{2.129}$$

Either increasing the number of bits in the phase shifters or increasing the aperture size reduces the beam pointing error.

When two or more elements have the same quantized phase shift, then the error is correlated and quantization lobes form. This situation occurs when the beam is steered to a small angle off boresite. The maximum phase shift across an aperture is given by

$$\psi_T = (N-1)kdu_s \qquad (2.130)$$

The total number of elements that receive the same quantized phase shift is then

$$N_Q = \frac{\Delta_p}{\psi_T}N = \frac{N}{(N-1)2^{N_{bp}}(d/\lambda)u_s} \qquad (2.131)$$

This means that there are N/N_Q subarrays of N_Q elements that receive the same phase shift. The grating lobes due to these subarrays occur at [21]

$$u_m = u_s \pm \frac{m\lambda}{N_Q d_e} = u_s\left[1\pm\frac{m(N-1)2^{N_{bp}}}{N}\right] \simeq u_s\left(1\pm m2^{N_{bp}}\right) \qquad (2.132)$$

The approximation in (2.132) assumes the array has many elements. For large scan angles, quantization lobes do not form, because the element-to-element phase difference appears random. The relative peaks of the quantization lobes are given by [21]

$$AF_N^{QL} = \frac{1}{2^{N_p}}\sqrt{\frac{\sqrt{1-u^2}}{\sqrt{1-u_s^2}}} \qquad (2.133)$$

Figure 2.60 shows a low-sidelobe array factor for a 10-element, $d = 0.5\lambda$ array with its beam steered to $u = 0.1$ when the phase shifters have 3, 4, and 5 bits.

Example. Find the location and heights of the quantization lobes for a 10-element array with $d = 0.5\lambda$ and the beam steered to $u_s = 0.05$ when the phase shifters have 3, 4, and 5 bits.

The location and heights of the grating lobes are calculated from (2.132) and (2.133).

m	−2	−1	1	2
u_m	−.71	−.33	.43	.81
3	−19.6 dB	−18.3 dB	−18.5 dB	−20.4 dB

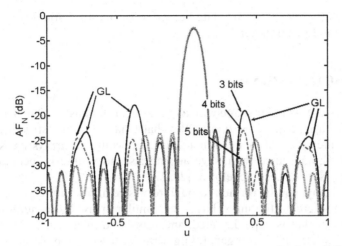

Figure 2.60. Array factors steered to $u = 0.1$ when there are 3-, 4-, and 5-bit phase shifters.

m	-1	1
u_m	$-.71$	$.81$
4	$-25.6\,\text{dB}$	$-26.4\,\text{dB}$

There are no grating lobes for the 5-bit phase shifters. The location of the grating lobes and their heights are only approximate. The actual values calculated from the array factor are given by

m	-2	-1	1	2
u_m	$-.72$	$-.3764$	$.4148$	$.8705$
3	$-23.2\,\text{dB}$	$-17.9\,\text{dB}$	$-19.1\,\text{dB}$	$-24.3\,\text{dB}$

m	-1	1
u_m	$-.7795$	$.8196$
4	$-24.4\,\text{dB}$	$-25.9\,\text{dB}$

Phase dithering reduces the size of the quantization lobes by adding a random phase to each phase shifter [23]. For large arrays, this random phase has very little impact on the main beam-pointing angle. Two other similar approaches that breakup correlated errors with small random errors include frequency and beam dithering [23].

2.13. FRACTAL ARRAYS

A fractal is a self-similar shape with a non-integer dimension [24]. Self-similar means that a magnified portion of the shape has the same structure as the whole geometry. Fractals were first applied to antenna arrays by Kim and Jaggard [25]; and since then, several analysis and synthesis techniques for fractal arrays have been developed [26, 27].

Fractal arrays can be formed through recursive application of a generating array used to create the much larger self similar array. The generating array has a pattern that is copied, scaled, and translated many times. There are a total of N_G elements in the generating array with N_1 of the elements having an amplitude of one and the rest having an amplitude of zero. An example is the Cantor array [28] that has a 3-element ($N_G = 3$) generating array with weights given by

$$w = [101] \tag{2.134}$$

The next scale is found by replacing a 1 with 101 and replacing a 0 with 000.

$$w = [101000101] \tag{2.135}$$

Applying the formula to obtain the next scale yields

$$w = [101000101000000000101000101] \tag{2.136}$$

Although this array does not have practical use, it does have some interesting properties. For instance, the array factor can be expressed as a product rather than a sum:

$$\mathrm{AF}_N(\psi) = \prod_{s=1}^{S} \mathrm{AF}_G\left(N_G^{s-1}\psi\right) \tag{2.137}$$

where AF_G is the array factor of the generating array. If $d = 0.25\lambda$, then the closest that two elements are spaced in the thinned aperture is 0.5λ, so the directivity is derived from (2.68)

$$D = 2^S \tag{2.138}$$

The generating array for Cantor arrays with every other element having an amplitude of zero may be expressed in the form

$$\mathrm{AF}_G(\psi) = \frac{\sin[N_1\psi]}{N_1\sin(\psi)} \qquad (2.139)$$

Other versions of the generating array are also possible. For instance, $N_G = 5$ has the weighting

$$w = [10101] \qquad (2.140)$$

The recursive building of larger arrays results in much different array weights. Proceeding to the second scale results in the weights

$$w = [101010000101010000010101] \qquad (2.141)$$

which differs from either (2.135) or (2.136).

The fractal array factor is based upon an iterated (recursive) function. Therefore, the array factor can be calculated via the product of S generating array factors rather than the N additions and multiplications in a normal Fourier series representation of the array factor.

$$\mathrm{AF}_N = \prod_{s=1}^{S} \frac{\sin[N_1 N_G^{s-1}\psi]}{N_1\sin[N_G^{s-1}\psi]} \qquad (2.142)$$

The fractal dimension D of these Cantor arrays are calculated using [28]

$$f_{\mathrm{dim}} = \frac{\log N_1}{\log N_G} \qquad (2.143)$$

which results in $f_{\mathrm{dim}} = 0.6309$ for $N_G = 3$, $f_{\mathrm{dim}} = 0.6826$ for $N_G = 5$, and $f_{\mathrm{dim}} = 0.7124$ for $N_G = 7$.

Example. Plot the array factors for the first four scales of the Cantor array when the generating array has $d = 0.25\lambda$.

The array factor of the three-element generating array is

$$\mathrm{AF}_G(\psi) = 2\cos(\psi) \qquad (2.144)$$

where $\psi = \pi u$. An expression for the normalized Cantor array factor given by

$$\mathrm{AF}_N(\psi) = \prod_{s=1}^{S} \mathrm{AF}_G(3^{m-1}\psi) = \prod_{s=1}^{S} \cos(3^{s-1}\psi) \qquad (2.145)$$

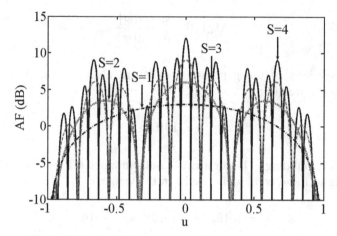

Figure 2.61. Array factors for $N_G = 3$ and $S = 1, 2, 3, 4$.

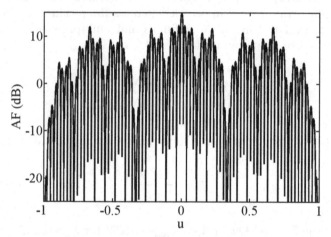

Figure 2.62. Array factors for $N_G = 3$ and $S = 5$.

The array factors for $S = 1, 2, 3$, and 4 are shown in Figure 2.61. You can see how similar the shape of the array factor are. Figure 2.62 shows the array factor at $S = 7$.

A Sierpinski carpet is a two-dimensional version of the Cantor set [29]. An example of a generating array on a square lattice is

$$
\begin{matrix}
1 & 1 & 1 \\
1 & 0 & 1 \\
1 & 1 & 1
\end{matrix}
$$

The normalized array factor associated with this generating subarray for $d_x = d_y = \lambda/2$ is given by

$$AF_G(u_x, u_y) = \frac{1}{4}[\cos(\pi u) + \cos(\pi v) + 2\cos(\pi u)\cos(\pi v)] \quad (2.146)$$

The expression for the fractal array factor at stage S is

$$AF_N = \frac{1}{4^S}\prod_{p=1}^{S}[\cos(3^{p-1}\pi u) + \cos(3^{p-1}\pi v) + 2\cos(3^{p-1}\pi u)\cos(3^{p-1}\pi v)] \quad (2.147)$$

The geometry for this Sierpinski carpet fractal array at progressive stages of growth appears in Figure 2.63 along with a plot of the corresponding array factors. The array factors look self-similar.

Linear polyfractal arrays can exhibit ultra-wideband characteristics when numerically optimized [30]. Polyfractal arrays are constructed from a set of multiple generatoring arrays rather than a single generating array. A 32-element linear polyfractal array was designed and built to have a wide bandwidth with suppressed grating lobes and relatively low sidelobes (-16.3 dB at f_0 and -5.39 dB at $4f_0$). The 32-element array was divided into 4 subarrays of

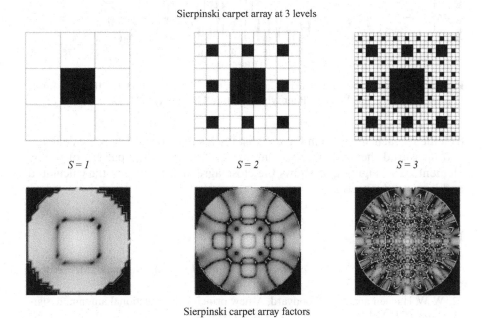

Sierpinski carpet array at 3 levels

$S = 1$ $S = 2$ $S = 3$

Sierpinski carpet array factors

Figure 2.63. Sierpinski carpet array and associated array factors for the first three stages.

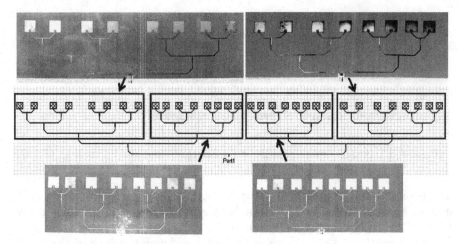

Figure 2.64. Sierpinski carpet array and associated array factors for the first three. (Courtesy of Douglas Werner, Pennsylvania State University.)

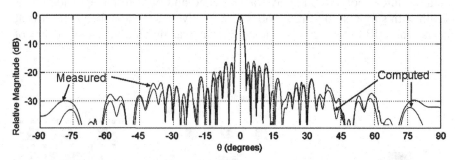

Figure 2.65. Sierpinski carpet array and associated array factors for the first three. (Courtesy of Douglas Werner, Pennsylvania State University.)

8 elements each as shown in Figure 2.64. Each subarray pattern was calculated and measured then coherently combined to find the array pattern of the 32 element array. Figure 2.65 shows the close agreement between the calculated and measured patterns.

REFERENCES

1. R. J. Mailloux, *Phased Array Antenna Handbook* 2nd ed., Norwood, MA: Artech House, 1994.
2. W. W. Hansen and J. R. Woodyard, A new principle in directional antenna design, *Proc. IRE*, Vol. 26, No. 3, March 1938, pp. 333–345.
3. W. K. Bartley, Near-field antenna focussing, Goddard Space Flight Center Report X-811-75-183, Greenbelt, MD, August 75.

4. A. J. Fenn, C. J. Diederich, and P. R. Stauffer, An adaptive-focusing algorithm for a microwave planar phased-array hyperthermia system, *Lincoln Lab. J.*, Vol. 6, No. 2, 1993, pp. 269–288.

5. L. J. Ricardi and R. C. Hansen, Comparison of Line and Square Source Near-Fields, *Trans. IEEE*, Vol. AP-11, November. 1963, pp. 711–712.

6. R. C. Hansen, Minimum spot size of focused apertures, *URSI Symposium on EM Theory*, Delft: Pergamon, Press, 1965, pp. 661–667.

7. E. Sharp, A triangular arrangement of planar-array elements that reduces the number needed, *IRE AP-S Trans.*, Vol. 9, No. 2, 1961, pp. 126–129.

8. J. Hsiao, Properties of a nonisosceles triangular grid planar phased array, *IEEE AP-S Trans.*, Vol. 20, No. 4, 1972, pp. 415–421.

9. R. J. Mailloux et al., Multiple mode control of grating lobes, RADC Technical Report 76–307, September 1976.

10. L. Corey, J. Weed, and T. Speake, Modeling triangularly packed array antennas using a hexagonal FFT, *IEEE AP-S Symp.*, June 1984, pp. 507–510.

11. S. Silver, Aperture illumination and antenna patterns, in *Microwave Antenna Theory and Design*, London: Peter Peregrinus, 1949, pp. 169–199.

12. S. A. Schelkunoff, A mathematical theory of linear arrays, *Bell Syst. Tech. J.*, Vol. 22, 1943, pp. 80–107.

13. C. A. Balanis, *Antenna Theory Analysis and Design*, 2nd ed., New York: John Wiley & Sons, 1997.

14. C. Stearns and A. Stewart, An investigation of concentric ring antennas with low sidelobes, *IEEE AP-S Trans.*, Vol. 13, No. 6, November 1965, pp. 856–863.

15. R. Das, Concentric ring array, *IEEE AP-S Trans.*, Vol. 14, No. 3, May 1966, pp. 398–400.

16. F. Adcock, Improvement in means for determining the direction of a distant source of electromagnetic radiation, British Patent 1304901919, 1917.

17. E. J. Baghdady, New Developments in Direction-of-Arrival Measurement Based on Adcock Antenna Clusters, *Proc. of the IEEE Aerospace and Electronics Conference*, Dayton, OH, May 22–26, 1989, pp. 1873–1879.

18. A. Manikas, A. Alexiou, and H. R. Karimi, Comparison of the ultimate direction-finding capabilities of a number of planar array geometries, *IEE Proc. Radar, Sonar Navig.*, Vol. 144, No. 6, December 1997, pp. 321–329.

19. A. Swindlehurst, and T. Kailath, Azimuth/elevation direction finding using regular array geometries, *IEEE AES Trans.*, Vol. 29, No. 1, January 1993, pp. 145–156.

20. R. C. Hansen, *Phased Array Antennas*, New York: John Wiley & Sons, 1998.

21. R. J. Mailloux, Array grating lobes due to periodic phase, amplitude, and time delay quantization, *IEEE AP-S Trans.*, Vol. 32, December 1984, pp.1364–1368.

22. E. Brookner, Antenna Array Fundamentals—Part 2, in *Practical Phased Array Antenna Systems*, E. Brookner, ed., Norwood, MA, Artech House, 1991.

23. R. H. Sahmel and R. Manasse, Spatial statistics of instrument-limited angular measurement errors in phased array radars, *IEEE Transactions on Antennas and Propagation*, Vol. 21, No. 4, July 1973, pp. 524–532.

24. B. B. Mandelbrot, *The Fractal Geometry of Nature*. New York: W. H. Freeman, 1983.

25. Y. Kim and D. L. Jaggard, The fractal random array, *Proc. IEEE*, Vol. 74, No. 9, pp. 1278–1280, 1986.

26. D. H. Werner, et al., The theory and design of fractal antenna arrays, in *Frontiers in Electromagnetics*, D. H. Werner and R. Mittra, eds., New York: John Wiley & Sons, 2000, pp. 94–203.

27. D. H. Werner, R. L. Haupt, and P. L. Werner, Fractal antenna engineering: The theory and design of fractal antenna arrays, *IEEE Antennas Propagat. Mag.*, Vol. 41, No. 5, October 99, pp. 1009–1015.

28. C. Puente Baliarda and R. Pous, Fractal design of multiband and low side-lobe arrays, *IEEE Trans. Antennas Propagat.*, Vol. 44, No. 5, pp. 730–739, May 1996.

29. H. O. Peitgen, H. Jurgens and D. Saupe, *Chaos and Fractals: New Frontiers of Science*, New York: Springer-Verlag, 1992.

30. J. S. Petko, and D. H. Werner, The Pareto optimization of ultrawideband poly-fractal arrays, *IEEE AP-S Trans.*, Vol. 56, No. 1, pp. 97–107, January 2008.

3 Linear and Planar Array Factor Synthesis

The array factor is a function of the amplitude and phase weights, the relative element positions, and the frequency. Values for these variables exist that yield a desirable array factor (as long as the laws of physics are obeyed). This chapter presents analytical, statistical, and numerical techniques to synthesize or optimize an array factor.

3.1. SYNTHESIS OF AMPLITUDE AND PHASE TAPERS

The techniques presented in this section are more academic than practical. They provide some insight into the design of low-sidelobe tapers, but the fact that amplitude and phase weighting are required and that the weights can significantly vary from element to element make them very difficult to implement with real hardware.

3.1.1. Fourier Synthesis

In Chapter 2, the array weights were shown to be coefficients of a Fourier series. These coefficients come from the inner product of the desired array factor with a single Fourier component.

$$w_m = \frac{d}{\lambda} \int_{-\lambda/2d}^{\lambda/2d} \mathrm{AF}(u) \cos\left[(2m-1)\pi\left(\frac{d}{\lambda}\right)u \right] du, \qquad m = 1, 2, \ldots, M \quad \textbf{even} \qquad (3.1)$$

$$w_m = \frac{d}{\lambda} \int_{-\lambda/2d}^{\lambda/2d} \mathrm{AF}(u) \cos\left[2m\pi\left(\frac{d}{\lambda}\right)u \right] du, \qquad m = 0, 1, \ldots, M \quad \textbf{odd} \qquad (3.2)$$

where m is one of the M harmonics. The number of elements in the array is $2M$ (even) or $2M + 1$ (odd). Note that u ranges from -1 to $+1$. The limits on the integral should also span this range. Thus,

---

Antenna Arrays: A Computational Approach, by Randy L. Haupt
Copyright © 2010 John Wiley & Sons, Inc.

$$\frac{\lambda}{2d} = 1 \Rightarrow d = \frac{\lambda}{2} \tag{3.3}$$

If $d > \lambda/2$, then the limits of integration do not cover the range of u. If $d < \lambda/2$, then the limits of integration cover a greater range than -1 to $+1$. Sine terms in the Fourier series expansion are included whenever the current weights (amplitude and phase) are not an even function with respect to the center of the array.

Equations (3.1) and (3.2) provide a method of synthesizing a desired array factor for a given number of elements. The steps needed in a Fourier synthesis technique are as follows:

1. Determine desired pattern AF(u).
2. Determine number of elements and element spacing.
3. Calculate limits of integration.
4. Find the w_m using (3.1) or (3.2).

Example. Design a 16-element equally spaced array to receive signals at a constant level over an angular range of $-0.5 \le u \le 0.5$ and zero elsewhere.

The specifications require that $M = 8$ and the array factor is represented by

$$\mathrm{AF}(u) = \begin{cases} 0, & u < -0.5 \\ 1, & -0.5 \le u \le 0.5 \\ 0, & 0.5 < u \end{cases}$$

The weights derived using the Fourier series synthesis method are shown in Figure 3.1 and the corresponding array factor in Figure 3.2. Some of the weights are negative, so the amplitude weights also have to be accompanied by 180° phase shifts.

The amplitude taper for low sidelobes in a sum pattern will not result in the same low sidelobes in a difference pattern and vice versa. Since the difference array factor is an odd function, the weights for a difference array are the inner product of the desired array factor with a single Fourier sine component.

$$w_n = \frac{d}{\lambda} \int_{-\frac{\lambda}{2d}}^{\frac{\lambda}{2d}} \mathrm{DF}(u) \sin\left[(2n-1)\pi\left(\frac{d}{\lambda}\right)u\right] du \tag{3.4}$$

Example. Find the weights for a 20-element array with a difference pattern having the following specifications:

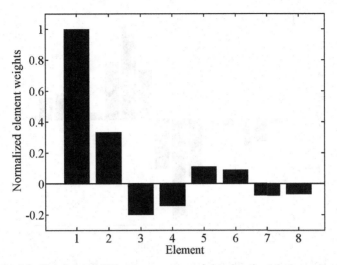

Figure 3.1. Fourier coefficients or array weights for the 16-element array.

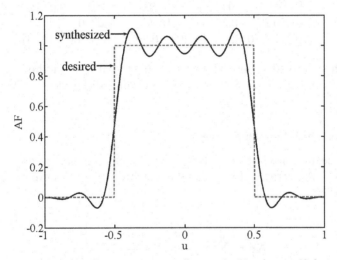

Figure 3.2. Array factor corresponding to the Fourier coefficients.

$$D(u) = \begin{cases} 0, & -1 \le u < -0.5 \\ -1, & -0.5 \le u < 0 \\ 0, & u = 0 \\ 1, & 0 < u \le 0.5 \\ 0, & 0.5 < u \le 1 \end{cases}$$

Substituting into (3.4) and solving yields

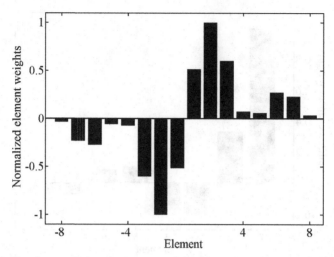

Figure 3.3. Fourier coefficients or array weights for the 16-element difference array.

$$w = [-0.0124 \quad -0.0836 \quad -0.0988 \quad -0.0207 \quad -0.0266$$
$$-0.2170 \quad -0.3620 \quad -0.1860 \quad 0.1860 \quad 0.3620$$
$$0.2170 \quad 0.0266 \quad 0.0207 \quad 0.0988 \quad 0.0836 \quad 0.0124]$$

The weights derived using the Fourier series synthesis method are shown normalized in Figure 3.3 and the corresponding difference array factor in Figure 3.4.

3.1.2. Woodward–Lawson Synthesis

Sines and cosines are not the only building blocks that can be used to create array factors. An array factor is also the weighted sum of steered linear array factors [1, 2].

$$\text{AF} = \sum_{m=-M}^{M} b_m \frac{\sin\left[\dfrac{N}{2}kd(u-u_m)\right]}{N\sin\left[\dfrac{1}{2}kd(u-u_m)\right]} \tag{3.5}$$

where the coefficients are the samples of the desired array factor given by

$$b_m = \text{AF}_{\text{desired}}(u_m) \tag{3.6}$$

The samples are taken at points where the maximum of one beam is at a zero of all the other beams (the beams are orthogonal).

$$u_m = \frac{m\lambda}{Nd} \tag{3.7}$$

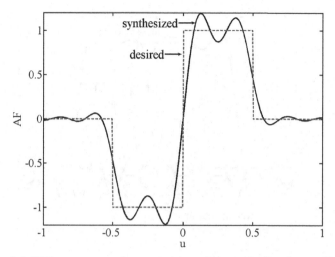

Figure 3.4. Difference array factor corresponding to the Fourier coefficients.

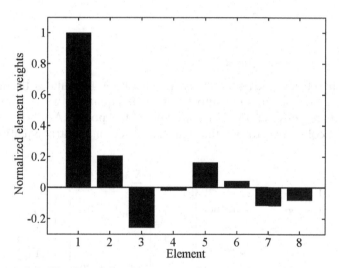

Figure 3.5. Woodward–Lawson array weights for the 16-element array.

With this approach, the amplitude weights for the array are given by

$$w_n = \frac{1}{N} \sum_{m=-M}^{M} b_m e^{-jk(n-1)du_m} \qquad (3.8)$$

Example. Repeat the Fourier series array synthesis example using the Woodward–Lawson synthesis technique.

Figure 3.5 is a plot of the resulting array weights and Figure 3.6 is the corresponding array factor with the beams and desired array factor superimposed.

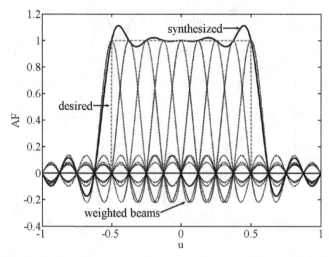

Figure 3.6. Array factor corresponding to the Woodward–Lawson array weights.

3.1.3. Least Squares Synthesis

A more direct approach to array factor synthesis formulates the linear or planar array factor equation into a set of M equations with N unknowns. Samples of the desired array factor are taken at M points $(\mathrm{AF}_1, \mathrm{AF}_2, \ldots, \mathrm{AF}_M)$ and form a column vector for the right-hand-side of a matrix equation.

$$
\begin{bmatrix}
1 & e^{j\psi_1} & \cdots & e^{j(N-1)\psi_1} \\
1 & e^{j\psi_2} & & e^{j(N-1)\psi_2} \\
\vdots & & \ddots & \vdots \\
1 & e^{j\psi_M} & \cdots & e^{j(N-1)\psi_M}
\end{bmatrix}
\begin{bmatrix}
w_1 \\ w_2 \\ \vdots \\ w_N
\end{bmatrix}
=
\begin{bmatrix}
\sum_{n=1}^{N} w_1 e^{j(n-1)\psi_1} \\
\sum_{n=1}^{N} w_1 e^{j(n-1)\psi_2} \\
\vdots \\
\sum_{n=1}^{N} w_1 e^{j(n-1)\psi_M}
\end{bmatrix}
=
\begin{bmatrix}
\mathrm{AF}_1 \\ \mathrm{AF}_2 \\ \vdots \\ \mathrm{AF}_M
\end{bmatrix}
\tag{3.9}
$$

Each row in (3.9) is the array factor with the w_n as unknowns. If $M = N$, then the weights are found using a direct matrix inversion. Otherwise, a least squares solution is necessary to solve the over or under determined system of equations.

Example. Repeat the Fourier series and Woodward–Lawson example using a least squares approach.

Taking 20 equally spaced samples of the array factor for the right-hand-side vector permits a direct matrix inversion to find the array weights. The weights

TABLE 3.1. Amplitude Weights for Half of the Array Synthesized to Produce the Desired Array Factor for a 16-Element Array

Element	Fourier Series	Woodward–Lawson	Least Mean Squares
1	0.45016	0.49291	0.46711
2	0.15005	0.10149	0.13343
3	−0.090032	−0.12688	−0.10769
4	−0.064308	−0.0096566	−0.047619
5	0.050018	0.080463	0.068812
6	0.040924	−0.020572	0.023733
7	−0.034627	−0.0576	−0.055073
8	−0.030011	0.039841	−0.011761

and array factor are very similar to those of the previous examples, so the synthesized weights are compared to those of the Fourier series and Woodward–Lawson techniques in Table 3.1.

Unlike Fourier and Woodward–Lawson synthesis methods, the least squares technique works for any element lattice and array shape. As with the other methods, the synthesized weights are complex and highly oscillatory. Also, the array factors are matched either exactly or in a least squares sense at M specified points. A much more desirable approach would be to place limits on the power pattern and the array weights. This can be done using a robust, global optimization technique like a genetic algorithm. Many examples of optimizing array factors using a genetic algorithm may be found in the literature [3, 4].

3.2. ANALYTICAL SYNTHESIS OF AMPLITUDE TAPERS

There are many different methods to synthesize amplitude weights that produce desirable sidelobe levels. This section presents several approaches to analytically calculate the array weights for linear and planar arrays. In general, the analystical synthesis approaches require linear or circular apertures. Weights for all other geometries must be numerically found.

3.2.1. Binomial Taper

If all the unit circle zeros of an array factor are at $\psi = 180°$, then the array factor has no sidelobes and is written as

$$\text{AF} = \sum_{n=1}^{N-1} \left(z - e^{j\psi_n} \right) = \left(z + 1 \right)^{N-1} \tag{3.10}$$

TABLE 3.2. List of Binomial Amplitude Weights for Arrays with 1 through 9 Elements

Number of Elements	Amplitude Weights	Taper Efficiency
1	1	1.000
2	1 1	1.000
3	1 2 1	0.889
4	1 3 3 1	0.800
5	1 4 6 4 1	0.731
6	1 5 10 10 5 1	0.667
7	1 6 15 20 15 6 1	0.633
8	1 7 21 35 35 21 7 1	0.597
9	1 8 28 56 70 56 28 8 1	0.566

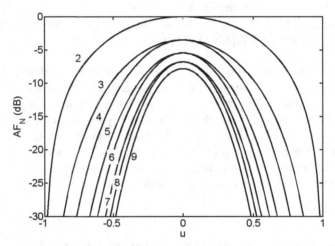

Figure 3.7. Array factors normalized to the number of elements in the array for several sizes of binomial arrays.

The corresponding amplitude weights are the binomial coefficients, hence this array is known as a binomial array [5]. Table 3.2 lists the binomial coefficients for up to nine element arrays, and Figure 3.7 shows some examples of binomial array factors normalized to a uniform array factor of the same size. The taper efficiency for binomial arrays is very low. The price paid for lowering sidelobe levels is a decrease in taper efficiency (Table 3.2) and directivity and a corresponding wide main beam. Errors are always present in an array, so even though the amplitude taper results in no sidelobes, the array factor has error sidelobes. Consequently, other more efficient sidelobe tapers with sidelobe levels comparable to the error sidelobes are more desirable.

3.2.2. Dolph–Chebyshev Taper

Rather than totally eliminating the sidelobes by using the binomial coefficients as the array weights, sidelobe levels can be set to a specified level by mapping the array factor to a Chebyshev polynomial [6]. The Chebyshev polynomials are represented by

$$T_m(x) = \begin{cases} \cos(m\cos^{-1}x), & -1 \le x \le 1 \\ \cosh(m\cosh^{-1}x), & |x| > 1 \end{cases} \tag{3.11}$$

When x is between one and minus one, these polynomials oscillate as a cosine function with a maximum amplitude of one (for $m = 0, 1, \ldots , 4$ graphed in Figure 3.8). Outside of that range, they quickly increase or decrease as described by the cosh function in (3.11).

Assume the maximum sidelobe level is 1.0, so that it equals the height of the ripples of the Chebyshev polynomial between $-1 \le x \le 1$. The number of sidelobes corresponds to the number of extrema in the polynomial. An N element array corresponds to a Chebyshev polynomial of order $N - 1$. If the sidelobes are to be sll (in dB) below the peak of the main beam, then the value of the Chebyshev polynomial at the peak of the main beam must equal

$$T_{N-1}(x_{mb}) = 10^{\text{sll}/20} \tag{3.12}$$

Setting (3.11) equal to (3.12) results in the peak of the main beam at

$$x_{mb} = \cosh\left[\frac{\pi A}{N-1}\right] \tag{3.13}$$

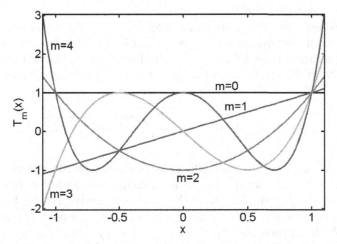

Figure 3.8. Graph of the first four Chebyshev polynomials for $0 \le m \le 4$.

where

$$A = \frac{1}{\pi} \cosh^{-1}(10^{sll/20})$$ (3.14)

Now, the main beam maps to the Chebyshev polynomial. Next, the array factor zeros (nulls) map to the zeros of the Chebyshev polynomial. The zeros of the Chebyshev polynomial are located at

$$x_n = \cos\left[\frac{\pi(n-0.5)}{N-1}\right]$$ (3.15)

Mapping the zeros of the array factor to the zeros of the Chebyshev polynomial is done through the following relation:

$$x_n = x_{mb} \cos\left(\frac{\psi_n}{2}\right)$$ (3.16)

The following equation provides the zeros of the array factor that correspond to a sidelobe level in dB of sll:

$$\psi_n = 2\cos^{-1}\left\{\frac{\cos\left(\frac{(n-0.5)\pi}{N-1}\right)}{\cosh\left(\frac{\pi A}{N-1}\right)}\right\}$$ (3.17)

When the number of elements and sidelobe level are specified, the null locations on the unit circle are easily determined. Once the ψ_n are known, the polynomial in factored form easily follows. Multiplying all the factored terms results in a polynomial of degree $N-1$. The polynomial coefficients are the amplitude weights for the array elements.

It turns out that for a specified sidelobe level, the Dolph–Chebyshev taper has the minimum null-to-null beam width. A reverse process is possible in which the lowest sidelobe level can be found for a specified null-to-null beamwidth [6]. The beam-broadening factor for a Chebyshev array is given by [7]

$$B_b = 1 + 0.636\left\{2 \times 10^{-sll/20} \cosh\sqrt{\left[\cosh^{-1}(10^{sll/20})\right]^2 - \pi^2}\right\}^2$$ (3.18)

Figure 3.9 is a plot of taper efficiency as a function of sidelobe level for Chebyshev arrays having 10, 20, and 40 elements. Smaller arrays have a peak efficiency at a higher sidelobe level than do larger arrays. The peak efficiency of the smaller arrays is higher than the peak efficiency of the larger arrays. As the sidelobe level decreases, the taper efficiency of the larger arrays surpasses that of the smaller arrays.

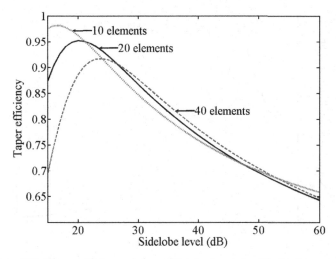

Figure 3.9. Taper efficiency versus sidelobe level for Chebyshev arrays.

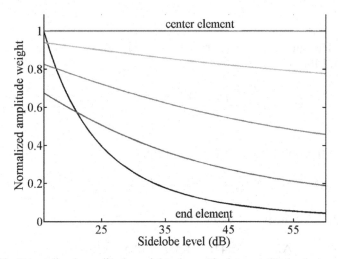

Figure 3.10. Normalized amplitude weights for a 10-element Chebyshev array versus sidelobe level.

Figure 3.10 are the normalized weights for a 10-element Chebyshev array as a function of sidelobe level. The end element does not have the smallest amplitude for sidelobe levels above –22 dB. At –22 dB and below, the weights monotonically decrease from the center to the edge. Similar behavior occurs for larger arrays too. The center amplitude weight for a 40-element array does not become larger than the edge element weight until the sidelobe level is –25 dB or below as shown in Figure 3.11.

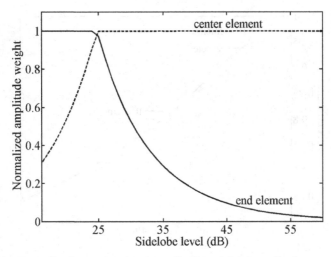

Figure 3.11. Normalized center and end amplitude weights for a 40-element Chebyshev array versus sidelobe level.

Example. Design a 6-element array ($d = 0.5\lambda$) with −20-dB sidelobes.
First, find the ψ_n:

$$\psi_m = 2\cos^{-1}\left[\frac{\pi(m-0.5)}{\cosh(0.1)}\right] = \pm73.2°, \pm120.5°, 180° \tag{3.19}$$

Next, write the polynomial in factored form:

$$AF = \left(z - e^{j73.2}\right)\left(z - e^{-j73.2}\right)\left(z - e^{j120.5}\right)\left(z - e^{-j120.5}\right)\left(z - e^{j180}\right) \tag{3.20}$$

Multiplying these factors together yields

$$AF = z^5 + 1.44z^4 + 1.85z^3 + 1.85z^2 + 1.44z + 1 \tag{3.21}$$

The coefficients of AF are the amplitude weights for the 6-element array pattern in Figure 3.12 with $\eta_T = 0.944$.

Tseng and Cheng developed a Chebyshev synthesis technique for rectangular arrays with an equal number of elements in the x and y directions [8]. They applied the Baklanov transformation [9] to represent the array factor (a function of two angle variables: u and v) as a polynomial of one variable, t.

$$t = t_0 \cos u \cos v \tag{3.22}$$

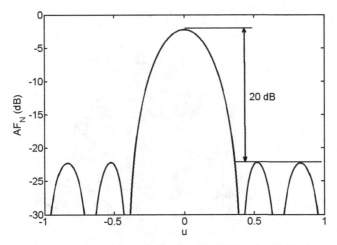

Figure 3.12. Array factor for a 6-element 20-dB Dolph–Chebyshev array.

With some manipulation, the weights for the fourth quadrant of a rectangular array are given by

$$w_{mn} = \sum_{s=\begin{cases} m \text{ if } m \geq n \\ n \text{ if } m < n \end{cases}}^{N} \frac{(-1)^{N-s}(2N-1)}{2(N+s-1)} \binom{N+s-1}{2s-1}\binom{2s-1}{s-m}\binom{2s-1}{s-n}\left(\frac{t}{2}\right)^{2s-1} \qquad (3.23)$$

for $m, n = 1, 2, \dots, N$. The element spacing in the x and y directions can be different, but the number of elements in those directions must be the same.

Example. Calculate the Chebyshev weights for a 30-dB square array with 256 elements. Plot the array factor. The element spacing is square with $d_x = d_y = 0.5\lambda$.

The weights are calculated and shown in Figure 3.13. The array factor appears in Figure 3.14. The weights are real but do not monotonically decrease from the center to the edges.

3.2.3. Taylor Taper

The Chebyshev weighting is practical for a small linear array; but as N becomes large, the amplitude weights at the edge of the aperture increase. Increasing the edge taper presents problems with edge effects and mutual coupling. Other amplitude tapers are better suited for large, low-sidelobe arrays. One such taper was developed by Taylor [10]. The Taylor taper is a continuous taper for line sources that can be sampled for application to antenna arrays.

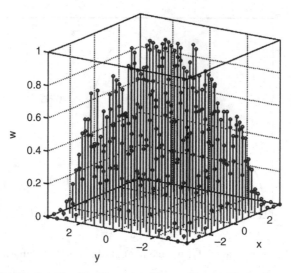

Figure 3.13. Weights for the 30-dB square Tseng–Cheng–Chebyshev array with 256 elements arranged in a square grid having $d_x = d_y = 0.5\lambda$.

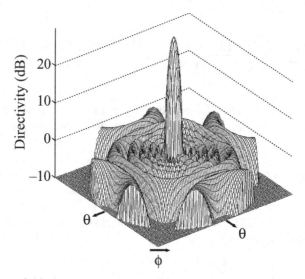

Figure 3.14. Array factor for the Tseng–Cheng–Chebyshev array.

The Taylor taper is similar to the Chebyshev taper in that the maximum sidelobe level can be specified. The difference is that the Taylor taper only has the first $\bar{n} - 1$ sidelobes on either side of the main beam at a specified height. All remaining sidelobes decrease at the same rate as the corresponding sidelobes in a uniform array.

The Taylor taper moves the first $\bar{n} - 1$ nulls on either side of the main beam away from the main beam. This equates to moving the corresponding zeros on the unit circle closer to the negative real axis. Since the other zeros remain untouched, the outer sidelobes and nulls stay in the same locations as those for a uniform array. The null locations for the Taylor array factor are calculated from

$$u_n = \frac{\pm\lambda}{Nd} \begin{cases} \bar{n}\sqrt{\dfrac{A^2 + (n-0.5)^2}{A^2 + (\bar{n}-0.5)^2}}, & n < \bar{n} \\ n, & n \geq \bar{n} \end{cases} \tag{3.24}$$

with A given by (3.14). These null locations translate to zeros on the unit circle by

$$\psi_n = kdu_n \tag{3.25}$$

Next, the factored form of AF is found. Multiplying the terms together gives a polynomial whose coefficients are the Taylor weights.

Example. Find the amplitude weights and array factor for a 20-element Taylor taper with $\bar{n} - 5$ and sidelobes 20 dB below the peak of the main beam.

$$A = 0.95$$

$$u_n = \pm.117, \pm.193, \pm.291, \pm.394, \pm.5, \pm.6, \pm.7, \pm.8, \pm.9, 1$$

Converting u_n into ψ_n produces

$$\pm 0.367, \pm 0.607, \pm 0.914, \pm 1.239, \pm 1.571, \pm 1.885, \pm 2.199, \pm 2.513, \pm 2.827, 3.142$$

and forming the polynomial equation yields

$$\begin{aligned} \text{AF} = {} & 0.667z^{19} + 0.621z^{18} + 0.589z^{17} + 0.624z^{16} + 0.718z^{15} + 0.818z^{14} \\ & + 0.888z^{13} + 0.933z^{12} + 0.972z^{11} + z^{10} + z^9 + 0.972z^8 + 0.933z^7 \\ & + 0.888z^6 + 0.818z^5 + 0.718z^4 + 0.624z^3 + 0.589z^2 + 0.621z + 0.667 \end{aligned}$$

Figure 3.15 and Figure 3.16 show graphs of the amplitude weights and array factors, respectively. The efficiency is given by $\eta_T = 0.965$.

Notice that the first four sidelobes in Figure 3.16 slowly drop off rather than stay constant as expected. The peak sidelobe level is predicted accurately, though.

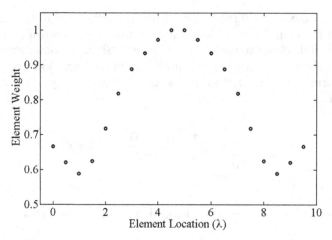

Figure 3.15. Amplitude weights for Taylor 20-dB, $\bar{n} = 5$ taper.

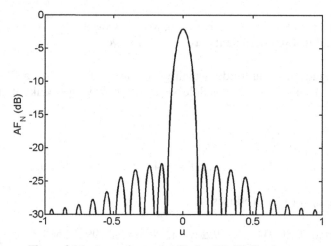

Figure 3.16. Array factor for Taylor 20-dB, $\bar{n} = 5$ taper.

The previous method of obtaining the Taylor amplitude weights is numerically inefficient for a large number of elements due to the many convolutions with the polynomial zeros to obtain the full polynomial. The weights become unsymmetrical for arrays larger than 50 elements. A more direct and numerically stable method is given by the formula

$$w_n = 1 + 2\sum_{m=1}^{\bar{n}-1} \frac{\left[(\bar{n}-1)!\right]^2 \cos\left[m\pi\left(n-1-\dfrac{N-1}{N}\right)\right]}{(\bar{n}-1+m)!(\bar{n}-1-m)!} \prod_{i=1}^{\bar{n}-1}\left[1-\left(\frac{m^2\left[A^2+(\bar{n}-0.5)^2\right]}{\bar{n}^2\left[A^2+(i-0.5)^2\right]}\right)^2\right]$$

$$(3.26)$$

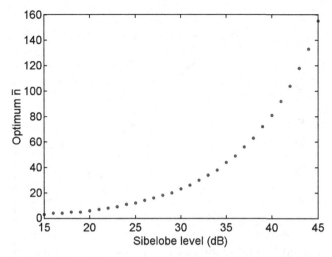

Figure 3.17. Plot of the $\bar{n}$ that yields the most efficient taper for a given sidelobe level.

This formula is easy to program and finds the weights for very large linear arrays but becomes unstable for large values of $\bar{n}$. The formula in (3.26) can be used to calculate the taper up to a sidelobe level of 40 dB and $\bar{n} = 81$ using MATLAB. An $\bar{n}$ above 86 requires higher precision and/or putting (3.26) in a format even less prone to numerical errors. It is recommended to compute the product in (3.26) using logarithms and then converting back by raising the result to the base.

For a specified sidelobe level, there is an $\bar{n}$ that results in a maximum directivity. As $\bar{n}$ increases, more energy goes into the sidelobes until a point is reached where the directivity decreases. As $\bar{n}$ decreases, the beamwidth increases, so power is robbed from the peak of the main beam in order to increase the width of the main beam. The graph in Figure 3.17 indicates the $\bar{n}$ (sidelobe levels between 15 and 45 dB) that results in the largest η_T. Although a large η_T is desirable, it is not the only consideration when implementing an amplitude taper. For small $\bar{n}$, the amplitude taper monotonically decreases from the center to the edge. Above a certain $\bar{n}$ for a given sidelobe level, however, the amplitude taper increases at the edges. Consider the amplitude tapers for a 30 dB Taylor taper with $\bar{n} = 7$ and $\bar{n} = 23$ shown in Figure 3.18. The $\bar{n} = 7$ amplitude taper is the most efficient taper while still having a monotonically decreasing amplitude from the center of the aperture to the edge. The $\bar{n} = 23$ taper has the highest efficiency. In most arrays, a monotonically decreasing amplitude taper is desirable, because the feed network is easier to build and the contribution from the edge elements are minimized. A plot of the $\bar{n}$ that yields the most efficient taper for a given sidelobe level while still having a monotonically decreasing amplitude is shown in Figure 3.19. The $\bar{n}$

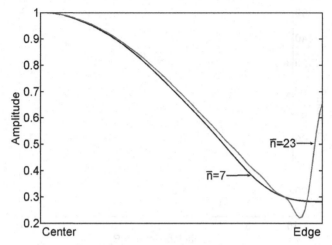

Figure 3.18. Taylor 30-dB amplitude tapers for $\bar{n} = 7$ and $\bar{n} = 23$.

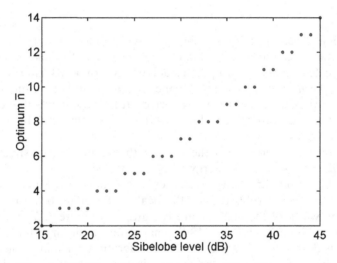

Figure 3.19. Plot of the $\bar{n}$ that yields the highest η_T while still having a monotonically decreasing amplitude taper.

needed for the most efficient taper dramatically increases as the sidelobe level decreases. In general, computing such a high $\bar{n}$ is not necessary, since the monotonically decreasing tapers are more desirable and not that less efficient (see Figure 3.20).

Taylor developed a similar taper for a desired sidelobe level without having to specify $\bar{n}$ [11]. Weights for the one parameter taper are given by

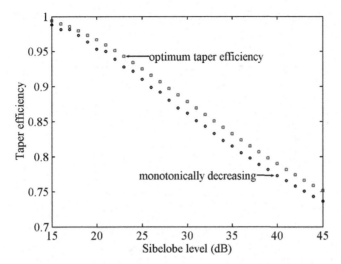

Figure 3.20. Graph of η_T versus sidelobe level for Taylor tapers that have an optimum η_T and the highest η_T while still having a monotonically decreasing amplitude taper.

$$w_n = I_0\left(\pi B\sqrt{1-(r_n/r_{max})^2}\right) \tag{3.27}$$

where I_0 is the zeroth-order modified Bessel function of the first kind, B is the Taylor parameter, $r_n = \sqrt{x_n^2 + y_n^2}$ is the distance of element from origin, and r_{max} is the array radius. The Taylor one parameter is related to the relative sidelobe level of the array and can be found by solving

$$13.26 + 20\log_{10}\frac{\sinh(\pi B)}{\pi B} - \text{sll}_{dB} = 0 \tag{3.28}$$

where sll_{dB} is the maximum relative sidelobe level in decibels. Figure 3.21 is a plot of the efficiency at a particular sidelobe level for the Taylor one parameter taper. This taper is not as efficient as those displayed in Figure 3.20.

Taylor extended his amplitude taper for a continuous line source to a continuous circular aperture [12]. The following formula calculates the amplitude weights for a discrete Taylor taper for a circular aperture:

$$w(x,y) = 1 - \sum_{m=1}^{\bar{n}-1} \frac{J_0\left(\dfrac{r\mu_m}{r_{max}}\right)}{J_0(\mu_m)} \frac{\displaystyle\prod_{n=1}^{\bar{n}-1}\left[1-\dfrac{\mu_m^2}{U_n^2}\right]}{\displaystyle\prod_{\substack{n=1\\n\neq m}}^{\bar{n}-1}\left(1-\dfrac{\mu_m^2}{\mu_n^2}\right)} \tag{3.29}$$

where $r = \sqrt{x^2 + y^2}$ is the distance of element from array center; r_{max} is the radius of circular array; $J_1(\mu_m) = 0$, $m = 1, \dots, \bar{n}$; $\mu_m = $ is the zeros of J_1; and

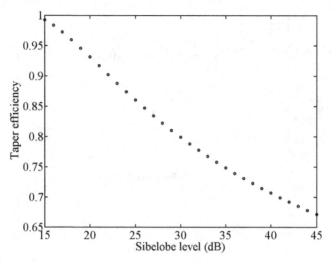

Figure 3.21. Efficiency versus sidelobe level for the Taylor one parameter amplitude taper.

$U_n^2 = \mu_{\bar{n}}^2 \dfrac{A^2 + (n - .5)^2}{A^2 + (\bar{n} - .5)^2}$. As with the continuous line source, samples of this taper are taken at the normalized element locations.

Example. Calculate the 30-dB, $\bar{n} = 5$ Taylor weights for a circular array with 284 elements. Plot the array factor when the elements are on a square grid with $d_x = d_y = 0.5\lambda$.

First, calculate $A = 1.32$. Next, find the Bessel function zeros and the null movements.

n	μ_n	U_n
1	3.8317	4.9574
2	7.0156	7.0176
3	10.1735	9.9291
4	13.3237	13.1377
5	16.4706	16.4706

Finally the weights are calculated and shown in Figure 3.22. Weight values are difficult to read from Figure 3.22, so a plot of the weights as a function of normalized distance from the center of the array is shown in Figure 3.23. The corresponding far field pattern is shown in Figure 3.24 with a cut in the $\phi = 0°$ plane shown in Figure 3.25.

It is possible to find the $\bar{n}$ for a given sidelobe level that results in the highest taper efficiency while maintaining a monotonically decreasing amplitude taper

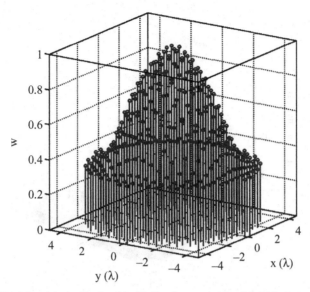

Figure 3.22. Weights for the Taylor 30-dB, $\bar{n} = 5$ array with with 284 elements arranged in a square grid having $d_x = d_y = 0.5\lambda$.

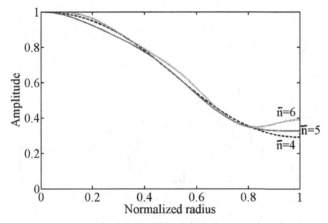

Figure 3.23. Taylor 30-dB, $\bar{n} = 5$ weights.

(Figure 3.26). The previous example selected $\bar{n} = 5$ for a 30-dB sidelobe level. If $\bar{n} = 4$ were selected, then the dashed line in Figure 3.23 results. It has less area under the curve, so it is less efficient. If $\bar{n} = 6$ were selected, then the dotted line in Figure 3.23 results. Note that this taper is more efficient but is not monotonically decreasing. Figure 3.27 is a graph of the taper efficiency versus sidelobe level when the Taylor taper has the highest η_T while still having a monotonically decreasing amplitude taper.

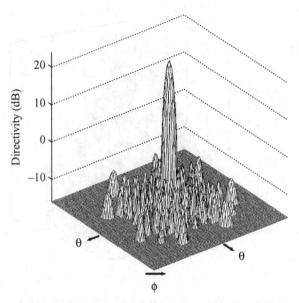

Figure 3.24. Array factor for the weights in Figure 3.22.

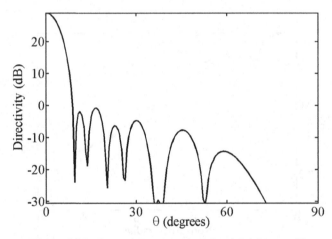

Figure 3.25. Array factor as a function of θ for $\phi = 0°$.

Hansen developed a low-sidelobe distribution for circular apertures that has a desired maximum sidelobe level and is specified by a single parameter [13]. His approach follows that of Taylor for the Taylor one-parameter taper. The amplitude distribution is given by

$$w_n = I_0\left(\pi H \sqrt{1 - (r_n/r_{\max})^2}\right) \qquad (3.30)$$

where H is the Hansen parameter found by solving

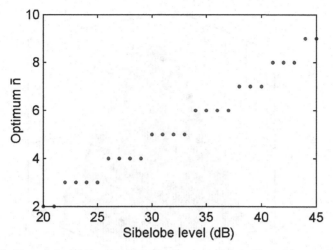

Figure 3.26. Plot of the $\bar{n}$ that yields the highest η_T while still having a monotonically decreasing circular Taylor amplitude taper.

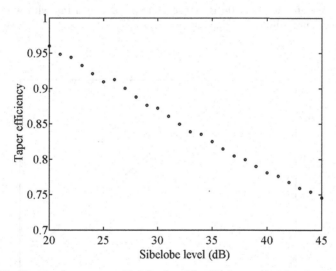

Figure 3.27. Graph of η_T versus sidelobe level for Taylor tapers that have the highest η_T while still having a monotonically decreasing amplitude taper.

$$17.57 + 20\log_{10}\frac{2I_1(\pi H)}{\pi H} - \text{sll}_{dB} = 0 \qquad (3.31)$$

and I_1 is the first-order modified Bessel function of the first kind.

Example. Calculate the 30-dB Hansen one-parameter weights for a circular array with 284 elements. Plot the array factor. The square element lattice has $d_x = d_y = 0.5\lambda$.

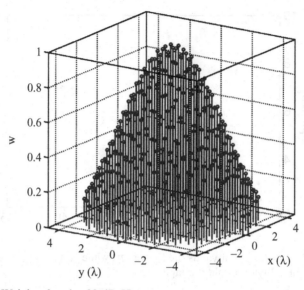

Figure 3.28. Weights for the 30-dB Hansen one-parameter array with with 284 elements arranged in a square grid having $d_x = d_y = 0.5\lambda$.

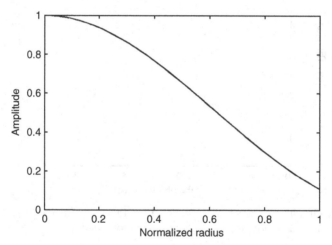

Figure 3.29. The 30-dB Hansen one-parameter weights.

First, find $H = 1.1977$. Next, the weights are calculated and shown in Figure 3.28. Weight values are difficult to read from Figure 3.28, so a plot of the weights as a function of normalized distance from the center of the array is shown in Figure 3.29. The corresponding far field pattern is shown in Figure 3.30 with a cut in the $\phi = 0°$ plane shown in Figure 3.31.

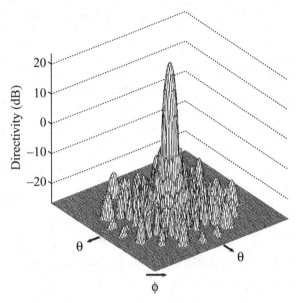

Figure 3.30. Array factor for the weights in Figure 3.28.

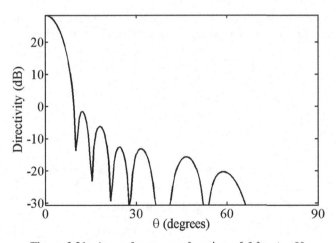

Figure 3.31. Array factor as a function of θ for $\phi = 0°$.

Figure 3.32 is a graph of the efficiency versus sidelobe level for the Hansen one-parameter amplitude taper. In general, the sidelobe levels drop off dramatically which produces a much less efficient Taper than is possible with the Taylor weights.

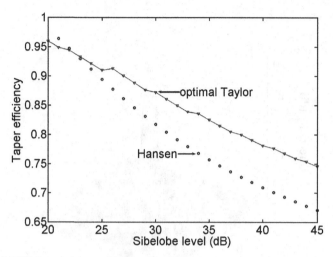

Figure 3.32. Efficiency versus sidelobe level for the Hansen one-parameter amplitude taper.

3.2.4. Bickmore–Spellmire Taper

The Bickmore–Spellmire taper for a linear or circular array encompasses some of the amplitude tapers already covered and can be written as [14]

$$w_n = \left[1 - \left(\frac{r_n}{r_{\max}}\right)^2\right]^q \Lambda_q\left(j\pi A\sqrt{1 - \left(\frac{r_n}{r_{\max}}\right)^2}\right) \tag{3.32}$$

where

$$\Lambda_v(z) = \Gamma(v+1)\left(\frac{2}{z}\right)^v J_v(z) = \text{lambda function}$$

$$\Gamma(n) = (n-1)! = \text{gamma function}$$

$$q = \begin{cases} v - 1/2 & \text{for a line source} \\ v - 1 & \text{for a circular aperture} \end{cases}$$

with A given by (14). Two special cases of interest occur when $v = 1/2$ (Taylor's one-parameter) and $v = 1$ (Hansen's one-parameter). For a given v, A can be varied to effect a tradeoff between beamwidth and peak sidelobe level. Picking A holds the beamwidth constant, so v governs a tradeoff between the peak sidelobe level near the main beam and the height of the more remote sidelobes.

3.2.5. Bayliss Taper

Bayliss developed a low sidelobe taper for difference patterns that is analogous to the Taylor taper for sum patterns [15]. The Bayliss taper has $\bar{n} - 1$ equally high sidelobes on either side of the main beam, while the rest decrease away from the main beam as with a uniform difference pattern. Null locations are given by

$$u_n = \frac{\lambda}{Nd}\begin{cases} 0, & n = 0 \\ (\bar{n}+0.5)\sqrt{\dfrac{q_n}{B^2+\bar{n}^2}}, & 1 \leq \bar{n} \leq 4 \\ (n+0.5)\sqrt{\dfrac{B^2+n^2}{B^2+\bar{n}^2}}, & 5 \leq \bar{n} \leq \bar{n}-1 \\ n, & n \geq \bar{n} \end{cases} \qquad (3.33)$$

where

$$B = 0.3038753 + \text{sll}\{0.05042922 + \text{sll}[-2.7989 \times 10^{-4} \\ + \text{sll}(3.43 \times 10^{-6} - 2 \times 10^{-8}\,sll)]\}$$

$$q_1 = 0.9858302 + \text{sll}\{0.0333885 + \text{sll}[1.4064 \times 10^{-4} + \text{sll}(-1.9 \times 10^{-6} + 1 \times 10^{-8}\,sll)]\}$$

$$q_2 = 2.00337487 + \text{sll}\{0.01141548 + \text{sll}[4.159 \times 10^{-4} \\ + \text{sll}(-3.73 \times 10^{-6} + 1 \times 10^{-8}\,sll)]\}$$

$$q_3 = 3.00636321 + \text{sll}\{0.00683394 + \text{sll}[2.9281 \times 10^{-4} + \text{sll}(-1.61 \times 10^{-6})]\}$$

$$q_4 = 4.00518423 + \text{sll}\{0.00501795 + \text{sll}[2.1735 \times 10^{-4} + \text{sll}(-8.8 \times 10^{-7})]\}$$

$$(3.34)$$

The peak of the difference array factor is located at

$$0.4797212 + \text{sll}\{0.01456692 + \text{sll}[-1.8739 \times 10^{-4} + \text{sll}(2.18 \times 10^{-6} - 1 \times 10^{-8}\text{sll})]\}$$

$$(3.35)$$

Example. Find the amplitude weights and array factor for a 20-element Bayliss taper with $\bar{n} = 5$ and -20-dB sidelobes.

$$B = 0.95$$

$$u_n = \pm.117, \pm.193, \pm.291, \pm.394, \pm.5, \pm.6, \pm.7, \pm.8, \pm.9, 1$$

Converting u_n into ψ_n produces

$$\pm 0.367, \pm 0.607, \pm 0.914, \pm 1.239, \pm 1.571, \pm 1.885, \pm 2.199, \pm 2.513, \pm 2.827, 3.142$$

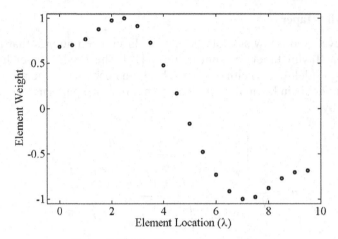

Figure 3.33. Bayliss amplitude weights.

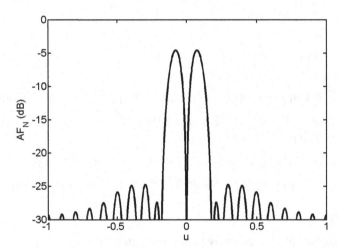

Figure 3.34. Bayliss 20-element array factor.

and forming the polynomial equation yields

$$\begin{aligned} AF = {} & 0.667z^{19} + 0.621z^{18} + 0.589z^{17} + 0.624z^{16} + 0.718z^{15} + 0.818z^{14} \\ & + 0.888z^{13} + 0.933z^{12} + 0.972z^{11} + z^{10} - z^{9} - 0.972z^{8} - 0.933z^{7} \\ & - 0.888z^{6} - 0.818z^{5} - 0.718z^{4} - 0.624z^{3} - 0.589z^{2} - 0.621z - 0.667 \end{aligned}$$

The resulting amplitude weights and array factor are shown in Figures 3.33 and 3.34.

In the same paper, Bayliss developed his taper for a circular difference array by moving the zeros of the first-order Bessel function. Half of the aperture receives a 180° phase shift. The weights are given by the formula

$$w(x, y) = \cos\phi \sum_{m=0}^{\bar{n}-1} \frac{J_1\left(\dfrac{r\mu'_m}{r_{max}}\right)}{J_1(\mu'_m)} \frac{\displaystyle\prod_{n=1}^{\bar{n}-1}\left[1 - \dfrac{\mu'^2_m}{U_n^2}\right]}{\displaystyle\prod_{\substack{i=0\\i\neq m}}^{\bar{n}-1}\left(1 - \dfrac{\mu'^2_m}{\mu'^2_n}\right)} \tag{3.36}$$

where

$$J'_1(\mu'_m) = 0, \qquad m = 0, 1, \ldots, \bar{n}$$

$$\mu'_m = \text{zeros of } J'_1$$

$$U_n^2 = \mu'^2_{\bar{n}}\frac{A^2 + n^2}{A^2 + \bar{n}^2}$$

with A given by (3.14). The $\cos\phi$ term forces the weights in half of the aperture $(x < 0)$ to be negative. The null occurs at $\phi = 90°$ or $270°$ for all θ.

Example. Calculate the 30-dB $\bar{n} = 5$ Bayliss weights for a circular array with 284 elements. Plot the array factor. The element spacing is square with $d_x = d_y = 0.5\lambda$.
    First, calculate $A = 1.32$. Next, find the zeros of $J_1(r)$ and the null movements.

n	μ'_n	U_n
0	1.8412	—
1	5.3314	7.0894
2	8.5363	8.9951
3	11.706	11.747
4	14.864	14.815
5	18.016	—

    Finally, the weights are calculated and shown in Figure 3.35. Weight values are difficult to read from Figure 3.35, so a plot of the weights as a function of normalized distance from the center of the array is shown in Figure 3.36. The corresponding far field pattern is shown in Figure 3.37 with a cut in the $\phi = 0°$ plane shown in Figure 3.38.

3.2.6. Unit Circle Synthesis of Arbitrary Linear Array Factors

Sometimes the standard amplitude tapers are not sufficient to meet design specifications. This section presents several examples of using the unit circle to create desirable linear array factors [16]. The consequence of most of these designs is that the element weights are complex rather than real. If each array polynomial root has a complex conjugate pair, then the weights are real.

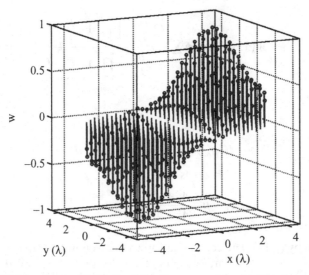

Figure 3.35. Weights for the 30-dB, $\bar{n} = 5$ Bayliss array with with 284 elements arranged in a square grid having $d_x = d_y = 0.5\lambda$.

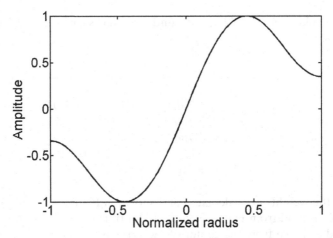

Figure 3.36. Bayliss 30-dB, $\bar{n} = 5$ weights.

Otherwise, they are complex. Elliot presents numerous examples of synthesizing modified Taylor patterns and null free array factors in his book [17]. These topics are presented as examples here. Their practical value is limited and numerical synthesis methods provide superior results today.

Example. In this example, very low sidelobes are needed for $u < 0$ while only low sidelobes are needed for $u > 0$. The desired array factor has the following characteristics:

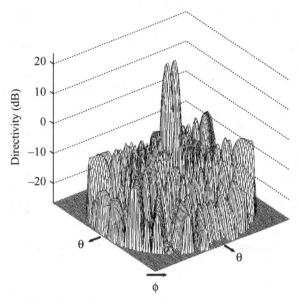

Figure 3.37. Array factor for the weights in Figure 3.35.

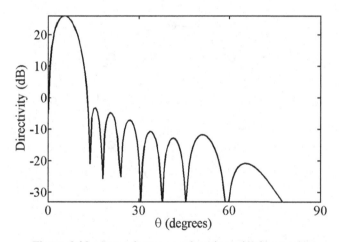

Figure 3.38. Array factor as a function of θ for $\phi = 0°$.

1. Taylor $\bar{n} = 5$ sll = 25 dB for $u < 0$.
2. Taylor $\bar{n} = 3$ sll = 15 dB for $u > 0$.

This is a two-step process. First, the roots are found for the 25-dB taper and the 15-dB taper. Next, only the roots from the 15-dB taper that have an imaginary part less than zero are kept, then the roots from the 25-dB taper that have an imaginary part greater than or equal to zero are kept. These roots are

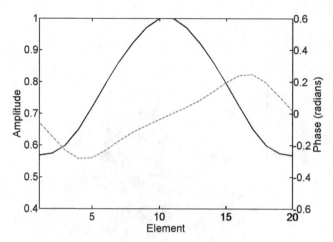

Figure 3.39. Synthesized amplitude weights (solid line) and phase weights (dashed line).

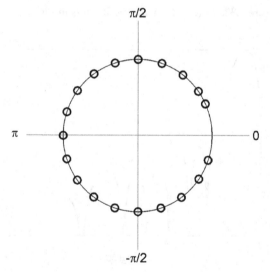

Figure 3.40. Unit circle of asymmetrical Taylor weights.

then placed in factored form and multiplied to get the array polynomial. The complex weights of this polynomial are the weights shown in Figure 3.39. The corresponding unit circle is in Figure 3.40 and the array factor in Figure 3.41.

Example. Start with a Taylor $\bar{n} = 4$ sll = 25-dB taper and place a null at $u = 0.25$ when $d = 0.5\lambda$. Do not allow complex weights.

First, the roots of the Taylor taper are found:

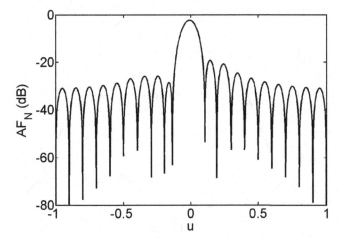

Figure 3.41. Array factor of asymmetrical Taylor weights.

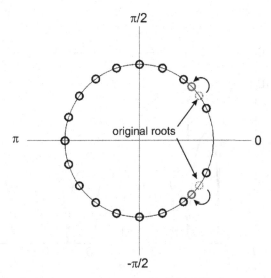

Figure 3.42. Unit circle representation of the initial Taylor taper and the null at $u = 0.25$.

$-1.0000, -0.9509 \pm j0.3094, -0.8085 \pm j0.5885, -0.5867 \pm j0.8098, -0.3072 \pm j0.9516, 0.0026 \pm j1.0000, 0.3124 \pm j0.9499, 0.5977 \pm j0.8017, 0.8092 \pm j0.5875, 0.9066 \pm j0.4220$

For a null at $u = 0.25$, $r_{null} = \cos(.25\,kd) + j\sin(.25\,kd) = 0.7071 + j0.7071$.

To keep the weights real, then the roots should be complex conjugate pairs. Replace the roots $0.8092 \pm j0.5875$ with the roots $0.7071 \pm j0.7071$. The new roots are shown on the unit circle in Figure 3.42. Weights from the Taylor

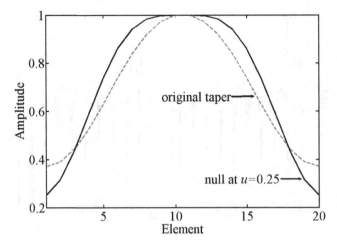

Figure 3.43. Amplitude weights for the Taylor taper and the null at $u = 0.25$.

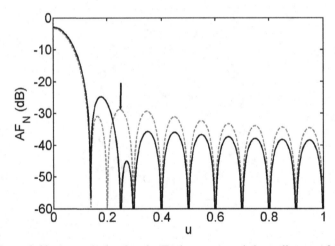

Figure 3.44. Array factor for the Taylor taper and the null at $u = 0.25$.

taper and the nulled taper are in Figure 3.43. The two array factors are shown in Figure 3.44.

Example. Modify a 20-element uniform array with $d = 0.5\lambda$ so that it has a flat top from $-0.1 < u < 0.1$ and sidelobes 20 dB below the main beam.

This problem could be done best using numerical methods. Here, a good guess is used instead. A zero placed on the real axis and outside the unit circle

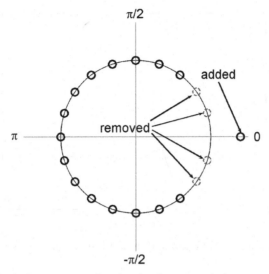

Figure 3.45. Unit circle representation for the flat top main beam and uniform arrays.

in order to flatten the main beam. Also, the first two zeros on either side of the main beam are removed to give the main beam room to expand. Adding one zero and removing 4 zeros while maintaining $d = 0.5\lambda$ necessitates that there are only 17 elements in the new array. The resulting weights are

$$w = [0.026 \quad 0.081 \quad 0.144 \quad 0.174 \quad 0.132 \quad -0.003 \quad -0.224 \quad -0.493 \quad -0.750$$
$$-0.934 \quad -1.0 \quad -0.936 \quad -0.766 \quad -0.539 \quad -0.314 \quad -0.139 \quad -0.037]$$

The uniform array and the flat top main beam array are compared via the unit circle (Figure 3.45), the amplitude weights (Figure 3.46), and the array factors (Figure 3.47).

3.2.7. Partially Tapered Arrays

A partial array taper uses only a subset of all the element weights to control the array factor [18]. Center elements have an amplitude of 1 and a phase of 0. Edge element weights are synthesized using an appropriate numerical method. The partially tapered array factor is

$$\text{AF} = \sum_{n=1}^{N_1} w_n e^{jkx_n u} + \sum_{n=N_1+1}^{N_1+N_{cen}} e^{jkx_n u} + \sum_{n=N_1+N_{cen}+1}^{N_{cen}+N_1+N_2} w_n e^{jkx_n u} \qquad (3.37)$$

where N_{cen} is the number of uniformly weighted elements in the center and N_1 and N_2 are the number of weighted edge elements. The center Fourier

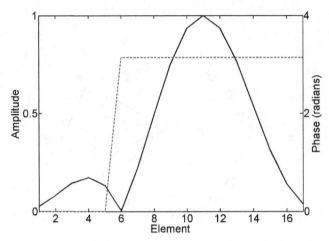

Figure 3.46. Amplitude (solid line) and phase (dashed line) weights for the flat top main beam array.

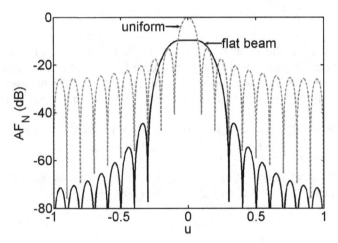

Figure 3.47. Array factors for the flat top main beam and uniform arrays.

coefficients are all one. If the array has an even number of equally spaced elements and is symmetric about the center of the array, then $N_1 = N_2 = (N - N_{cen})/2$ and the array factor is written as

$$\text{AF} = 2 \sum_{n=N_{cen}/2+1}^{N} w_n \cos(kx_n u) + \frac{\sin[N_{cen}\psi/2]}{\sin(\psi/2)} \tag{3.38}$$

Equation (3.38) may be rewritten in matrix form

$$
\begin{bmatrix}
\cos\left(kx_{N_{cen}/2+1}u_1\right) & \cos\left(kx_{N_{cen}/2+2}u_1\right) & \cdots & \cos\left(kx_{N/2}u_1\right) \\
\cos\left(kx_{N_{cen}/2+1}u_2\right) & \cos\left(kx_{N_{cen}/2+2}u_2\right) & \cdots & \cos\left(kx_{N/2}u_2\right) \\
\vdots & \vdots & \ddots & \vdots \\
\cos\left(kx_{N_{cen}/2+1}u_M\right) & \cos\left(kx_{N_{cen}/2+2}u_M\right) & \cdots & \cos\left(kx_{N/2}u_M\right)
\end{bmatrix}
\begin{bmatrix}
w_{N_{cen}/2+1} \\
w_{N_{cen}/2+2} \\
\vdots \\
w_{N/2}
\end{bmatrix}
$$

$$
=
\begin{bmatrix}
AF(u_1) - \dfrac{\sin\left[N_{cen}kdu_1/2\right]}{\sin\left(kdu_1/2\right)} \\[2ex]
AF(u_2) - \dfrac{\sin\left[N_{cen}kdu_2/2\right]}{\sin\left(kdu_2/2\right)} \\[2ex]
\vdots \\[1ex]
AF(u_M) - \dfrac{\sin\left[N_{cen}kdu_M/2\right]}{\sin\left(kdu_M/2\right)}
\end{bmatrix}
\tag{3.39}
$$

The weights in (3.39) can be found using least squares described earlier or one of the numerical methods described in the next section.

3.3. NUMERICAL SYNTHESIS OF LOW-SIDELOBE TAPERS

Analytical approaches to finding optimum array amplitude weights are still used today. They work well because the unknown array weights are coefficients of a complex Fourier series. If the unknowns are the element spacings or element phases, then they appear in the complex exponent and are not easily found. Checking all combinations of values of the array variables is not realistic unless the number of variables is small. Optimizing one variable at a time does not work nearly as well as following the gradient vector downhill. The steepest descent method, invented in the 1800s, is based on this concept and is still widely used today [19]. Newton's method uses second-derivative information in the form of the Hessian matrix to find the minimum. Although more powerful than steepest descent, calculating the second derivative of the cost function may be too difficult.

In order to avoid the calculation of derivatives, Nelder and Mead introduced the downhill simplex method in 1965 [20]. This technique has become widely used by commercial computing software like MATLAB because it is very stable. A simplex has $n + 1$ sides in n-dimensional space. Each iteration generates a new vertex for the simplex. A new point replaces the worst vertex when it is better. In this way the simplex gets smaller and the solution becomes more accurate.

Successive line minimization methods were developed in the 1960s [21]. A successive line minimization algorithm starts at a random point and then moves in a predetermined direction until the cost function increases. Then, it tries a new direction. A conjugate direction does not interfere with the minimization of the prior direction. Powell developed a technique in which changes to the gradient of the cost function remain perpendicular to the previous conjugate directions [22]. The BFGS algorithm [23–26] approximates the Hessian matrix (square matrix of second-order partial derivatives) in order to calculate the next search point. This algorithm is "quasi-Newton" in that it is equivalent to Newton's method for prescribing the next best point to use for the iteration, yet it doesn't use an exact Hessian matrix. Quadratic programming is a technique that assumes that the cost function is quadratic (variables are squared) and that the constraints are linear. It is based upon Lagrange multipliers and requires derivatives or approximations to derivatives [19].

Numerical optimization has been used to find nonuniform element spacings, complex weights, and phase tapers that resulted in desired antenna patterns. Some examples of nonuniform spacing synthesis include dynamic programming [27], Nelder Mead downhill simplex algorithm [28], steepest descent [29], and simulated annealing [30]. Numerical methods were used to iteratively shape the main beam while constraining sidelobe levels for planar arrays [31–33]. Linear programming [34] and the Fletcher–Powell method [35] were applied to optimizing the footprint pattern of a satellite planar array antenna. Quadratic programming was used to optimize aperture tapers for various planar array configurations [36, 37]. Numerical optimization was used to find phase tapers that maximized the array directivity [38], and a steepest descent algorithm was used to find the optimum phase taper to minimize sidelobe levels [39].

The cost function returns the values of an attribute of an array antenna that are to be minimized. As an example, consider finding the minimum maximum sidelobe level of a 6-element array by adjusting either the amplitude weights, element spacing, or phase weights of a linear array that lies along the x axis [40]. The spacing, amplitude weights, and phase weights are symmetric with respect to the center of the array, so only the right half of the array needs to be specified. In order to visualize the cost surface, only two variables can be used. The center two elements have an amplitude of one and a phase of zero. Figure 3.48 is the cost function when the amplitude weights are the optimization variables with limits $0.1 \leq a_{2,3} \leq 1.0$, and $\delta_{1,2} = 0$ and $x_1 = 0.25\lambda$, $x_2 = 0.75\lambda$, and $x_3 = 1.25\lambda$. Figure 3.49 is the cost function when $a_{2,3} = 1.0$, $0 \leq \delta_{1,2} \leq \pi$, and $x_1 = 0.25\lambda$, $x_2 = 0.75\lambda$, and $x_3 = 1.25\lambda$. Figure 3.50 is the cost function when $a_{2,3} = 1.0$ and $\delta_{1,2} = 0$, and the element spacings are bound by $x_1 = 0.25\lambda$, $x_2 = 0.25\lambda + \Delta_2$, and $x_3 = 0.25\lambda + \Delta_2 + \Delta_3$.

All the cost functions in these figures have ridges, narrow valleys, and dramatic variations in slope. The cost surface variations slows the convergence of local minimization algorithms. Speed of convergence and quality of the minimum depends upon the starting point. For the 6-element case, the local minimization algorithms find the true minimum most of the time. On the other hand, adding more array variables increases the cost surface complexity by introducing many other local minima that fool local optimizers.

Figure 3.51 is a graph of the maximum sidelobe level in decibels versus the thinning configuration for a 32-element array. Elements in the array are either

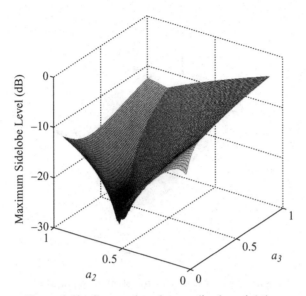

Figure 3.48. Cost surface for amplitude weighting.

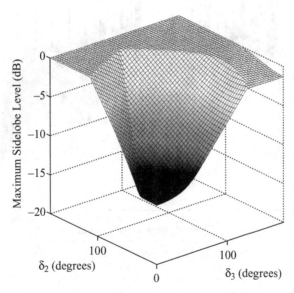

Figure 3.49. Cost surface for phase weighting.

turned on with an amplitude of 1 or turned off with an amplitude of 0. The end elements are always on, and the array is assumed to be symmetric. Values along the x axis are the decimal versions of the 15-bit binary thinning configuration. As an example, one of the thinned array configurations is

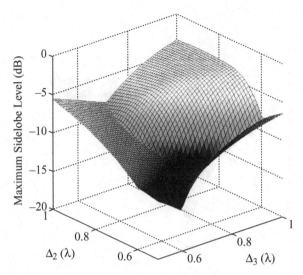

Figure 3.50. Cost surface for element separation.

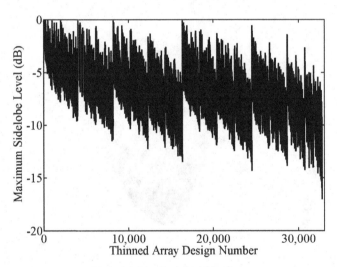

Figure 3.51. Cost for thinning.

$$101111010011010\underbrace{1010110010111101}_{=22110}1 \qquad (3.40)$$

There are a total of 2^{15} possible thinning configurations. Not only is the cost surface riddled with local minima, but the variable values are discrete.

A minimum seeking algorithm cannot find the global minimum in Figure 3.51 unless it has a lucky first guess at an initial starting point. In order to get out of a local minimum, the optimization algorithm incorporates random components so it can jump to different positions on the cost surface. Trying many different random starting points for a local optimizer helps but is less powerful than global optimization methods that have been developed. Simulated annealing is an excellent global optimization approach [41, 42] modeled after the annealing. Annealing is the process that heats a substance above its melting temperature, then gradually cools it to produce a crystalline lattice that has a minimum energy probability distribution. A genetic algorithm is part of a larger field of evolutionary computations. This approach models genetics and natural selection in order to optimize a design [43, 44]. Other natural optimization algorithm offshoots have also proven useful.

The genetic algorithm begins with a random set of array configurations (rows of a matrix called the population) consisting of variables such as element amplitude, phase, and spacing. Each array configuration is evaluated by the cost function that returns a value like maximum sidelobe level. Array configurations have either binary or continuous values. Array configurations with high costs are discarded, while array configurations with low costs form a mating pool. Two parents are randomly selected from the mating pool. Selection is inversely proportional to the cost. Offspring result from a combination of the parents. The offspring replace the discarded array configurations. Next, random array configurations in the population are randomly modified or mutated. Finally, the new array configuration are evaluated for their costs and the process repeats. A flow chart of a genetic algorithm appears in Figure 3.52.

Since its introduction, the genetic algorithm has become a dominant numerical optimization algorithm in many disciplines. Details on implementing a genetic algorithm can be found in reference 45, and a variety of applications to electromagnetics are reported in reference 46. Some of the advantages of a genetic algorithm include:

- Optimizing continuous or discrete variables
- Avoiding calculation of derivatives
- Handling a large number of variables
- Suitability for parallel computing
- Jumping out of a local minimum
- Providing a list of optimum variables, not just a single solution
- Encoding the variables so that the optimization is done with the encoded variables
- Working with numerically generated data, experimental data, or analytical functions

Other, similar algorithms modeled after nature's methods of optimization have also been applied to the optimization of arrays.

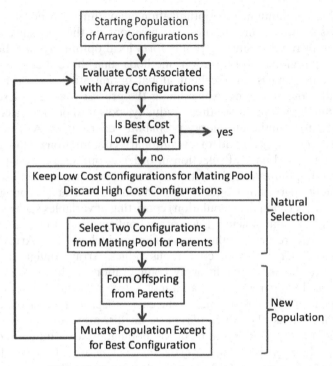

Figure 3.52. Flow chart of a genetic algorithm.

3.4. APERIODIC ARRAYS

So far, only arrays with equally spaced elements have been considered. Nonperiodic element spacing is another approach to synthesizing low sidelobes. Areas of the aperture with a high element density have a higher effective amplitude than do areas with a low element density. Two approaches considered in this section are thinned and nonuniformly spaced arrays.

3.4.1. Thinned Arrays

A thinned or density tapered array turns elements off in a uniform array with a periodic lattice in order to obtain a spatial taper that results in low sidelobes [47]. The elements that are turned off are connected to a matched load and deliver no signal to form a beam. The elements that are turned off are not removed from the aperture, so the element lattice is not disturbed.

3.4.1.1. Statistical Density Tapering. The normalized desired amplitude taper serves as a probability density function for a uniform array that is to be thinned [48]. In a thinned array, the elements either have an amplitude of one or zero. Elements with an amplitude of one are connected to the feed network, while elements with an amplitude of zero are connected to a matched load and do not contribute to a signal to the array output. Elements that correspond to a high amplitude have a greater probability of being turned on than those that correspond to a low-sidelobe amplitude taper. This type of taper has some advantages including:

- A cheap method to implement an amplitude taper. Designing, building, and testing low-sidelobe feed networks is expensive. Thinned arrays use cheap uniform feed networks.
- A narrow beamwidth for a small number of active elements. Active elements are more expensive, especially if each element has a transmitter and/or receiver.
- The mutual coupling is more well-defined than for an array with variable spacing between elements. Knowing the mutual coupling effects makes the array more predictable and easier to design.

Thinning works best for large arrays, since the statistics are more reliable for a large number of elements.

Use the following steps to design a statistically thinned array:

1. Normalize the desired amplitude taper, a_n^{desired}.
2. This normalized amplitude taper looks like a probability density function. The probability that an element is on is equal to its desired normalized amplitude.
3. Generate a uniform random number, r, $0 \leq r \leq 1$.
4. $w_n = \begin{cases} 1 & \text{if } r \leq w_n^{\text{desired}} \\ 0 & \text{if } r > w_n^{\text{desired}} \end{cases}$
5. This process repeats for each element in the array.

The resulting thinned array is not symmetric unless done for only half the array, and the other half is a mirror image.

The total number of elements in the array is the sum of the elements turned on and the elements turned off.

$$N = N_1 + N_0 \qquad (3.41)$$

where N is the total number of elements, N_1 is the number of active elements, and N_0 is the number of inactive elements. The directivity and sidelobe level depend upon the number of active elements. As an example, the directivity of a thinned array with half-wavelength spacing is

$$D_{\text{thin}} = N_1 \qquad (3.42)$$

A high taper efficiency is desirable, so η_t is a merit factor when designing thinned arrays. Comparing the directivity of the thinned array to that of a uniform array with $d = 0.5\lambda$ yields a taper efficiency given by

$$\eta_t = \frac{N_1}{N} \tag{3.43}$$

An expression for the rms sidelobe level of a thinned array is given by [49]

$$\overline{\text{sll}^2} = \frac{1}{N_1} \tag{3.44}$$

where $\overline{\text{sll}^2}$ is the power level of the average sidelobe level. An expression for the peak sidelobe level is found by assuming all the sidelobes are within three standard deviations of the rms sidelobe level and has the form [49]

$$P\left(\text{all sidelobes} < \text{sll}_p^2\right) \simeq \left(1 - e^{-\text{sll}_p^2/\overline{\text{sll}^2}}\right)^{N/2} \tag{3.45}$$

for linear and planar arrays having half-wavelength spacing.

Example. A 5000-element linear array has a taper efficiency of 50% and elements spaced a half-wavelength apart. What are the average and peak sidelobe levels of this array?

$$\overline{\text{sll}^2} = \frac{1}{2500} = 0.0004$$
$$= 10\log_{10}(0.0004) = -34\,\text{dB}$$

The peak sidelobe level is found by manipulating (3.45).

$$\text{sll}_p^2 = -\frac{1}{N_{on}} \ln\left(1 - P^{2/N}\right)$$

The peak sidelobe level of this array at which 90% of the sidelobes fall below is given by

$$\text{sll}_p^2 = -\frac{1}{2500} \ln\left(1 - 0.9^{2/5000}\right)$$
$$= 4.0298 \times 10^{-3} = -24\,\text{dB}$$

Increasing the probability to 0.99 yields a sidelobe level of

$$-\frac{1}{2500} \ln\left(1 - 0.99^{2/5000}\right) = -23\,\text{dB}$$

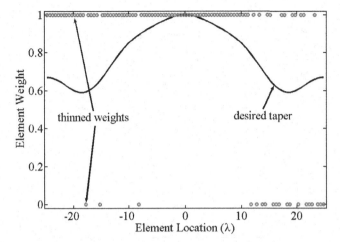

Figure 3.53. The desired Taylor amplitude taper is shown as a solid line. The thinned array amplitude weights are shown as small circles.

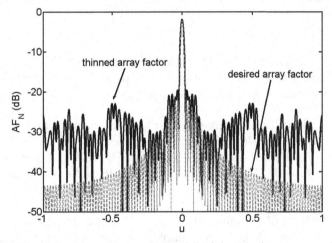

Figure 3.54. The desired array factor (dashed line) is superimposed on the thinned array factor (solid line).

Example. Start with a 100-element array with element spacing $d = 0.5\lambda$, and thin to a 20-dB, $\bar{n} = 5$ Taylor taper.

The normalized amplitude taper shown as a solid line in Figure 3.53 serves as the probability cutoff for selecting elements turned on or off. The small circles at the top or bottom are the amplitude weights of the thinned array. The thinned aperture is not symmetric. The desired and thinned array factors appear in Figure 3.54. The thinned array factor has relatively constant sidelobe levels, while the sidelobes of the Taylor array decrease away from the main beam.

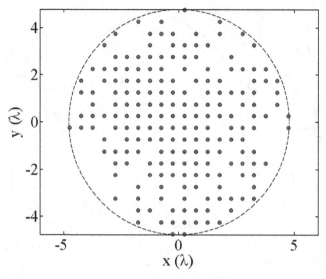

Figure 3.55. Thinned array based upon a 20-dB $\bar{n} = 2$ Taylor circular amplitude distribution.

Statistical thinning can also be applied to planar arrays [50]. The amplitude taper serves as a two-dimensional proability distribution function. This concepts works for any shape or element spacing.

Example. Statistically thin a circular array with 284 elements and $d_x = d_y = 0.5\lambda$ to get a 20-dB $\bar{n} = 2$ Taylor amplitude distribution.

The resulting thinned aperture is shown in Figure 3.55. Figure 3.56 is the resulting array factor. It has a directivity of 27.2 dB and a peak sidelobe level of 20 dB.

The z transform of a thinned linear array is a polynomial with coefficients of either zero or one [51]. As an example, consider a 20-element linear array of isotropic elements. Figure 3.57 is the unit circle representation when the elements are equally weighted and spaced ($d = 0.5\lambda$). The nulls in the pattern are equally spaced in phase, $\Psi = kdu$, but not equally spaced in angle, u. The roots of the uniform array polynomial are given by

$$\Psi = \pm 18°, \pm 36°, \pm 54°, \pm 72°, \pm 90°, \pm 108°, \pm 126°, \pm 144°, \pm 162°, 180°$$

If this same array is thinned to obtain the array factor with the minimum maximum sidelobe level, then the second element from each end of the array is turned off ($a_2 = 0$ and $a_{19} = 0$). This configuration yields a far-field pattern that has a maximum sidelobe level of −15.74 dB. The unit circle representation and array factor are displayed in Figure 3.58 and Figure 3.59. Note that the

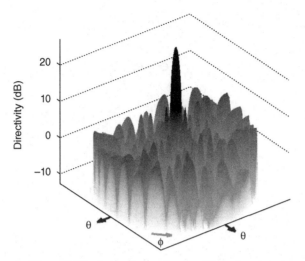

Figure 3.56. Array factor for the thinned array.

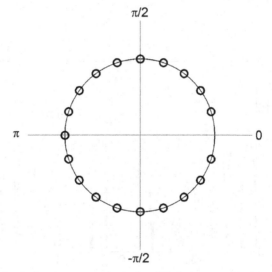

Figure 3.57. Unit circle representation of a 20-element uniform array factor with $d = 0.5\lambda$.

zeros are still equally spaced on the unit circle. This pattern appears to have one less zero on the unit circle than the uniform array, because there are double roots at ±60°. The zeros of the thinned array have phases given by

$$\Psi = \pm20°, \pm40°, \pm60°, \pm60°, \pm80°, \pm100°, \pm120°, \pm140°, \pm160°, 180°$$

Since the roots have a magnitude of one, they lie on the unit circle.

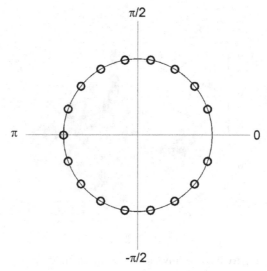

Figure 3.58. Unit circle representation of a 20-element array factor when $a_2 = 0$ and $a_{19} = 0$.

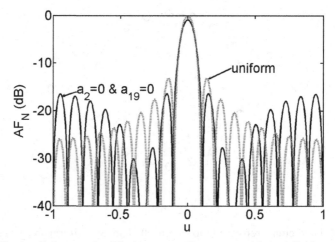

Figure 3.59. Array factor of a 20-element uniform array (dotted line) and an array with $a_2 = 0$ and $a_{19} = 0$ (solid line) for $d = 0.5\lambda$.

Not all thinned arrays have their roots on the unit circle though. In fact, for the 20-element array, only when one of the element pairs 1 & 20, 2 & 19, 7 & 14, or 10 & 11 are zero will all the roots lie on the unit circle. In all other cases of turning off symmetric pairs of elements, at least one pair of roots will be off the unit circle. Figures 3.60 and 3.61 show an example of the unit circle

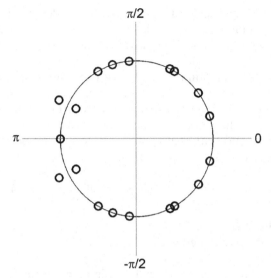

Figure 3.60. Unit circle representation of a 20-element uniform array with $a_8 = 0$ and $a_{13} = 0$.

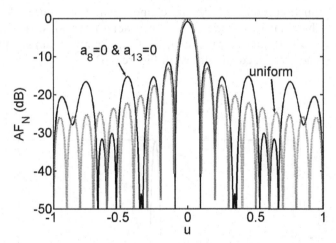

Figure 3.61. Array factor of a 20-element uniform array with $d = 0.5\lambda$ having $a_8 = 0$ and $a_{13} = 0$.

and far-field pattern of a thinned array with zeros off the unit circle. The array has 20 elements with elements 8 and 13 turned off. The zeros whose magnitudes don't equal one correspond to the pattern minimum at $u = 0.85$.

3.4.1.2. Thinning Using Numerical Optimization. One way to find the thinned array that results in the lowest sidelobe level is to check every possible array thinning combination. Finding the maximum sidelobe level for 2^N array factors becomes an unreasonable task as N get large. Numerical optimization techniques offer more feasible alternatives to finding the best (or more likely, a very good) solution without resorting to an exhaustive search. Most approaches to numerical optimization require continuous variables and do not adapt well to the on/off condition of thinning. The introduction of genetic algorithms changed the ability to thin arrays [52]. Arrays thinned with genetic algorithms have a higher taper efficiency and lower sidelobe levels than do previous numerical and statistical approaches. Numerical optimization offers several advantages over statistical thinning:

1. The maximum sidelobe level can be specified (as long as it is physically possible).
2. Higher efficiency.
3. Works for small and medium-sized arrays.

The genetic algorithm can also be used to thin planar arrays [53, 54].

Example. Use a genetic algorithm to thin a 50-element linear array with $d = 0.5\lambda$ to obtain the lowest sidelobe level.

The cost function requires both end elements and both center elements of the array to have an amplitude of one in order to minimize the maximum sidelobe level. The thinned array has $\eta_T = 0.80$ and a maximum sidelobe level of $-17.6\,\text{dB}$. The array factor is shown in Figure 3.62.

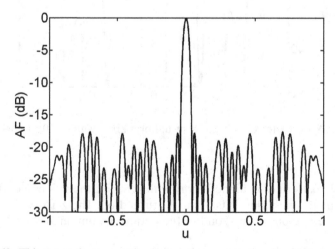

Figure 3.62. This array factor results from using a genetic algorithm to thin a 50-element array.

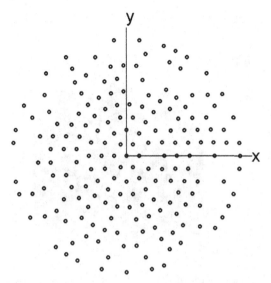

Figure 3.63. Thinned uniform concentric ring array having 9 rings.

Example. Use a genetic algorithm to thin a uniform concentric ring array with 270 elements layed out in nine rings spaced $\lambda/2$ apart and having $d_n = 0.5\lambda$ to obtain the lowest sidelobe level.

The thinned aperture is shown in Figure 3.63. It only has 66.3% of the 279 elements in the fully populated array. Turning off 94 of the elements reduced the directivity to 27.17 dB and the maximum relative sidelobe level to 22.44 dB. Since the thinning is not symmetric, the array factor is not symmetric either (Figure 3.64). The number of elements in each ring is shown in Table 3.3. The inner five rings have between 0% and 13% of the element removed, while the outer four rings have between 30% and 60% of the elements removed.

3.4.2. Nonuniformly Spaced Arrays

Thinned arrays have a large but finite number of possible active element locations. In contrast, nonuniformly spaced or aperiodic arrays have an infinite number of possible element locations. All elements in nonuniformly spaced arrays are active. Initial attempts at nonuniformly spaced arrays were based upon trial and error. Thinned arrays are more desirable than nonuniformly spaced arrays, because the feed network for the thinned arrays is much easier to design. Also, mutual coupling effects are easier to characterize for periodic array spacing than for aperiodic spacing. Also, implementing nonuniform spacing on planar arrays is extremely difficult.

3.4.2.1. Density Tapering. Density tapering for a nonuniformly spaced linear array differs from density tapering for thinned arrays in that the element

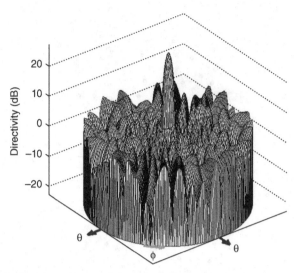

Figure 3.64. Array factor of the thinned uniform concentric ring array.

TABLE 3.3. Ring Radius and Number of Elements per Ring for a Nine-Ring Uniform Thinned Concentric Ring Array

n:	1	2	3	4	5	6	7	8	9
$r_n(\lambda)$:	0.5	1.0	1.5	2.0	2.5	3.0	3.5	4.0	4.5
N_n:	6	11	18	22	27	26	24	20.	30

spacing is related to the desired amplitude taper rather than the probability of whether an element in an equally spaced array is on or off. The idea is to divide the area under the desired amplitude taper into N equal sub-areas. An element is placed at the point that equally splits each sub-area.

Example. Find the density taper for a 15-element array that mimics a 20-dB, $\bar{n} = 3$ Taylor taper. Figure 3.65 shows the desired amplitude taper as a dashed line. The dotted vertical lines divide the area under the amplitude taper into 15 equal sub-areas. An element is placed at the point that evenly divides a sub-area (small circles). Figure 3.66 is the resulting array factor due to the nonuniform spacing. The pattern accurately reproduces the desired Taylor pattern out to about $u = \pm 0.5$. This technique does not work well for very low sidelobes. Even a 20-dB Taylor taper cannot be accurately reproduced.

3.4.2.2. Fractional z-Transform Synthesis. Earlier in this chapter, the unit circle representation of thinned arrays was introduced. For the more general case of nonuniformly spaced arrays, the far-field pattern is not represented by

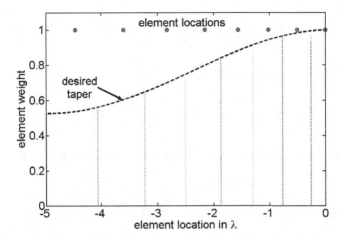

Figure 3.65. Taylor 20-dB taper is divided into equal areas, and an element is placed at the center of each area.

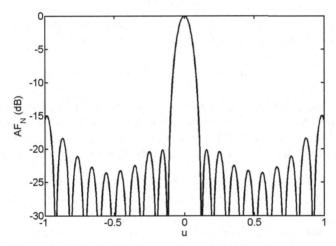

Figure 3.66. Taylor 20-dB array factor from density tapering.

a polynomial, so polynomial root finding techniques cannot be applied [51]. An approach to adapting the unit circle representation of a nonuniformly spaced array can be found by examining roots of a 4-element array that is symmetric about its center. It has an array factor given by

$$AF(u) = \cos[ks_1u] + \cos[ks_2u] \tag{3.46}$$

where s_1 and s_2 are the distances of the center and edge elements from the center of the array. Using a trigonometric identity and substituting $\psi = kdu$ ($d = 2s_1$) allows the array factor to be written as

$$\text{AF}(u) = 2\cos\left[\frac{\psi}{4}\left(\frac{s_2}{s_1}+1\right)\right]\cos\left[\frac{\psi}{4}\left(\frac{s_2}{s_1}-1\right)\right] \qquad (3.47)$$

This equation is zero when either

$$\frac{\psi}{4}\left(\frac{s_2}{s_1}+1\right) = \frac{(2n+1)\pi}{2} \quad \text{or} \quad \frac{\psi}{4}\left(\frac{s_2}{s_1}-1\right) = \frac{(2n+1)\pi}{2} \qquad (3.48)$$

Thus, the location of the roots on the unit circle are a function of the spacings and are given by

$$\psi_{n+} = \frac{2(2n+1)\pi}{\left(\frac{s_2}{s_1}+1\right)} \quad \text{and} \quad \psi_{n-} = \frac{2(2n+1)\pi}{\left(\frac{s_2}{s_1}-1\right)}, \qquad n = 0, 1, 2, \ldots \qquad (3.49)$$

The zeros of larger arrays are more difficult to find and are not always on the unit circle.

Consider an N element linear array of equally weighted, nonuniformly spaced point sources. Its array factor is represented by

$$\text{AF}(u) = 1 + e^{jks_1u} + e^{jks_2u} + \cdots + e^{jks_{N-1}u} \qquad (3.50)$$

where s_n is the distance of element n from the first element. The phase terms of this equation are not harmonics in a Fourier series. In order to write (3.50) in terms of a z transform, let d_{min} = minimum spacing between any two adjacent elements [55]. Next find $\psi = kd_{min}u$. Equation (3.50) is then written as

$$\text{AF}(\psi) = 1 + e^{j\frac{s_1}{d_{min}}\psi} + e^{j\frac{s_2}{d_{min}}\psi} + \cdots + e^{j\frac{s_{N-1}}{d_{min}}\psi} \qquad (3.51)$$

Making the substitution $z = e^{j\psi}$ yields

$$\text{AF}(z) = 1 + z^{\frac{s_1}{d_{min}}} + z^{\frac{s_2}{d_{min}}} + \cdots + z^{\frac{s_{N-1}}{d_{min}}} \qquad (3.52)$$

Equation (3.52) is a polynomial when s_n/d_{min} are integers. When z is raised to a noninteger power, this representation is known as a fractional z transform [55]. Only the roots for which $|u| \leq 1$ are valid. All others are ignored.

Example. Find the roots on the unit circle corresponding to an array with the element spacings given by

$$s_n = 0, 1, 1.9, 2.7, 3.4, 4.0, 4.5, 5.0, 5.5, 6.0, 6.5,$$
$$7.0, 7.5, 8.0, 8.5, 9.1, 9.8, 10.6, 11.5, 12.5\lambda$$

The array factor with $z = e^{jk(0.5)u}$ is written as

$$\text{AF}(z) = 1 + z^2 + z^{3.8} + z^{5.4} + z^{6.8} + z^8 + z^9 + z^{10} + z^{11} + z^{12} + z^{13} + z^{14} + \qquad (3.53)$$
$$z^{15} + z^{16} + z^{17} + z^{18.2} + z^{19.6} + z^{21.2} + z^{23} + z^{25}$$

This equation can be converted to a polynomial by letting $z = e^{jk(0.1)u}$.

$$\text{AF}(z) = 1 + z^{10} + z^{19} + z^{27} + z^{34} + z^{40} + z^{45} + z^{50} + z^{55} + z^{60} + z^{65} + z^{70} + \qquad (3.54)$$
$$z^{75} + z^{80} + z^{85} + z^{91} + z^{98} + z^{106} + z^{115} + z^{125}$$

Factoring this polynomial yields 125 roots. Only the roots for $|u| \leq 1$ exist and are given by

$$u = \pm 0.0953, \pm 0.143, \pm 0.216, \pm 0.304, \pm 0.366, \pm 0.440, \pm 0.521,$$
$$\pm 0.819, \pm 0.901, \pm 0.988$$

Plots of the far-field pattern and unit circle representation are shown in Figures 3.67 and 3.68, respectively. Note that there are 26 zeros to (3.54) even though there are only 20 elements. Since the average separation between element is 0.66λ, there are more nulls.

If d is very small, then the maximum power of z can be quite large for even a small array. One way to increase the size of s is to first round or quantize the value of s_n. The value of s will quantize the spacing of the array elements

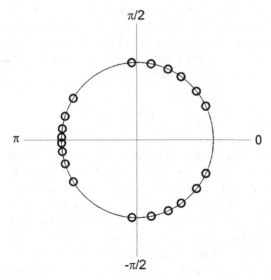

Figure 3.67. Unit circle representation of the 20-element nonuniformly spaced array.

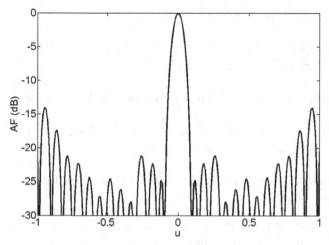

Figure 3.68. Array factor of the 20-element nonuniformly spaced array.

and produce an error in the array factor. This quantization error must produce error sidelobe levels less than the desired sidelobe level of the array.

3.4.2.3. Minimum Redundancy Arrays.

Large, sparsely populated arrays are used to observe faint extragalactic radio sources, because they have a narrow main beam and are economical (use only a few elements). A minimum redundancy array is a sparse array that maximizes the distance between grating lobes while maintaining the sidelobes in between the grating lobes at a relatively constant level. Its narrow beamwidth scans distant sections of space while keeping the grating lobes out of those sections. The distance between the first grating lobe and the main beam can be maximized while minimizing the number of elements by including all of the integer multiples of the fundamental spatial frequencies only once. The fundamental spatial frequency is given by one over the minimum element spacing

$$f_s = \frac{1}{d} \tag{3.55}$$

The other spatial frequencies are just multiples (harmonics) of f_s. A 5-element uniform array has $N - 1 = 4$ representations of f_s, 3 of $2f_s$, 2 of $3f_s$, and 1 of $4f_s$. The highest spatial frequency, f_s, is due to the total aperture length. Higher resolution is possible when the number of redundant spacings are reduced. The number of distinct elements which can be resolved in a linear array is approximately equal to $(N - 1)d$, and this is obtained when the angular width of the source is equal to the separation between the grating lobes. When the size of the source is known (e.g., the sun), the grating lobe spacing should be matched to the source size. Figure 3.69 shows a 4-element minimum redun-

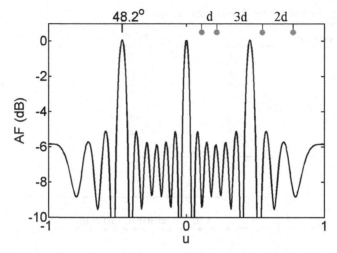

Figure 3.69. Array factor for a 4-element minimum redundancy array. The array spacing is shown in the upper right corner.

dancy array and its array factor when $d = 1.5\lambda$ and the first grating lobe occurs at 48.2°. A completely nonredundant linear array must have less than 5 elements. Looking at the separation distances between all possible combinations of two elements in the array reveals that each integer multiple of d up to the length of the array $d, 2d, \ldots, 6d$ occurs once and only once. Arrays with more than 4 elements will have more than one sample at some spatial frequencies. The redundancy factor is given by [56]

$$R = \frac{N(N-1)}{2N_{max}} \tag{3.56}$$

where N is the number of elements in the array and $(N_{max} - 1) \times d$ is the length of the array. Ideally, R should be as close to one as possible (only arrays with less than five elements can have $R = 1$). The lower limit for R on large arrays is 4/3. Global optimization techniques, such as simulated annealing, are needed to find large minimum redundancy arrays [57].

Example. List the number of possible spatial harmonics associated with a 4-element uniform array, a 7-element uniform array, and a four element minimum redundancy array. Table 3.4 lists the spatial frequencies associated with the arrays. Note that dc refers to a single element.

3.4.2.4. Numerical Optimization of Array Spacing. Finding the element spacings of a uniformly weighted array that produced the lowest possible sidelobe levels was investigated using a genetic algorithm [58]. The cost function is given by

TABLE 3.4. List of Spatial Frequencies in the Arrays

Spatial Frequency	OOOO	OOOOOOO	OO	O	O
dc	4	7	4		
1/d	3	6	1		
2/d	2	5	1		
3/d	1	4	1		
4/d	0	3	1		
5/d	0	2	1		
6/d	0	1	1		

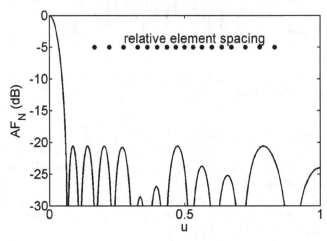

Figure 3.70. Array factor for an optimized element spacing of a 19-element array.

$$sll = \max\left\{1 + 2\sum_{n=1}^{N}\cos\left[ku(x_{n-1}+d_0+\Delta_n)\right]\right\} \qquad \text{for } u > u_{MB} \qquad (3.57)$$

where $2N + 1$ is the number of elements, x_n is the location of element n, $x_0 = 0$, $d_0 = \lambda/2$, Δ_n is the variable spacing between 0 and $\lambda/2$, and u_{MB} is the first null next to the main beam. This cost function requires a minimum spacing of d_0, so if all $\Delta_n = 0$, then a uniform array with spacing d_0 results. Since the array is uniformly weighted, the value of u_{MB} can be approximated by $0.5\lambda/x_N$.

Example. Find the optimum spacing for a 19-element array with $d_0 = 0.5\lambda$ that yields the lowest sidelobe level.

The genetic algorithm had a population size of 8 and mutation rate of 0.15. After 1000 generations, a Nelder–Mead algorithm polished the results. Figure 3.70 shows the resulting array factor with a maximum sidelobe level of −20.6 dB. The 19 round dots in the figure represent the relative element spacings of the array elements.

3.4.2.5. Nonuniformly Spaced Concentric Ring Array. Nonuniformly spacing the rings in a concentric ring array can also produce low sidelobe levels [54]. First, the rings are assumed to be separated by a minimum distance of $\lambda/2$ (not necessary but practical). An additional separation of Δ_n is added to the radius of ring $n - 1$, so that ring n has a radius of

$$r_n = r_{n-1} + \lambda/2 + \Delta_n \qquad (3.58)$$

where $0 \leq \Delta_n \leq \lambda$. A hybrid genetic algorithm that combines a continuous genetic algorithm with a local optimizer is used to find the r_n that result in the minimum maximum sidelobe level in the array factor. Table 3.5 displays the optimized ring spacing and the number of elements in each ring (keeping $d \simeq \lambda/2$). This new array has only 201 elements. The resulting directivity is 27.38 dB and the maximum relative sidelobe level is 22.94 dB. Figure 3.71 is a diagram of the array. The first and last rings have the largest Δ_n. Since the ring

TABLE 3.5. Ring Radius and Number of Elements per Ring for a Six-Ring Nonuniformly Spaced Concentric Ring Array with Optimized r_n

n:	1	2	3	4	5	6
$r_n(\lambda)$:	1.001	1.587	2.137	2.875	3.659	4.977
N_n:	12	19	26	36	45	62

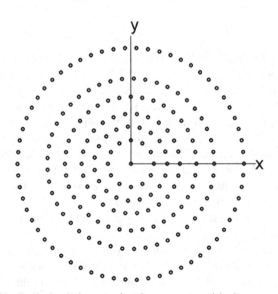

Figure 3.71. Optimized ring spacing for an array with six concentric rings.

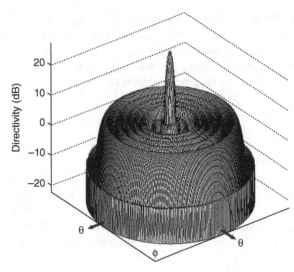

Figure 3.72. Array factor associated with optimized ring spacing for an array with six concentric rings.

TABLE 3.6. Ring Radius and Number of Elements per Ring for a Nine-Ring Uniform Concentric Ring Array with Optimized N_n

n:	1	2	3	4	5	6	7	8	9
$r_n(\lambda)$:	0.5	1.0	1.5	2.0	2.5	3.0	3.5	4.0	4.5
N_n:	6	12	18	25	17	23	22	27	32

separation remains close to $\lambda/2$, the sidelobes have sufficient sampling in ϕ, and they maintain symmetry in ϕ. Figure 3.72 shows the corresponding array factor. All the sidelobes have nearly the same height.

Another approach to reducing the sidelobe level is to let $r_n = n\lambda/2$ and optimize the N_n in order to minimize the maximum sidelobe level. A genetic algorithm is used to find the optimum N_n listed in Table 3.6. The first four rings have the same number of elements as a uniform concentric ring array with $\lambda/2$ spacing. The optimized array has 183 elements arranged as shown in Figure 3.73. This array has only 65.6% of the elements of the uniform array. The directivity is 28.5 dB and the maximum sidelobe level is 25.58 dB. The array factor shown in Figure 3.74 is nearly circularly symmetric for small θ but not for large θ.

An even more powerful approach is to optimize both the ring spacing and the number of elements in each ring. The maximum radius is slightly less than the maximum radius found by optimizing the r_n alone. Table 3.7 has the ring radii and the number of elements in each ring. This array has 142 elements or

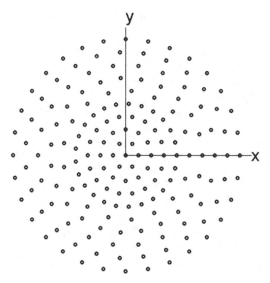

Figure 3.73. Optimized number of elements in nine concentric rings with equal ring spacing.

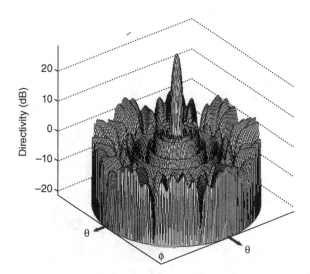

Figure 3.74. Array factor associated with optimized number of elements in nine equally spaced concentric rings.

TABLE 3.7. Ring Radius and Number of Elements per Ring for a Nine-Ring Uniform Concentric Ring Array with Optimized r_n and N_n

n:	1	2	3	4	5	6
$r_n(\lambda)$:	0.758	1.355	2.090	2.989	3.783	4.702
N_n:	9	17	25	31	26	33

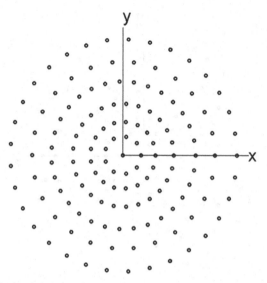

Figure 3.75. Optimized ring spacing and number of elements in each ring for an array with six concentric rings.

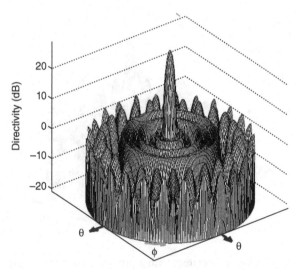

Figure 3.76. Array factor associated with optimized ring spacing and number of elements in each ring for array with six concentric rings.

50.9% of the nine-ring uniform concentric array. It has a directivity of 28.89 dB and a maximum relative sidelobe level of 27.82 dB. The optimized array appears in Figure 3.75 with the corresponding array factor in Figure 3.76. All the sidelobes are at nearly the same height.

TABLE 3.8. Sidelobe Level Increase as a Result of a 5% Increase in Frequency

Array	Sidelobe Level Increase (dB)
Optimized r_n	0.1
Optimized N_n	3.1
Optimized r_n and N_n	11.6

The peak sidelobe level of the optimized concentric ring arrays will not change if the frequency decreases by 10%. The sidelobes do not increase because the spatial sampling gets better as the frequency goes down. Increasing the frequency by 5%, however, results in significant sidelobe level increases. Table 3.8 summarizes the sidelobe level increases with a 5% frequency increase. The optimized r_n and N_n case had the greatest increase in sidelobe level, while the optimized ring spacing case has almost no increase. As a result, the array should be optimized at the highest frequency, then the lower frequencies will have low sidelobes.

The concentric ring arrays were optimized without taking scanning into account. When the array beams are scanned, the sidelobe level increases. The nonuniform ring spacing example did not have a sidelobe level increase until the beam was scanned beyond $\theta = 22.5°$. All the other examples experienced increasingly worse sidelobe levels with increasing scan in θ. As an example, steering the main beam in Figure 3.74 to 30° increases the maximum sidelobe level by 8.5 dB. Better performance can be obtained by optimizing over all the scan angles.

3.5. LOW-SIDELOBE PHASE TAPER

A phase taper is another approach to lowering sidelobes in the array factor [59]. The synthesis of a phase taper for an array with uniform amplitude weights and spacing is formulated as a minimization of the maximum sidelobe level [60]. One possible cost function for the numerical optimization of the phase, δ_n, in an N-element array is

$$\text{sll} = \max\left\{2\sum_{n=1}^{N/2} e^{j\delta_n}\cos[(n-.5)kdu]\right\} \qquad \text{for } u > u_{\text{mb}} \qquad (3.59)$$

where u_{mb} is the maximum angle of the main beam. The advantage of a phase taper is that low sidelobe levels can be obtained through adjustments to the beam-steering phase shifters rather than by any amplitude weighting via the feed network. These tapers have a modest ability to lower sidelobes and tend to be less efficient than amplitude tapers.

Example. Find a phase taper that results in the lowest possible sidelobe level for a 20-element array with $d = 0.5\lambda$.

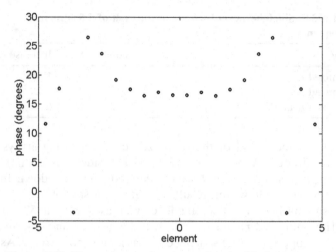

Figure 3.77. The optimum phase taper for the 20-element array.

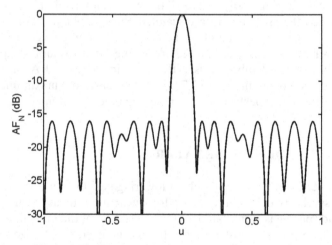

Figure 3.78. The array factor due to the phase taper.

A hybrid genetic/Nelder–Meade algorithm was used to find the phase weights. The resulting weights are shown in Figure 3.77 and the associated array factor appears in Figure 3.78. The maximum sidelobe level is 16.1 dB below the peak of the main beam with $\eta_T = 0.84$. The resulting phase taper for the right-hand side of the array is

$$2\pi(0.145\ 0.149\ 0.144\ 0.153\ 0.168\ 0.207\ 0.231 - 0.0312\ 0.155\ 0.101)$$

For comparison, a 20-element Chebyshev array with a 16.1-dB sidelobe level has $\eta_T = 0.91$; and a 16.1-dB, $\bar{n} = 2$ Taylor array has $\eta_T = 0.98$.

3.6. SUPPRESSING GRATING LOBES DUE TO SUBARRAY WEIGHTING

3.6.1. Subarray Tapers

For most applications, placing the weights at the subarray ports alone produces unacceptable grating lobes in the array factor. Minimum sidelobe level in the array factor cannot be less than the highest grating lobe. This is demonstrated in Figure 3.79. The array has 128 elements, 16 subarrays, and an element spacing of 0.5λ. A genetic algorithm optimizes the amplitude weights at the subarray ports while the element weights are uniform. The sidelobe levels next to the main beam are at the same level as the first grating lobe. In order to further reduce the peak sidelobe level, the grating lobes must be dealt with at the element level.

A tradeoff exists between attaining the desired low-sidelobe performance and simplicity of design. One approach places a low-sidelobe amplitude taper at the subarray outputs while the element weights are the same for each subarray [61]. The linear subarray model in Figure 3.80 has amplitude weights at the elements as well as at the subarray ports. All subarrays have identical amplitude tapers across the elements. The elements are assumed to be equally spaced and have symmetric weights about the center of the array. Based upon these assumptions, the array factor for a linear array along the x axis is given by

$$AF = 2\sum_{q=1}^{N_s} b_q \sum_{n=1}^{N_e} a_m \cos\{kd\sin\theta[n - 0.5 + (q-1)N_e]\} \qquad (3.60)$$

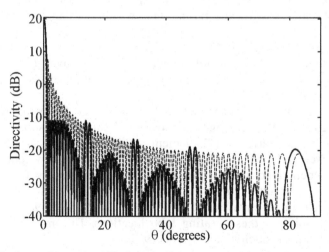

Figure 3.79. Array factor for 128-element linear array with 16 subarrays. The amplitude weights at the subarray ports are optimized to give the lowest maximum sidelobe level.

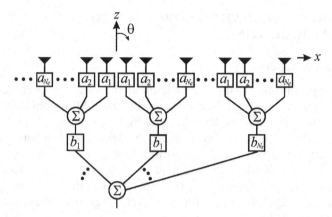

Figure 3.80. Subarrayed antenna with subarray weights and identical element weights in each subarray.

where a_m represents element amplitude weights, b_q represents subarray amplitude weights, $2N_s$ is the number of subarrays, N_e is the number of elements in a subarray, and θ is the angle from boresight. The effective weights are then represented by a $2N_s \times N_e$ row vector.

$$\mathbf{w} = [\underbrace{b_{N_s}a_{N_e}, \cdots b_{N_s}a_1}_{N_e} \cdots \underbrace{b_1a_{N_e}, \cdots, b_1a_1}_{N_e} \underbrace{b_1a_1, \cdots, b_1a_{N_e}}_{N_e} \cdots \underbrace{b_{N_s}a_1, \cdots b_{N_s}a_{N_e}}_{N_e}] \qquad (3.61)$$

Optimizing the subarray and element weights results in a low-sidelobe array factor with a simplified array design. This approach produces identical corporate feeds for all subarrays and T/R modules weight the signals at the subarray ports.

Example. Find the subarray and element amplitude tapers (same for each subarray) for a 144-element array with ($d = 0.5\lambda$) that produces the minimum maximum sidelobe level.

Assume that the array is symmetric about its center. The array has 12 subarrays of 12 elements. Breaking the array into 12 equal-sized subarrays results in 16 optimization variables (11 element amplitude weights and 5 subarray amplitude weights), since $a_1 = 1$ and $b_1 = 1$ and the subarray weights are symmetric. A genetic algorithm is used to find the weightings in Figure 3.81 to Figure 3.83. The resulting array factor is shown in Figure 3.84. Grating lobes normally occur in this array at 9.6°, 19.5°, 30.0°, 41.8°, 56.4°, 90.0°. Arrows in Figure 3.84 point to these grating lobe locations. The array directivity is 20.2 dB with a peak sidelobe level of −35.9 dB.

The array factor for a planar array in the x–y plane that is symmetric about the x and y axes is given by [62]

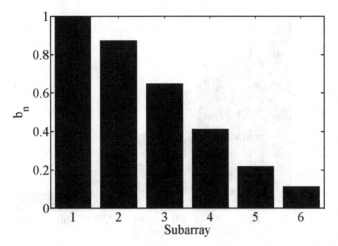

Figure 3.81. Optimized subarray weights when the element weights are identical for all subarrays.

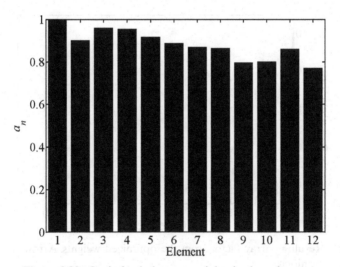

Figure 3.82. Optimized element weights in the subarrays.

$$AF = 4\sum_{p=1}^{N_{sy}}\sum_{q=1}^{N_{sx}} b_{pq} \sum_{m=1}^{N_{ey}}\sum_{n=1}^{N_{ex}} a_{mn} \cos\{kd_x \sin\theta\cos\phi[n-0.5+(q-1)N_{ex}]\}$$
$$\times \cos\{kd_y \sin\theta\sin\phi[m-0.5+(p-1)N_{ey}]\} \tag{3.62}$$

where a_{mn} represents element weights, b_{pq} represents subarray weights, $2N_{sx}$ is the number of subarrays in x direction, $2N_{sy}$ is the number of subarrays in y direction, N_{ex} is the number of elements in a subarray in x direction, N_{ey} is the number of elements in a subarray in x direction, d_x is the element spacing in

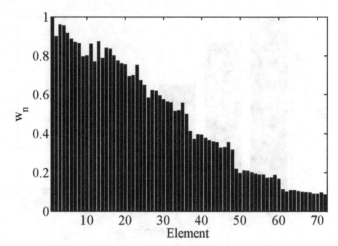

Figure 3.83. Effective element weights resulting from the product of the element weight times its corresponding subarray weight.

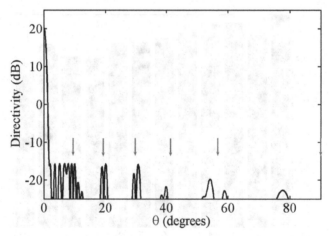

Figure 3.84. Resulting array factor from the optimized weights. Arrows point at the grating lobe locations.

x direction, and d_y is the element spacing in y direction. A diagram of this array is shown in Figure 3.85. The array is divided into 6×6 subarrays each having 5×5 elements. A dark color indicates a high-amplitude weight at the element (a_{mn}). Note that the shade of the dots for each element is the same for every subarray. The tint of the subarray squares is proportional to the subarray weight (b_{pq}).

Example. Find the subarray and element amplitude tapers (same for each subarray) for a planar array with with 8×8 subarrays each having 8×8 ele-

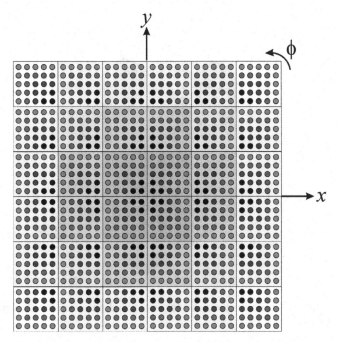

Figure 3.85. Planar array with 6 × 6 subarrays each having 5 × 5 elements. All element weights are the same for all subarays. Darker dots have higher amplitudes than lighter dots. Darker subarrays have higher amplitudes than lighter subarrays.

ments ($d_x = d_y = 0.5\lambda$) that produces the minimum maximum sidelobe level. Assume that the array is symmetric about its center.

Since the array is symmetric, there are 15 unknown subarray weights and 63 unknown element weights for a total of 78 unknown variables. A genetic algorithm is used to find these variables. The optimized subarray weights appear in Figure 3.86 with the optimized element weights in Figure 3.87. Figure 3.88 shows the product of the element weights times the subarray weights. The resulting array factor is shown in Figure 3.89. This array factor has a directivity of 34.7 dB with a maximum relative sidelobe level of 23.6 dB.

3.6.2. Thinned Subarrays

Figure 3.90 is a diagram of a 256-element square array divided into 16, 4 × 4 element thinned subarrays [63]. Dark circles indicate that the element has an amplitude of one, and white circles indicate that the element is terminated in a matched load and receives an amplitude of zero. Note that the subarrays are either identical or are mirror images.

A genetic algorithm is used to simultaneously optimize the subarray amplitude weights and the thinning for the subarrays. The cost is the maximum relative sidelobe level and is calculated from the array factor given by

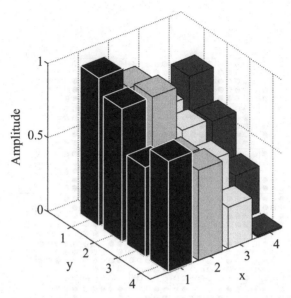

Figure 3.86. Optimized subarray weights for one quandrant when the element weights are identical for all subarrays.

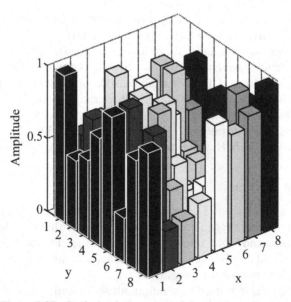

Figure 3.87. Optimized element weights in the subarrays.

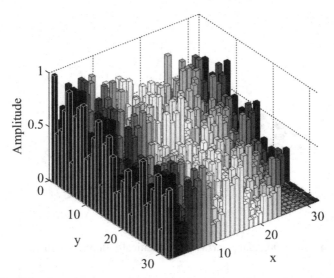

Figure 3.88. Effective element weights resulting from the product of the element weight times its corresponding subarray weight.

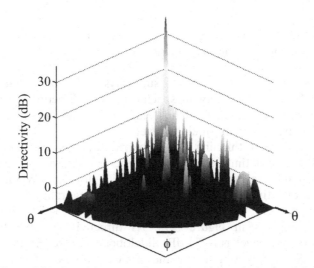

Figure 3.89. Resulting array factor from the optimized weights.

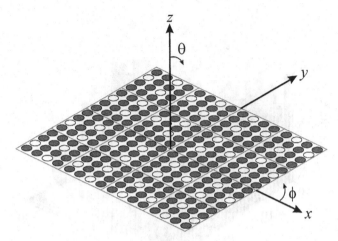

Figure 3.90. Planar array with identically thinned subarrays.

$$\text{cost} = \sum_{n=1}^{N} w_n \cos[kx_n \sin\theta\cos\varphi]\cos[ky_n \sin\theta\sin\varphi] \qquad (3.63)$$

where w_n is the product of element weight and subarray weight for element n, and (x_n, y_n) is the location of element n.

Example. An array has $6 \times 6 = 36$ subarrays with each subarray having $5 \times 5 = 25$ elements. The element spacing in both directions of the square lattice is one-half wavelength. Find the thinned subarrays and subarray weights that minimize the maximum sidelobe level.

When the element thinning and subarray weights are simultaneously optimized, then the optimized element weights are shown in Figure 3.91 and the optimized subarray weights are shown in Figure 3.92. When multiplied together, the effective element weights for one quadrant of the array are shown in Figure 3.93. This taper has an efficiency of 54.5%. The resulting array factor appears in Figure 3.94. Its directivity is 26.9 dB and the maximum sidelobe level is –22.9 dB. Sidelobes that are 4.5 dB lower than the optimized subarray weighting come at a cost in taper efficiency and loss in directivity.

One way to eliminate the grating lobes is to create overlapped subarrays. Although difficult to do with corporate array feeds, they can be realized using a Butler matrix or lens feeds that will be discussed in a later chapter. Some examples of implementing a partially overlapped constrained feed network are found in references 64 and 65. Several approaches to amplitude weighting the subarray outputs of identical subarrays exist. Other approaches have tried randomizing the number of elements within a given subarray [66], rotating

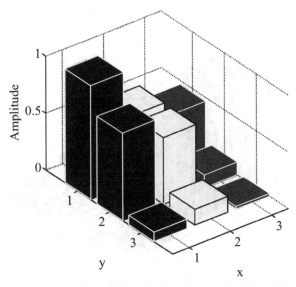

Figure 3.91. Optimized subarray weights for one quadrant of the array.

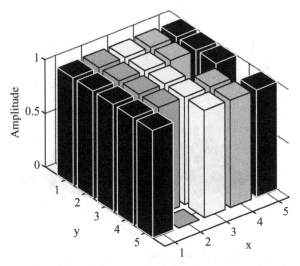

Figure 3.92. Optimized element weights in one subarray.

the subarray orientation [67], and nonuniformly spacing the elements within the subarrays [68, 69]. These methods are designed to break the periodicity and redistribute the energy in the grating lobes throughout the far-field pattern.

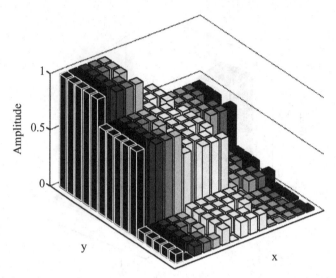

Figure 3.93. Optimized effective element weights in one quadrant of the array.

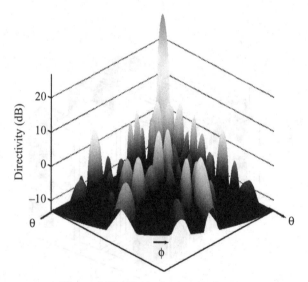

Figure 3.94. Optimized array factor.

If an array has subarrays of different sizes and has amplitude weights at the subarray ports when all element weights are one (Figure 3.95), then the effective element weights are given by [70]

$$\Big[\underbrace{b_1 b_1 \cdots b_1}_{N_e(1)} \underbrace{b_2 b_2 \cdots b_2}_{N_e(2)} \cdots \underbrace{b_{N_s} b_{N_s} \cdots b_{N_s}}_{N_e(N_s)}\Big] \tag{3.64}$$

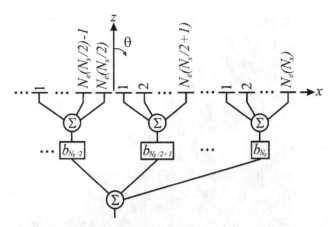

Figure 3.95. Diagram of an antenna with unequal sized subarrays and amplitude weights at the subarray ports.

N_e is a vector that is N_s (number of subarrays) long. The total number of elements in the array is given by

$$N_T = \sum_{m=1}^{Ns} N_e(m) \tag{3.65}$$

A genetic algorithm chromosome or input vector to the cost function consists of $2 \times (N_s - 1)$ values

$$[b_2 \cdots b_{N_s} N_e(1) N_e(1) \cdots N_e(N_s - 1)] \tag{3.66}$$

Since $b_1 = 1$ and $N_e(N_s) = N_T - \sum_{r=1}^{N_s-1} N_e(r)$. The minimum number of elements that are allowed in a subarray is represented by N_{emin}. Each subarray starts by having N_{emin} elements. The remaining $N_T - N_s \times N_{emin}$ elements are distributed among the subarrays. Although not necessary, the array is assumed to have an even number of subarrays, and the subarrays are symmetric about the center of the array. Selecting N_{emin} is arbitrary unless certain design constraints exist. A design constraint may have to do with the way the elements are fed via a corporate feed. For example, all subarrays may have the number of elements equal to a power of two. N_{emin} has to be at least greater than or equal to two, otherwise the subarrays consist of only a single element. In order to take advantage of the efficiencies offered by subarrays, it will be assumed that the minimum number of elements in a subarray is greater than or equal to four.

Example. Assume a symmetric array ($d = 0.5\lambda$) has $N_T = 128$ elements with $N_{emin} = 4$ and $N_s = 16$. Optimize the subarray weights and sizes to get the minimum maximum sidelobe level.

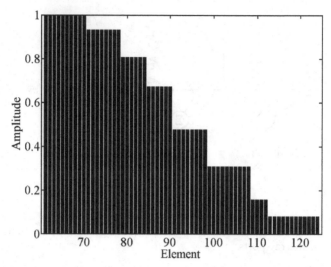

Figure 3.96. Effective element weights when the subarray weights and number of elements in the subarray are optimized for $N_{\text{emin}} = 4$ *and* $N_s = 16$.

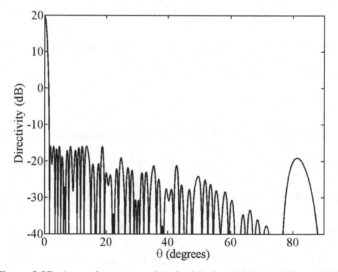

Figure 3.97. Array factor associated with the weights in Figure 3.96.

A genetic algorithm is used to find the weights and sizes shown in Figure 3.96. The resulting array factor is shown in Figure 3.97. The maximum sidelobe level is 35.9 dB below the peak of the main beam, and the directivity is 19.89 dB.

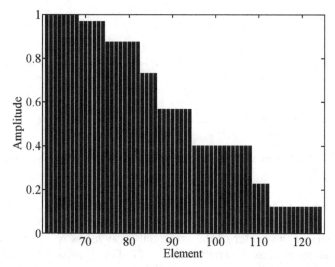

Figure 3.98. Effective element weights when the subarray weights and number of elements in the subarray are optimized for $N_{\text{emin}} = 6$ *and* $N_s = 16$.

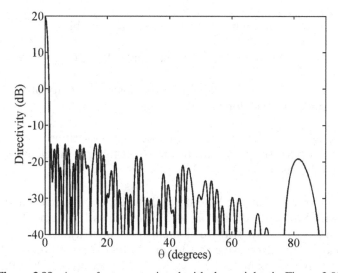

Figure 3.99. Array factor associated with the weights in Figure 3.98.

Example. Assume a symmetric array ($d = 0.5\lambda$) has $N_T = 128$ elements with $N_{\text{emin}} = 6$ and $N_s = 16$. Optimize the subarray weights and sizes to get the minimum maximum sidelobe level.

A genetic algorithm is used to find the weights and sizes shown in Figure 3.98. The resulting array factor is shown in Figure 3.99. The maximum sidelobe

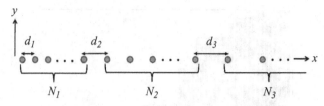

Figure 3.100. Array factor associated with the weights in Figure 3.98.

level is 35.1 dB below the peak of the main beam, and the directivity is 19.93 dB.

Another way to lower sidelobe levels with uniformly weighted elements ($a_n = 1$) and subarrays ($b_n = 1$) breaks the array into N_s subarrays that are symmetric about the center of the array. Subarray n has N_n elements with an element spacing of d_n as shown in Figure 3.100. This approach has been shown to be useful for linear and planar arrays [71].

Example. Assume that a linear array has 64 elements and is divided into three subarrays. Also, $d_3 > d_2 > d_1 > 0.5\lambda$ and a subarray must have at least one element.

A genetic algorithm found the following array variables:

$$d_1 = 0.501\lambda, d_2 = 0.906\lambda, d_3 = 1.890\lambda \quad \text{and} \quad N_1 = 21, N_1 = 9, N_1 = 2$$

The resulting array factor in Figure 3.101 has a directivity of 18.05 dB and a −22.4-dB maximum sidelobe level.

Optimizing a phase taper on an array with a low-sidelobe amplitude taper at the subarray ports (b_n) and uniform amplitude taper at the elements reduces the grating lobe levels ($a_n = 1$) [72].

Example. Find the element phase taper that best approximates a Taylor distribution with SLL = −30 dB and $\bar{n} = 4$. The array has 128 elements spaced $\lambda/2$ apart and has 8 subarrays with 16 elements per subarray.

The array is optimized using the iterative projection method (IPM) [73] in order to closely approximate the Taylor taper. By exploiting the efficiency in computing discrete Fourier transformations, the IPM is based on an iterative algorithm where the array illumination function is projected onto the corresponding space of far field patterns and then vice versa, until a desired pattern is synthesized. The amplitude taper appears in Figure 3.102 and the optimized phase taper in Figure 3.103. Figure 3.104 shows the peak sidelobe level is reduces by nearly 4 dB. The array without the phase taper has a directivity of 20.3 dB and a 3-dB beamwidth of 1.01°, while with the phase taper it has a directivity of 20.1 dB and a 3-dB beamwidth of 1.05°.

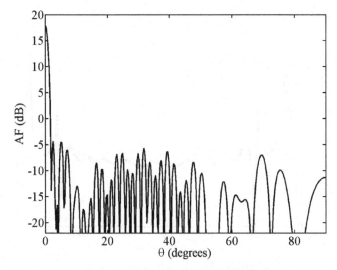

Figure 3.101. Array factor for the optimized subarray spacings.

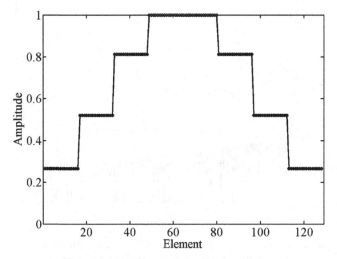

Figure 3.102. Element amplitude weights.

3.7. PLANE WAVE PROJECTION

Measuring far-field antenna patterns requires separating the transmit antenna and the antenna under test (AUT) by a large distance in order to minimize the amplitude and phase variations across the test aperture. When the AUT is many wavelengths across, the far-field distance can become quite large (recall IEEE definition of far field from Chapter 1). Thus, in order to measure

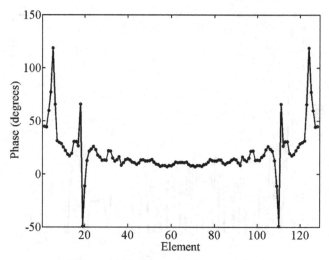

Figure 3.103. Element phase weights.

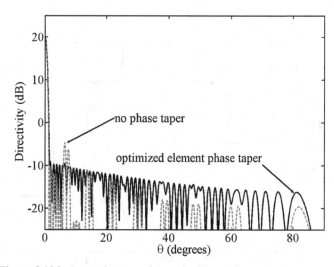

Figure 3.104. Array factors with and without element phase tapers.

the antenna pattern of an electrically large aperture indoors, techniques like compact ranges, near-field scanning, and antenna focusing are necessary. These approaches allow for accurate prediction of the far field pattern even though the measurements are taken in the near field.

A number of techniques have been developed to project a plane wave in the near field of an array by synthesizing appropriate complex element weights. Hill used a constrained least squares approach developed to find the weights that approximate a plane wave in the near field [74]. The results show a flat amplitude response for small linear arrays of line sources. Unfortunately, the phase response, which determines the far field criterion, is not shown. This approach was experimentally applied to a seven element array of Yagi-Uda antennas at 500 MHz [75]. In another study, a ring array was synthesized to create a plane wave inside the ring [76]. For further background information on creating a plane wave in the near field, the reader is referred to reference 77. A more recent method was developed to create a plane wave in the near field from a linear array of line sources [78] using a genetic algorithm. This approach combines the ideas of array focusing and a compact range. The location and weights of an array of line sources were found that approximate a plane wave at a desired location in space. The optimized amplitude and phase ripples across the AUT are much less than those created by a uniform array with the same number of elements. The resulting plane wave is robust in bandwidth and physical depth. This method also proved successful in experimental testing as reported in references 79 and 80. The idea behind these papers is to find the amplitude, phase, and position of the elements that create a relatively constant field amplitude and phase over a desired area in the near field. A linear array can only control the field in the plane of the array but not the field in the orthogonal plane.

A linear system of equations based on the near field array factor of the array is given by

$$
\begin{bmatrix} \dfrac{e^{jkR_{11}}}{R_{11}} & \cdots & \dfrac{e^{jkR_{1N}}}{R_{1N}} \\ \vdots & \ddots & \vdots \\ \dfrac{e^{jkR_{M1}}}{R_{M1}} & \cdots & \dfrac{e^{jkR_{MN}}}{R_{MN}} \end{bmatrix} \begin{bmatrix} w_1 \\ \vdots \\ w_N \end{bmatrix} = \begin{bmatrix} 1 \\ \vdots \\ 1 \end{bmatrix}
\tag{3.67}
$$

Figure 3.105 is a diagram of the plane wave projection layout. When $M = N$, then a direct solution to (3.67) is possible. Otherwise, when $M > N$ (more field points than elements), a least squares solution is found. Since the exact field values at the plane wave is unknown, the amplitude is assumed to be one and the phase zero. Once the weights are found, then they are normalized.

Solving (3.67) yields weights that produce a very flat amplitude and phase field distribution over the plane wave region. In fact, using $M = 81$ sample points in the first quadrant (Figure 3.105) results in weights that produce a maximum amplitude variation across the plane wave region of 0.003 dB and a phase variation of 0.02°. These amazing results are at a cost, however. Unfortunately, the very flat amplitude response in the desired plane wave region, but the level is at about −70 dB compared to the +5 dB for a uniform array. The amplitude of this plane wave would be in the noise level of an experimental setup, making this approach impractical.

The plane wave appears at such a low level because half the array weights are out of phase with the other half; in other words, it is a difference pattern with a null in the main beam. In the far field, the $1/R_{mn}$ terms are replaced by

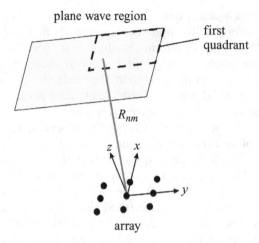

Figure 3.105. Plane wave projection.

a constant $1/R$ that can be factored out of the summation. In addition, symmetry can be imposed on both the amplitude and phase of the array weights in the form of combining symmetric phase terms using Euler's identity for cosine. In the near field, however, R_{mn} is not a constant and the amplitude of symmetric phase terms are not the same, so symmetry cannot be imposed. Consequently, numerical optimization must be used to find an acceptable solution.

The least squares constraint requires that

$$\sum_{n=1}^{N} |w_n|^2 \leq C \tag{3.68}$$

where C is a constant. Unfortunately, the Green's function matrix has a large condition number (on the order of 10^{10}), so when the Hermitan matrix is formed from the Green's function matrix, the condition number is even higher (on the order of 10^{17}). This condition number invalidates any results obtained with double precision arithmetic. Thus, another approach is needed for this planar array configuration.

Consider a 6×6 element transmitting array with elements arranged in a square lattice with spacings $d_x = 1.0\lambda$ and $d_y = 1.0\lambda$. These large spacings reduce mutual coupling effects, making the point source model a better approximation to reality. Optimized spacings [81] were on the order of a wavelength or larger. The desired $4\lambda \times 4\lambda$ plane wave region is $z_0 = 10\lambda$ away from the transmitting array. This distance is much less than the far field at $R_{ff} = 64\lambda$. Moving this plane wave region to $R_{ff} = 64\lambda$ would result in a phase variation of $22.5°$ and amplitude variation of $0.33\,\text{dB}$. The plane wave region at $z_0 = 10\lambda$ has a maximum phase deviation of $60°$ and a maximum amplitude variation of

8.6 dB. These variations exceed the IEEE standard and are unacceptable for most measurements.

The cost function used for the genetic algorithm optimization of the plane wave generator is given by

$$f = \left|\frac{\max \angle E_m - \min \angle E_m}{\pi}\right| + \frac{\max|E_m| - \min|E_m|}{\max|E_m|} \qquad (3.69)$$

where E_m are the complex field samples from the desired plane wave region. The first term in the cost function is the maximum phase deviation normalized to π. As long as the plane wave area is not too large, this term stays less than one. The second term is the normalized maximum field amplitude. These terms can be weighted to achieve a desired emphasis on either the field amplitude or phase. In this example, they are weighted equally.

The continuous parameter genetic algorithm used a population size of 80 with 1% mutation rate, 50% crossover rate, and single-point crossover. The amplitude and phase weights for the elements in one quadrant of the array are shown in Table 3.9. The maximum field amplitude variation across the desired plane wave region is 0.75 dB as seen in the contour plot of Figure 3.106. The maximum field phase variation across the desired plane wave region is 32.4° as seen in the contour plot of Figure 3.107. These field variations are a significant improvement over the same field variations due to a uniform array.

3.8. INTERLEAVED ARRAYS

Placing multiple antennas in the same area is important when space is limited as on a satellite. Antenna arrays that occupy the same area are called shared or common apertures. In these arrays, groups of elements dedicated to different frequencies or functions are interleaved, interlaced, or interspersed within the same aperture.

Interleaving elements dedicated to different feed networks have been successful in many applications. Random interleaving of elements in adjacent subarrays disrupt the grating lobes that result when only one phase shifter per subarray is used to steer the beam. Randomly assigning elements to subarrays reduces the peak sidelobe level. Interleaved arrays of waveguide radiators operating at different frequencies have been built [82]. An aperture with three arrays of interleaved elements operating in the L, S, and C bands was successfully built and measured to demonstrate the feasibility of interleaving three arrays on the same aperture [83]. A dual-band phased array using interleaved waveguides (C band) and printed dipoles on a high dielectric substrate (S band) has been developed. The element spacings for both arrays were equal, so no grating lobes formed in either band [84]. A phased array antenna with interleaved waveguide elements and wideband tapered elements was shown to operate over at least three frequency bands [85]. A dual-band/

TABLE 3.9. The Optimized Weights Found by the Genetic Algorithm

x Element Spacing →	0.5λ		1.5λ		2.5λ	
y Element Spacing ↓	Normalized Amplitude	Phase (Radians)	Normalized Amplitude	Phase (Radians)	Normalized Amplitude	Phase (Radians)
0.5λ	1.00	0.78π	0.29	0.94π	0.67	0.68π
1.5λ	0.61	0.80π	0.34	0.00	0.59	1.10π
2.5λ	0.39	0.86π	0.61	0.80π	0.29	1.38π

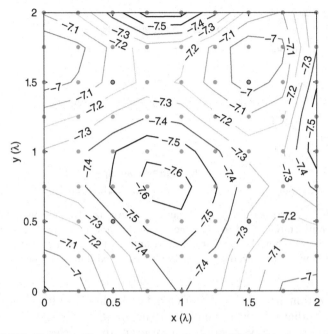

Figure 3.106. Field amplitude distribution (in decibels) at the plane wave area.

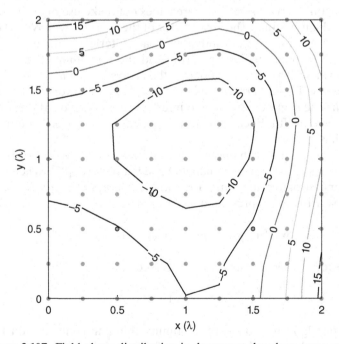

Figure 3.107. Field phase distribution in degrees at the plane wave area.

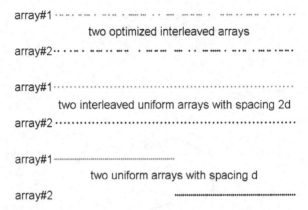

Figure 3.108. Two arrays occupying the same aperture size. The bottom arrays are two 30-element uniformly spaced arrays placed side by side ($d = 0.5\lambda$). The center two arrays have every other element turned on resulting in two arrays with $d = 0.5\lambda$. The top two arrays were optimized to create the lowest maximum sidelobe.

dual-polarization array had an interleaved cross-dipole radiator and a cavity-backed disk radiator in the same lattice structure [86]. Most interleaved arrays consist of multiple arrays operating at different frequencies.

There are several ways to have two arrays operating at the same frequency share the same aperture. Placing one array on the left side of the area and the other array on the right side of the area does not make use of any interleaving (bottom of Figure 3.108). The array factor of these arrays with $N = 30$ is shown in Figure 3.109 for $d = 0.5\lambda$. Another approach interleaves the two arrays with every other element feeding one array and the remaining elements feeding the second array. The pattern of one of the arrays having $d = 1.0\lambda$ is in Figure 3.109. The side-by-side arrays have the advantage of lower average sidelobe levels while the every other element interleaved arrays have the advantage of narrower beamwidth. Each array in the aperture has $\eta_t = 0.50$, but the whole aperture has $\eta_t = 1.00$.

A genetic algorithm can optimize the interleaving of two antisymmetrically thinned arrays to minimize the maximum sidelobe level [87]. When one array has an element on/off, the other array has it off/on, so the amplitude weights for the first array are represented as

$$\mathbf{w} = \left[w_1, w_2, \ldots, w_N, 1 - w_N, \ldots, 1 - w_2, 1 - w_1 \right] \qquad (3.70)$$

and those for the second array are expressed as

$$\mathbf{w}' = 1 - \mathbf{w} \qquad (3.71)$$

The genetic algorithm found the optimum thinning configuration for a 60-element array shown at the top of Figure 3.108, and the pattern of one of these

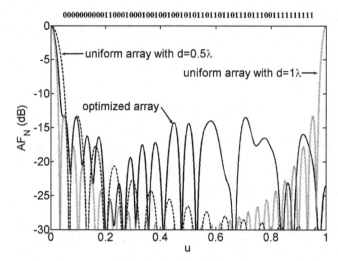

Figure 3.109. Array factor for an optimized fully interleaved array is compared to the array factors for a side-by-side array ($d = 0.5\lambda$) and every other element interleaved array ($d = 1.0\lambda$) when there are 60 total elements with 30 turned on.

arrays is the solid line in Figure 3.109. Adding **w** to **w**' yields a vector of all ones which implies $\eta_{ap} = 100\%$ with each array having $\eta_{ar} = 50\%$. The symmetry associated with **w** and **w**' ensures that 50% of the elements will be turned on in each array. The arrays have nearly the narrow beamwidth of the full aperture and has the same peak sidelobe level of a uniform array. The resulting arrays are a compromise between the side-by-side uniform array and the every-other element interleaved array.

Example. Find the interleaved configuration that yields the lowest maximum sidelobe level for two 10-element arrays with $d = 0.5\lambda$.

A genetic algorithm optimized the interleaved arrays to get the following weights:

$$\mathbf{w} = [11111111101000000000]$$

The maximum sidelobe level of the optimized array factor in Figure 3.110 is $-14\,\text{dB}$.

A thinned sum array has the highest density of elements turned on in the center corresponding to the desired low sidelobe amplitude tapers. A low-sidelobe thinned difference array has few elements turned on in the center. Both arrays have less "on" elements near the edges. Using the elements turned "off" in an optimally thinned sum array to create a thinned difference array does not result in an acceptable difference pattern [87].

1 1 1 1 1 1 1 1 1 0 1 0 0 0 0 0 0 0 0 0

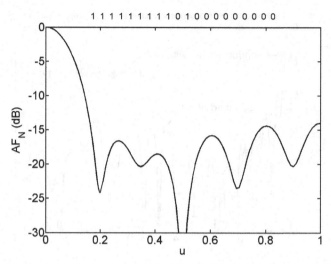

Figure 3.110. Array factor associated with two 10-element interleaved arrays.

1 0 0 0 0 1 0 1 0 0 0 0 1 1 0 1 0 0 1 1 0 0 0 0 0 0 1 0 1 1 1 0 0 0 0 0 1 0 1 1 1 1 1 1 0 1 1 1 1 1
1 1 1 1 1 0 1 1 1 1 1 1 0 1 0 0 0 0 0 1 1 1 0 1 0 0 0 0 0 0 1 1 0 0 1 0 1 1 0 0 0 0 1 0 1 0 0 0 0 1

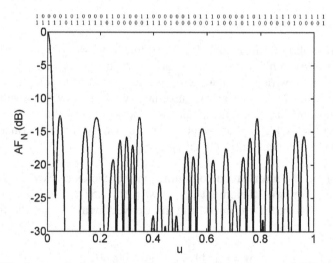

Figure 3.111. Sum array factor due to interleaved thinned sum and difference arrays.

If the thinned difference array has elements turned off then the sum array or $b_n = 1 - a_n$. A genetic algorithm found the thinning arrangement that results in the minimum maximum sidelobe level for both array factors results in an array having the patterns shown in Figure 3.111 and Figure 3.112. The patterns are normalized to their respective peaks. The peak sidelobe level of the sum pattern is 12.68 dB below its main beam, and the peak sidelobe level of the difference pattern is 12.5 dB below its main beam. The sum array has $\eta_{ar} = 0.48$

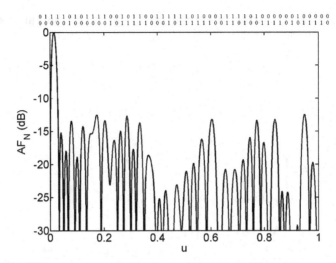

Figure 3.112. Difference array factor due to interleaved thinned sum and difference arrays.

and the difference array has $\eta_{ar} = 0.52$. The entire aperture is 100% efficient, so optimum use is made of the available space.

3.9. NULL SYNTHESIS

Earlier in this chapter, the z transform was used to move nulls to get low sidelobes and place nulls in the array factor at desired angles. That approach does not take into account the distortion to the array factor caused by the null placement. In this section, the null placement is done with minimal distortion to the array factor [88].

Element weights can be found that place nulls at M locations in the array factor. Assume that the new weights are given by

$$w_n = a_n + \Delta_n \tag{3.72}$$

where a_n is the amplitude taper and Δ_n is the complex weight perturbation that causes the nulls. Substituting (3.72) into the equation for a linear array factor results in a nulled array factor that is the sum of the quiescent array factor and a cancellation array factor.

$$\sum_{n=1}^{N} w_n e^{jkx_n u} = \underbrace{\sum_{n=1}^{N} a_n e^{jkx_n u}}_{\text{quiescent pattern}} + \underbrace{\sum_{n=1}^{N} \Delta_n e^{jkx_n u}}_{\text{cancellation pattern}} \tag{3.73}$$

There are an infinite number of combinations of w_n that result in nulls in the M desired directions. Writing (3.73) in matrix form results in

$$AW = B \tag{3.74}$$

where

$$A = \begin{bmatrix} a_1 e^{jkx_1 u_1} & a_2 e^{jkx_2 u_1} & \cdots & a_N e^{jkx_N u_1} \\ a_1 e^{jkx_1 u_2} & a_2 e^{jkx_2 u_2} & \cdots & a_N e^{jkx_N u_2} \\ \vdots & \vdots & \ddots & \vdots \\ a_1 e^{jkx_1 u_M} & a_2 e^{jkx_2 u_M} & \cdots & a_N e^{jkx_N u_M} \end{bmatrix}$$

$$W = \begin{bmatrix} \Delta_1 & \Delta_2 & \cdots & \Delta_N \end{bmatrix}^T$$

$$B = -\begin{bmatrix} \sum_{n=1}^{N} a_n e^{jkx_n u_1} & \sum_{n=1}^{N} a_n e^{jkx_n u_2} & \cdots & \sum_{n=1}^{N} a_n e^{jkx_n u_M} \end{bmatrix}^T$$

Since (3.74) has more columns than rows, finding the weights requires a least squares solution in the form [89]

$$W = A^{\dagger}(AA^{\dagger})^{-1}B \tag{3.75}$$

Example. A 40-element array of isotropic point sources spaced $\lambda/2$ apart has a 30-dB, $\bar{n} = 7$ Taylor taper. Plot the array factors and cancellation beam for a null at 61°. The adapted array factor is very close to the quiescent pattern (Figure 3.113), except near the nulls. Note that the cancellation pattern equals the height of the quiescent patterns where the nulls occur. The cancellation beam has the same Taylor taper as the quiescent pattern. The low sidelobes of the cancellation beam ensures that the pattern away from the null is minimally perturbed.

Example. Repeat the last example for nulls at 13° and 61°.
  Figure 3.114 shows that both nulls are placed with little perturbation to the pattern except for the sidelobes on either side of the nulls.

  Ideally, the quiescent pattern should be perturbed as little as possible when placing the nulls. The weights that produce the nulls are [90]

$$w_n = a_n - \sum_{m=1}^{M} \gamma_m c_n e^{-jnkdu_m} \tag{3.76}$$

where γ_m is the sidelobe level of the quiescent pattern at u_m and c_n is the amplitude taper of the cancellation beam. When $c_n = a_n$, the cancellation

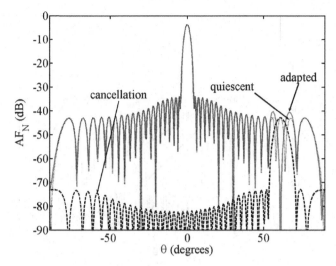

Figure 3.113. Adapted array factor with quiescent and cancellation pattern for a null at 61°.

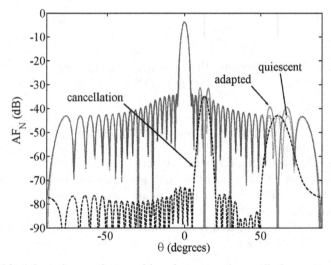

Figure 3.114. Adapted array factor with quiescent and cancellation pattern for nulls at 13° and 61°.

pattern looks the same as the quiescent pattern. When $c_n = 1.0$, the cancellation pattern is a uniform array factor and is the constrained least mean square approximation to the quiescent pattern over one period of the pattern. The nulled array factor can now be written as

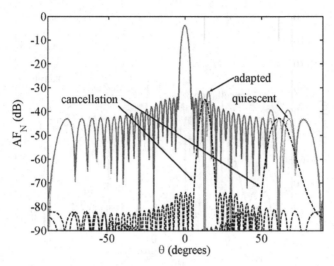

Figure 3.115. Adapted array factor with quiescent and weighted cancellation patterns for nulls at 13° and 61°.

$$\sum_{n=1}^{N} w_n e^{jkx_n u} = \sum_{n=1}^{N} a_n e^{jkx_n u} - \sum_{m=1}^{M} \gamma_m \sum_{n=1}^{N} a_n e^{jkx_n(u-u_m)} \tag{3.77}$$

Equation (3.77) works because the cancellation beams are orthogonal: When cancellation pattern m has a peak at u_m, all the other cancellation patterns have nulls.

Example. Repeat the previous example using both uniform and weighted cancellation beams.

The weighted cancellation beam (Figure 3.115) causes the sidelobes to become higher on either side of the null more than the uniform cancellation beam, because the uniform cancellation beam has a narrower beamwidth. The uniform cancellation beam (Figure 3.116) raises the sidelobe level in other places more than the weighted cancellation beam, because it has higher sidelobe levels (look at ±90°).

This procedure extends to phase-only nulling. When the phase shifts are small, then $e^{j\phi_n} \approx 1 + j\phi_n$, and the array factor can be written as [91]

$$\sum_{n=1}^{N} w_n e^{jkx_n u} = \sum_{n=1}^{N} a_n e^{j\delta_n} e^{jkx_n u} \approx \sum_{n=1}^{N} a_n (1 + j\delta_n) e^{jkx_n u} = \sum_{n=1}^{N} a_n e^{jkx_n u} + j \sum_{n=1}^{N} a_n \delta_n e^{jkx_n u}$$

$$\tag{3.78}$$

Using Euler's identity, (3.78) is written as

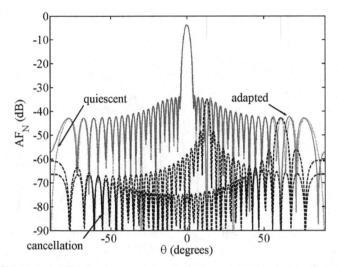

Figure 3.116. Adapted array factor with quiescent and uniform cancellation patterns for nulls at 13° and 61°.

$$\sum_{n=1}^{N} a_n \delta_n [\cos(kx_n u_m) + j\sin(kx_n u_m)] e^{jkx_n u} = j\sum_{n=1}^{N} a_n [\cos(kx_n u_m) + j\sin(kx_n u_m)]$$

(3.79)

Equating the real and imaginary parts of (3.79) leads to

$$\sum_{n=1}^{N} a_n \delta_n \sin(kx_n u_m) e^{jkx_n u} = \sum_{n=1}^{N} a_n \cos(kx_n u_m)$$

(3.80)

$$\sum_{n=1}^{N} a_n \delta_n \cos(kx_n u_m) e^{jkx_n u} = -\sum_{n=1}^{N} a_n \sin(kx_n u_m) = 0$$

(3.81)

Now, (3.80) is in the form that can be solved via (3.75).

Example. A 40-element array of isotropic point sources spaced $\lambda/2$ apart has a 40-dB, $\bar{n} = 7$ Taylor taper. Plot the array factors and cancellation beams for nulls at 13° and 61°. Use phase-only nulling.

Figure 3.117 is a plot of the quiescent, nulled, and cancellation patterns. Phase-only nulling with minimal weight perturbations results in cancellation beams with peaks at symmetric locations about the main beam [92]. A null is placed in the pattern at the desired location, while the pattern increases at the symmetric location. Note the higher sidelobes at −13° and −61°. Consequently, this approach cannot place nulls in symmetric locations in the array pattern.

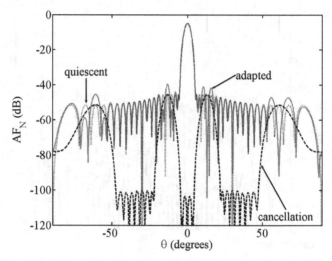

Figure 3.117. Adapted array factor with quiescent and uniform cancellation patterns for nulls at 13° and 61°.

Larger phase shifts that cause greater pattern distortion are required to place symmetric nulls in the array patterns [93].

A monopulse phased array has separate sum and difference channels. If each channel has a low-sidelobe taper as shown in Figure 3.118, then the shared phase shifters (and amplitude weights) can be used to simultaneously place nulls in the sum and difference patterns. The matrixes in (3.74) are now written as [94]

$$
A = \begin{bmatrix}
a_1 e^{jkx_1 u_1} & \cdots & a_N e^{jkx_N u_1} \\
\vdots & \ddots & \vdots \\
a_1 e^{jkx_1 u_M} & \cdots & a_N e^{jkx_N u_M} \\
b_1 e^{jkx_1 u_1} & \cdots & b_N e^{jkx_N u_1} \\
\vdots & \ddots & \vdots \\
b_1 e^{jkx_1 u_M} & \cdots & b_N e^{jkx_N u_M}
\end{bmatrix}
$$

$$
W = \begin{bmatrix} \Delta_1 & \Delta_2 & \cdots & \Delta_N \end{bmatrix}^T
$$

$$
B = -j \left[\sum_{n=1}^{N} a_n e^{jkx_n u_1} \quad \cdots \quad \sum_{n=1}^{N} a_n e^{jkx_n u_M} \quad \sum_{n=1}^{N} b_n e^{jkx_n u_1} \quad \cdots \quad \sum_{n=1}^{N} b_n e^{jkx_n u_M} \right]^T
$$

Phase-only nulling requires further simplification as noted in (3.79).

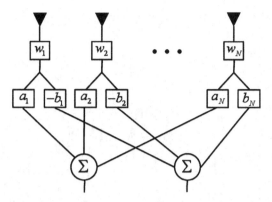

Figure 3.118. Diagram of an array with low-sidelobe sum and difference channels.

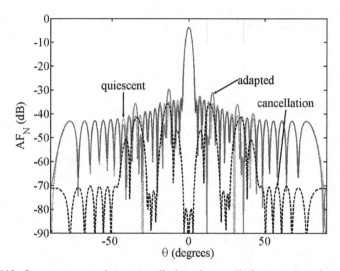

Figure 3.119. Sum pattern quiescent, nulled, and cancellation patterns for simultaneous nulling in the sum and difference patterns.

Example. A 40-element array of isotropic point sources spaced $\lambda/2$ apart has a 30-dB, $\bar{n} = 7$ Taylor taper and a 30-dB, $\bar{n} = 7$ Bayliss taper. Plot the array factors and cancellation beams for nulls at 13° and 61°. Use simultaneous phase-only nulling in the sum and difference patterns.

Figures 3.119 and 3.120 are the resulting patterns with deep nulls in the desired directions.

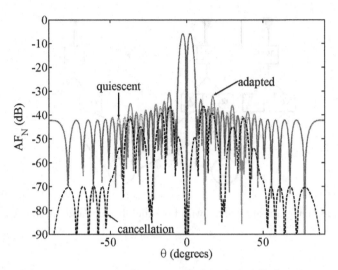

Figure 3.120. Difference pattern quiescent, nulled, and cancellation patterns for simultaneous nulling in the sum and difference patterns.

REFERENCES

1. P. M. Woodward, A method for calculating the field over a plane aperture required to produce a given polar diagram, *J. IEE*, Vol. 93, Pt. IIIA, No. 37, September 1946, pp. 1554–1558.

2. P. M. Woodward and J. D. Lawson, The theoretical precision with which an arbitrary radiation pattern may be obtained from a source of finite size, *J. IEE*, Vol. 95, Pt. III, No. 37, September 1948, pp. 363–370.

3. R. L. Haupt and D. H. Werner, *Genetic Algorithms in Electromagnetics*, New York: John Wiley & Sons, 2007.

4. R. L. Haupt, Genetic algorithm applications for phased arrays, *ACES J.*, Vol. 21, No. 3, November 2006, pp. 325–336.

5. J. S. Stone, US Patent 1,643,323, September 27, 1927.

6. C. L. Dolph, A current distribution for broadside arrays which optimizes the relationship between beam width and sidel-lobe level, *Proc. IRE*, Vol. 34, No. 6, June 1946, pp. 335–348.

7. R. S. Elliott, The theory of antenna arrays, in *Microwave Scanning Antennas*, R. C. Hansen, ed., New York: Academic Press, 1964, pp. 1–69.

8. F. I. Tseng and D. K. Cheng, Optimum scannable planar arrays with an invariant sidelobe level, *IEEE AP-S Trans.*, Vol. 56, No. 11, November 1968, pp. 1771–1778.

9. Y. V. Baklanov, Chebyshev distribution of currents for a plane array of radiators, *Radio Eng. Electron. Phys.*, Vol. 11, 1966, pp. 640–642.

10. T. T. Taylor, Design of line source antennas for narrow beamwidth and low side lobes, *IRE AP Trans.*, Vol. 3, 1955, pp. 16–28.

11. T. T. Taylor, One parameter family of line sources producing modified symmetry patterns, Report No. TM 324, Culver City, CA: Hughes Aircraft Co., 1953.

12. T. Taylor, Design of circular apertures for narrow beamwidth and low sidelobes, *IEEE AP-S Trans.*, Vol. 8, No. 1, January 1960, pp. 17–22.

13. R. C. Hansen, A one-parameter circular aperture distribution with narrow beamwidth and low sidelobes, Vol. 24, No. 4, July 1976, pp. 477–480.

14. W. D. White, Circular aperture distribution functions, *IEEE AP-S Trans.*, Vol. 25, No. 5, September 1977, pp. 714–716.

15. E. T. Bayliss, Design of monopulse antenna difference patterns with low sidelobes, *The Bell System Tech. J.*, Vol. 47, May–June 1968, pp. 623–650.

16. R. S. Elliott, Design of line source antennas for narrow beamwidth and asymmetric low sidelobes, *IEEE AP-S Trans.*, Vol. 23, No. 1, January 1975, pp. 100–107.

17. R. S. Elliott, *Antenna Theory and Design*, revised edition, Hoboken, NJ: John Wiley & Sons, 2003.

18. C. McCormack and R. Haupt, Antenna pattern synthesis using partially tapered arrays, *IEEE Trans. Magn.*, Vol. 27, No. 5, 1991, pp. 3902–3904.

19. W. H. Press et al., *Numerical Recipes in FORTRAN*, New York: Cambridge University Press, 1992.

20. J. A. Nelder and R. Mead, *Comput. J.*, Vol. 7, 1965, pp. 308–313.

21. D. G. Luenberger, *Linear and Nonlinear Programming*, Reading, MA: Addison-Wesley, 1984.

22. M. J. D. Powell, An efficient way for finding the minimum of a function of several variables without calculating derivatives, *Comput. J.*, 1964, pp. 155–162.

23. G. C. Broyden, A class of methods for solving nonlinear simultaneous equations, *Math. Comput.*, October 1965, pp. 577–593.

24. R. Fletcher, Generalized inverses for nonlinear equations and optimization, in *Nurnerical Methods for Nonlinear Algebraic Equations*, R. Rabinowitz, ed., London: Gordon and Breach, 1970, pp. 75–85.

25. D. Goldfarb and B. Lapidus, Conjugate gradient method for nonlinear programming problems with linear constraints, *I&EC Fundamentals*, February 1968, pp. 142–151.

26. D. F. Shanno, An accelerated gradient projection method for linearly constrained nonlinear estimation, *SIAM J. Appl. Math.*, March 1970, pp. 322–334.

27. M. I. Skolnik, G. Nemhauser, and J. W. Sherman, III Dynamic programming applied to unequally spaced arrays, *IEEE AP-S Trans.*, Vol. 12, January 1964, pp. 35–43.

28. N. Balakrishnan, P. K. Murthy, and S. Ramakrishna, Synthesis of antenna arrays with spatial and excitation constraints, *IEEE AP-S Trans.*, Vol. 27, No. 5, September 1979, pp. 690–696.

29. J. Perini, Note on antenna pattern synthesis using numerical iterative methods, *IEEE AP-S Trans.*, Vol. 12, July 1976, pp. 791–792.

30. C. S. Ruf, Numerical annealing of low-redundancy linear arrays, *IEEE Trans.*, Vol. 41, No. 1, January 1993, pp. 85–90.

31. W. L. Stutzman and E. L. Coffey, Radiation pattern synthesis of planar antennas using the iterative sampling method, *IEEE Trans.*, Vol. AP-23, No. 6, November 1975, pp. 764–769.

32. H. J. Orchard, R. S. Elliot, and G. J. Stern, Optimizing the synthesis of shaped beam antenna patterns, *IEE Proc.*, Vol. 132, No. 1, February 1985, pp. 63–68.

33. R. S. Elliot and G. J. Stearn, Shaped patterns from a continuous planar aperture distribution, *IEEE Proc.*, Vol. 135, No. 6, December 1988, pp. 366–370.

34. J. E. Richie and H. N. Kritikos, Linear program synthesis for direct broadcast satellite phased arrays, *IEEE AP-S Trans.*, Vol. 36, No. 3, March 1988, pp. 345–348.

35. F. Ares, R. S. Elliott, and E. Moreno, Design of planar arrays to obtain efficient footprint patterns with an arbitrary footprint boundary, *IEEE AP-S Trans.*, Vol. 42, No. 11, November 1994, pp. 1509–1514.

36. O. Einarsson, Optimization of planar arrays, *IEEE AP-S Trans.*, Vol. AP-27, No. 1, January 1979, pp. 86–92.

37. T. S. Ng, J. Yoo Chong Cheah, and F. J. Paoloni, Optimization with controlled null placement in antenna array pattern synthesis, *IEEE Trans.*, Vol. AP-33, No. 2, February 1985, pp. 215–217.

38. D. K. Cheng, Optimization techniques for antenna arrays, *Proc. IEEE*, Vol. 59, No. 12, December 1971, pp. 1664–1674.

39. J. F. DeFord and O. P. Gandhi, Phase-only synthesis of minimum peak sidelobe patterns for linear and planar arrays, *IEEE AP-S Trans.*, Vol. 36, No. 2, February 1988, pp. 191–201.

40. R. L. Haupt, Genetic algorithm applications for phased arrays, *ACES J.*, Vol. 21, No. 3, November 2006, pp. 325–336.

41. S. Kirkpatrick, C. D. Gelatt, and M. P. Vecchi, Optimization by simulated annealing, *Science*, Vol. 220, No. 4598, May 13, 1983, pp. 671–680.

42. N. Metropolis, A. Rosenbluth, and M. Rosenbluth, *J. Chem. Phys.*, Vol. 21, 1953, pp. 1087–1092.

43. J. H. Holland, 1975, *Adaptation in Natural and Artificial Systems*, Ann Arbor: The University of Michigan Press.

44. D. E. Goldberg, 1989, *Genetic Algorithms in Search, Optimization, and Machine Learning*, New York: Addison-Wesley.

45. R. L. Haupt and S. E. Haupt, *Practical Genetic Algorithms*, 2nd edition, New York: John Wiley & Sons, 2004.

46. R. L. Haupt and D. Werner, *Genetic Algorithms in Electromagnetics*, New York: John Wiley & Sons, 2007.

47. R. Willey, Space tapering of linear and planar arrays, *IEEE AP-S Trans.*, Vol. 10, 1962, pp. 369–377.

48. M. Skolnik, J. Sherman III, and F. Ogg, Jr., Statistically designed density-tapered arrays, *IEEE AP-S Trans.*, Vol. 12, No. 4, 1964, pp. 408–417.

49. E. Brookner, Antenna array fundamentals—part 1, in *Practical Phased Array Antenna Systems*, E. Brookner, ed., Norwood, MA: Artech House, 1991, pp. 2-1-2-37.

50. T. A. Milligan, Space-tapered circular (ring) array, *IEEE AP Mag.*, Vol. 46, No. 3, June 2004, pp. 70–73.

51. R. L. Haupt, Unit circle representation of aperiodic arrays, *IEEE AP-S Trans.*, Vol. 43, No. 9, October 1995, pp. 1152–1155.

52. R. L. Haupt, Thinned arrays using genetic algorithms, *IEEE AP-S Trans.*, Vol. 42, No. 7, July 1994, pp. 993–999.

53. R. L. Haupt, Thinned arrays using genetic algorithms, in *Electromagnetic Optimization by Genetic Algorithms*, Y. Rahmat-Samii and E. Michielssen, eds., New York: John Wiley & Sons, 1999, pp. 95–117.

54. R. L. Haupt, Optimized element spacing for low sidelobe concentric ring arrays, *IEEE AP-S Trans.*, Vol. 56, No. 1, January 2008, pp. 266–268.

55. V. Namias, The fractional order Fourier transform and its application to quantum mechanics, *J. Inst. Appl. Math.*, 25, 1980, pp. 241–265.

56. A. T. Moffet, Minimum-redundancy linear arrays, *IEEE AP-S Trans.*, Vol. 16, No. 2, March 1968, pp. 172–175.

57. C. S. Ruf, Numerical annealing of low-redundancy linear arrays, *IEEE AP-S Trans.*, Vol. 41, No. 1, January 1993, pp. 85–90.

58. R. L. Haupt, An introduction to genetic algorithms for electromagnetics, *IEEE Antennas Propagat. Mag.*, Vol. 37, No. 2, April 95, pp. 7–15.

59. J. F. DeFord and O. P. Gandhi, Phase-only synthesis of minimum peak sidelobe patterns for linear and planar arrays, *IEEE Trans. Antennas Propagat.*, Vol. 36, No. 2, February 1988, pp. 191–201.

60. R. L. Haupt, Optimum quantized low sidelobe phase tapers for arrays, *IEE Electron. Lett.*, Vol. 31, No. 14, July 95, pp. 1117–1118.

61. R. Haupt, Reducing grating lobes due to subarray amplitude tapering, *IEEE AP-S Trans.*, Vol. 33, No. 8, August 1985, pp. 846–850.

62. R. Haupt, Optimization of subarray amplitude tapers, *IEEE AP-S Symp.*, June 1995, pp. 1830–1833.

63. R. L. Haupt, Low sidelobes from identically thinned subarrays, in *IEEE AP-S Symposium*, Albuquerque, NM, 2006, pp. 4217–4220.

64. R. J. Mailloux, Constrained feed technique for subarrays of large phased arrays, *IEE Electron. Lett.*, Vol. 34, November 12, 1998, pp. 2191–2193.

65. R. J. Mailloux, A low-sidelobe partially overlapped constrained feed network for time-delayed subarrays, *IEEE AP-S Trans.*, Vol. 49, February 2001, pp. 280–291.

66. A. P. Goffer, M. Kam, and P. R. Herczfeld, Design of phased arrays in terms of random subarrays, *IEEE AP-S Trans.*, Vol. 42, June 1994, pp. 820–826.

67. P. S. Hall and M. S. Smith, Sequentially rotated arrays with reduced sidelobe levels, *IEE Proc. Microwave Antennas Propagat.*, Vol. 141, August 1994, pp. 321–325.

68. N. Toyama, Aperiodic array consisting of subarrays for use in small mobile earth stations, *IEEE AP-S Trans.*, Vol. 53, June 2005, pp. 2004–2010.

69. K. C. Kerby and J. T. Bernhard, Wideband periodic array of random subarrays, *IEEE AP-S Symp.*, June 2004, pp. 555–558.

70. R. L. Haupt, Optimized weighting of uniform subarrays of unequal sizes, *IEEE AP-S Trans.*, Vol. 56, No. 1, January 2008, pp. 1207–1210.

71. M. Alvarez-Folgueiras, J. Rodriguez-Gonzalez, and F. Ares-Pena, High-performance uniformly excited linear and planar arrays based on linear semiarrays

composed of subarrays with different uniform spacings, *IEEE AP-S Trans.*, Vol. 57, December 2009, pp. 4002–4006.

72. P. Rocca, R. L. Haupt, and A. Massa, Sidelobe reduction through element phase control in uniform subarrayed array antennas, *IEEE AWPL*, Vol. 8, 2009, pp. 437–440.

73. O. M. Bucci, G. Franceschetti, G. Mazzarella, and G. Panariello, Intersection approach to array pattern synthesis, *Proc. IEE, Part H*, Vol. 137, No. 6, December 1990, pp. 349–357.

74. D. A. Hill, A numerical method for near-field array synthesis, *IEEE EMC Trans.*, Vol. 27, No. 4, November 1985, pp. 201–211.

75. D. A. Hill, A near-field array of Yagi-Uda antennas for electromagnetic-susceptibility testing, *IEEE EMC Trans.*, Vol. 28, No. 4, August 86, pp. 170–178.

76. D. A. Hill, A circular array for plane-wave synthesis, *IEEE EMC Trans.*, Vol. 30, No. 1, February 1988, pp. 3–8.

77. J. E. Hansen, Plane-wave synthesis, in *Spherical Near-Field Antenna Measurements*, London: Peter Peregrinus Ltd., 1988, Chapter 7.

78. R. L. Haupt, Generating a plane wave with a linear array of line sources, *IEEE AP-S Trans.*, February 2002.

79. C. C. Courtney, D. E. Voss, R. Haupt, and L. LeDuc, The theory and architecture of a plane wave generator, 2002 AMTA Conference, Cleveland, OH, November 2002.

80. C. C. Courtney, D. E. Voss, R. Haupt, and L. LeDuc, The measured performance of a plane wave generator prototype, 2002 AMTA Conference, Cleveland, OH, November 2002.

81. R. Haupt, Generating a plane wave in the near field with a planar array antenna, *Microwave J.*, Vol. 63, August 2003, pp. 152–165.

82. J. Hsiao, Analysis of interleaved arrays of waveguide elements, *IEEE AP-S Trans.*, Vol. 19, No. 6 , November 1971, pp. 729–735.

83. J. Boyns and J. Provencher, Experimental results of a multifrequency array antenna, *IEEE AP-S Trans.*, Vol. 20, No. 1, November 1972, pp. 106–107.

84. K. Lee et al., A dual band phased array using interleaved waveguides and dipoles printed on high dielectric substrate, *IEEE AP-S Symp.*, Vol. 2, June 1984, pp. 886–889.

85. R. Chu, K. Lee, and A. Wang, Multiband phased-array antenna with interleaved tapered-elements and waveguide radiators, *IEEE AP-S Symp.*, Vol. 3, December 1996, pp. 21–26.

86. K. Lee, A. Wang, and R. Chu, Dual-band, dual-polarization, interleaved cross-dipole and cavity-backed disc elements phased array antenna, *IEEE AP-S Symp.*, Vol. 2, July 1997, pp. 694–697.

87. R. L. Haupt, Interleaved thinned linear arrays, *IEEE AP-S Trans.*, Vol. 53, September 2005, pp. 2858–2864.

88. H. Steyskal, Synthesis of antenna patterns with prescribed nulls, *IEEE Trans. Antennas Propagat.*, Vol. 30, No. 2, 1982, pp. 273–279.

89. R. Shore and H. Steyskal, Nulling in linear array patterns with minimization of weight perturbations, RADC-TR-82-32, February 1982.

90. H. Steyskal, R. Shore, and R. Haupt, Methods for null control and their effects on the radiation pattern, *IEEE Trans. Antennas Propagat.*, Vol. 34, No. 3, 1986, pp. 404–409.

91. H. Steyskal, Simple method for pattern nulling by phase perturbation, *IEEE Trans. Antennas Propagat.*, Vol. 31, No. 1, 1983, pp. 163–166.

92. R. Shore, A proof of the odd-symmetry of the phases for minimum weight perturbation phase-only null synthesis, *IEEE Trans. Antennas Propagat.*, Vol. 32, No. 5, 1984, pp. 528–530.

93. R. Shore, Nulling a symmetric pattern location with phase-only weight control, *IEEE Trans. Antennas Propagat.*, Vol. 32, No. 5, 1984, pp. 530–533.

94. R. Haupt, Simultaneous nulling in the sum and difference patterns of a monopulse antenna, *IEEE Trans. Antennas Propagat.*, Vol. 32, No. 5, 1984, pp. 486–493.

4 Array Factors and Element Patterns

So far, only isotropic point sources have been considered as elements in the arrays. This chapter takes a first step toward a more realistic antenna array through the introduction of an element pattern with its associated gain and polarization. Formulas for the normalized far-field element patterns and directivity of many common elements are given in this chapter for use with the array factor formulas already presented. More complicated antennas require the use of a numerical model to calculate the element pattern. Calculating the impedance of an element is important in order to determine the bandwidth.

4.1. PATTERN MULTIPLICATION

The elements in an actual array are not isotropic point sources. Instead, they have directionality that is proportional to their size. They also have frequency, impedance, and polarization properties not associated with point sources. Normally, the elements in an array are spaced relatively close together, so an element is typically no larger than $\lambda/2 \times \lambda/2$ in area in a square lattice. As such, the element pattern is too small to have sidelobes. A typical element pattern for an array in the x–y plane can be reasonably approximated by $\cos \theta$ or $\sin \phi$ or the change in the projected area of the element [1].

Element spacing in an array is determined by the distance between phase centers of adjacent elements. A point source actually represents the phase center of the antenna element, which is the center of a sphere of constant phase radiated by the antenna [2]. This phase center moves with frequency and angle; so in actuality, it only exists for a portion of a sphere at a given frequency.

The directivity of the array depends upon the directivity of the elements in the array. Consider a linear array of point sources that lies along the z axis. Assume that each point source is the phase center for an element pattern given by

$$e(\theta) = \sin \theta \qquad (4.1)$$

Antenna Arrays: A Computational Approach, by Randy L. Haupt
Copyright © 2010 John Wiley & Sons, Inc.

The far-field antenna pattern for this array is

$$\text{AP}(\theta) = w_1 e^{jkz_1 \cos\theta} \sin\theta + w_2 e^{jkz_2 \cos\theta} \sin\theta + w_3 e^{jkz_3 \cos\theta} \sin\theta + \cdots + w_N e^{jkz_N \cos\theta} \sin\theta$$
(4.2)

Since the element pattern is common to all the terms, it can be factored out to get

$$\text{AP} = \sin\theta \sum_{n=1}^{N} w_n e^{jkz_n \cos\theta}$$
(4.3)

which leads to general the formula for any element pattern and array factor:

$$\text{AP} = \text{Element pattern} \times \text{Array factor}$$
(4.4)

The directivity formula for an array with an element pattern, **e**, is

$$D = \frac{4\pi |\mathbf{e}(\theta, \phi) AF(\theta, \phi)|^2}{\displaystyle\int_0^{2\pi}\int_0^{\pi} |\mathbf{e}(\theta, \phi) AF(\theta, \phi)|^2 \sin\theta d\theta d\phi}$$
(4.5)

Note that **e** is a vector, because it has polarization, whereas the array factor does not. In some cases, (4.5) does not equal the directivity of the array factor times the directivity of the element pattern.

There are several important differences between the antenna pattern and the array factor. First, the antenna pattern has a polarization that is determined by the array elements. Usually, all the elements are oriented in the same direction, so the array polarization is the same as a single element. It is possible to orient the elements in a way that causes the array antenna pattern to have a different polarization from the element pattern. An example of this architecture is a 2×2 subarray of rectangular patch antennas where each patch receives a 90° rotation as shown in Figure 4.1 [3]. In addition, one set of diagonal patches that are horizontally polarized receive a 0° phase shift, while the other set of diagonal patches that are vertically polarized receive a 90° phase shift. A second difference is that the element pattern forms an envelope that contains the much faster oscillating array factor. The element pattern enhances the array factor in the direction of the element pattern peak and suppresses the array factor in direction of element pattern minima. A final difference is that element orientation is important, because the element directivity and polarization are a function of angle. If the peak of the element pattern points in the same direction as the array factor peak, then the array pattern main beam is enhanced. If an element pattern null points in the direction of the array factor peak, then the antenna pattern has a null in that direction.

Usually, the array elements have very broad patterns that cannot be steered. Thus, the element patterns remain fixed in space. When the peak of the array

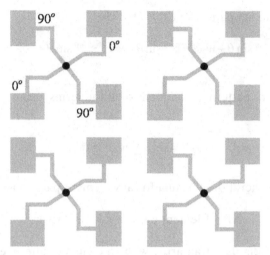

Figure 4.1. Circularly polarized 4 × 4 array made from linearly polarized elements. Four patches are combined to form a circularly polarized subarray.

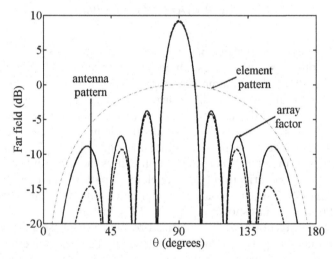

Figure 4.2. Element pattern superimposed on the array pattern and array factor.

factor and the peak of the element pattern align, then the main beam of the resulting antenna pattern is a maximum, while the sidelobes far from the main beam are reduced. Figure 4.2 shows the effects of pattern multiplication on the resulting normalized antenna pattern. The element pattern is given by $\sin \theta$. The array factor corresponds to an 8-element array ($d = 0.5\lambda$) along the z axis. When compared to the array factor directivity (9 dB), the directivity of the antenna pattern increases to 9.29 dB while its sidelobes decrease more quickly away from the main beam. When the array factor is steered, the

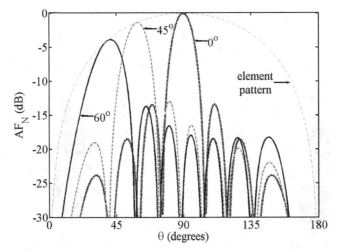

Figure 4.3. Element pattern superimposed on the array patterns when the array factor is steered to 90°, 60°, and 45°. The element pattern has a peak of 0 dB. The array patterns are normalized to the broadside peak.

element pattern remains stationary. Thus, the product of the array factor and main beam changes as the main beam is steered. The effects of the steering can be seen in Figure 4.3. Steering the main beam reduces the peak of the antenna pattern due to the decrease in the element pattern. The element pattern also causes a squint in the main beam away from broadside as can be seen in the beam steered to 45° with a standard steering phase of $\delta_n = -kx_n u_s$. Thus, a correction to the steering phase for the array factor is necessary to make sure that the peak of the antenna pattern points to the desired angle.

Hertzian dipoles are the simplest antenna elements outside of point sources. They are linearly polarized and have a pattern that is proportional to $\sin\theta$ when oriented in the z direction. Placing several of these dipoles in the vicinity of each other causes the dipoles to interact. In other words, the dipoles all radiate and receive time-varying fields from each other. This interaction is called mutual coupling. Although mutual coupling is very important in antenna design, we will ignore it until Chapter 6. A first approximation to the electric field of an array of Hertzian dipoles is

$$\mathbf{E} = \sum_{n=1}^{N} w_n j L_z Z I_z k \sin\theta \frac{e^{-jkR_n}}{4\pi R_n} \hat{\boldsymbol{\theta}} \qquad (4.6)$$

We are interested in the relative electric field as a function of angle. Thus, the constants can be ignored and the approximation $R_n \approx r$ is valid in the denominator. Also, the R_n in the phase term can be approximated as shown in Chapter 2. The resulting far-field pattern for an array of Hertzian dipoles is

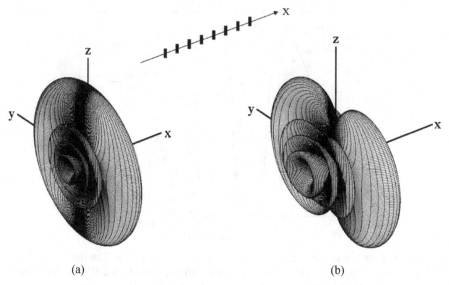

Figure 4.4. An 8-element array along the x axis with $d = \lambda/2$. **(a)** Isotropic point sources. **(b)** z-directed Hertzian dipoles.

$$\mathbf{E} = \sin\theta \sum_{n=1}^{N} w_n e^{jk(x_n\sin\theta\cos\phi + y_n\sin\theta\sin\phi + z_n\cos\theta)}\hat{\boldsymbol{\theta}} \qquad (4.7)$$

The $\sin\theta$ term is the relative shape of the far-field pattern of the Hertzian dipole, and the summation terms are the array factor.

An 8-element linear array lying along the x axis with $d = \lambda/2$ has the array factor shown in Figure 4.4a. This array factor has no polarization, because point sources have no polarization. Its directivity is 9.0 dB. The peak occurs at $\phi = 90°$ for all θ. Replacing the point sources with z-directed Hertzian dipoles having a directivity of 1.76 dB results in the array factor in Figure 4.4b. This antenna pattern is θ-polarized with no radiation in the z direction, because the element pattern has a null in that direction. The directivity of this array is 11.9 dB and can be calculated using numerical integration of the array factor times the element pattern or the formula [4]:

$$D = \frac{1.5N^2}{N + 3\sum_{n=1}^{N-1}(N-n)\left[\dfrac{\sin nkd}{nkd} + \dfrac{\cos nkd}{(nkd)^2} - \dfrac{\sin nkd}{(nkd)^3}\right]} \qquad (4.8)$$

If $d = \lambda/2$, then this formula reduces to

$$D = \frac{1.5N^2}{N + 3\sum_{n=1}^{N-1}\dfrac{(-1)^n(N-n)}{\pi^2 n^2}} \qquad (4.9)$$

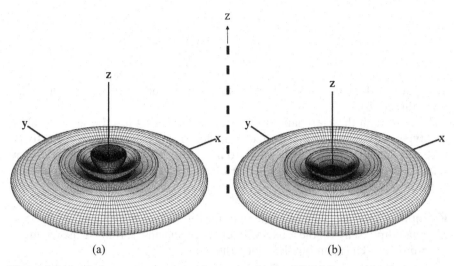

Figure 4.5. An 8-element array along the z axis with $d = \lambda/2$. **(a)** Isotropic point sources. **(b)** z-directed Hertzian dipoles.

The directivity of the Hertzian dipole array is 2.9 dB higher than the directivity of the array factor, because the Hertzian dipole enhances the array factor main beam at $\theta = 90°$ and produces a null at $\theta = 0°$. This example shows that the directivity of the array factor times the directivity of the element pattern does not have to equal the directivity of the array. In fact, in this case, the directivity of the element times the directivity of the array factor is less than the directivity of the array.

If the same z-directed Hertzian dipoles were placed along the z axis instead, would the directivity and antenna patterns be different? The answer is yes. An 8-element linear array lying along the z axis with $d = \lambda/2$ has the array factor shown in Figure 4.5a. Like the point source array along the x axis, this array has no polarization. Its directivity is also 9.0 dB. The peak occurs at $\theta = 90°$ for all ϕ. The array factor is identical to Figure 4.4a, except turned on its side. Replacing the point sources with z-directed Hertzian dipoles changes the array factor to look like Figure 4.4b. The antenna pattern is θ-polarized with no radiation in the z direction and has a directivity of 9.2 dB calculated using numerical integration or the formula [4]:

$$D = \frac{1.5N^2}{N - 6\sum_{n=1}^{N-1}(N-n)\left[\dfrac{\cos nkd}{(nkd)^2} - \dfrac{\sin nkd}{(nkd)^3}\right]} \tag{4.10}$$

If $d = \lambda/2$, then this formula reduces to

$$D = \frac{1.5N^2}{N - 6\sum_{n=1}^{N-1} \frac{(-1)^n (N-n)}{\pi^2 n^2}} \quad\quad (4.11)$$

This pattern does not differ much from its array factor, so it has nearly the same directivity.

Even though the Hertzian dipole array in Figure 4.4 has the same number and type of elements as the Hertzian dipole array in Figure 4.5, its directivity is 2.7 dB higher. Unlike with point sources, the orientation of the element relative to the array has a significant effect on the directivity and antenna pattern of the array.

Example. Calculate the antenna pattern for a 1×8 array of z-oriented Hertzian dipoles along the x axis with element spacing λ. Repeat the example for z-directed Hertzian dipoles along the z axis.

Figure 4.6a shows the antenna pattern for the array along the x axis. The main beam is narrow in the x–y plane. The grating lobes point along the x axis. A null occurs in the main beam along the z axis. The directivity is reduced to 10.4 dB from Figure 4.4 due to the formation of two grating lobes at $\theta = 90°$ and $\phi = 90°$ and $270°$. Figure 4.6b shows the antenna pattern for the array along the z axis. The main beam is omnidirectional in the x–y plane. Grating lobes do not appear at $\theta = 0°$ and $180°$, because the element pattern has nulls in the z direction. The directivity is 11.7 dB. In this case, the array along the z axis has a greater directivity than the array along the x axis, because the element pattern nulls cancel the grating lobes at $\theta = 0°$ and $180°$. Steering the

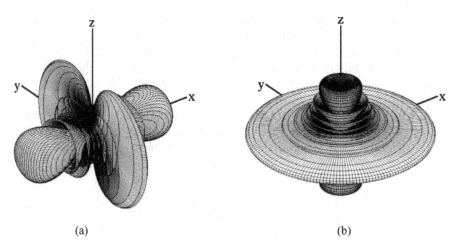

(a)   (b)

Figure 4.6. An 8-element array of z-directed Hertzian dipoles with $d = \lambda$. **(a)** Along the x axis. **(b)** Along the z axis.

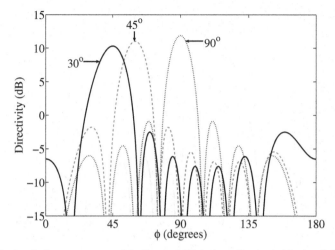

Figure 4.7. An 8-element array of z-directed Hertzian dipoles with $d = \lambda/2$ along the x axis. The beam is steered to $\phi = 90°$, $45°$, and $30°$.

main beam would cause the grating lobes to appear, because they would move out of the element pattern nulls.

The main beam of the array in Figure 4.4 can be steered to an angle ϕ_s in the x–y plane. Figure 4.7 shows the H-plane cut of the beam steered to $\phi = 90°$, $60°$, and $45°$. Even though the element pattern is omnidirectional in the x–y plane, the peak of the main beam decreases, because the projected area of the rectangular area containing the dipoles decreases. As a result, the directivity decreases from 11.9 dB at $\phi = 90°$ to 11.0 dB at $\phi = 60°$ and to 10.3 dB at $\phi = 45°$. Figure 4.8 shows the E plane cut of the main beam of the Hertzian dipole array in Figure 4.5 steered to $\phi = 90°$, $60°$, and $45°$. The element pattern directivity decreases with decreasing θ, but the antenna pattern decreases only from 9.2 dB at $\phi = 90°$ to 9.1 dB at $\phi = 60°$ and to 9.0 dB at $\phi = 45°$.

4.2. WIRE ANTENNAS

Hertzian dipoles and small loops serve as building blocks for larger wire antennas that have more directivity and are easier to match to the feed network.

4.2.1. Dipoles

Unlike Hertzian dipoles, long dipoles have current variations along the wire, so the constant current approximation is not valid. Since the dipole is narrowband and a standing wave exists along the wire, the current can be reason-

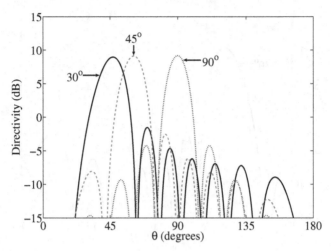

Figure 4.8. An 8-element array of z-directed Hertzian dipoles with $d = \lambda/2$ along the z axis. The beam is steered to $\theta = 90°, 45°$, and $30°$.

ably represented by a sinusoid. Substitute $I(z) = I_z \sin[k(L_z/2 \pm z')]$ into the equation for the magnetic vector potential to get

$$\mathbf{A} = \hat{\mathbf{z}} \int_{-a}^{a} \frac{I_z \sin[k(L_z/2 \pm z')]e^{-jkR}}{4\pi R} dz' \qquad (4.12)$$

Begin solving this integral by breaking it into two parts and approximating the denominator by $R \cong r$ and the numerator by $R \approx r - z' \cos\theta$.

$$A_z = \frac{I_z e^{-jkr}}{4\pi r} \left\{ \int_0^a \sin[k(L_z/2 - z')]e^{-jkz'\cos\theta} dz' + \int_{-a}^0 \sin[k(L_z/2 + z')]e^{-jkz'\cos\theta} dz' \right\}$$

$$(4.13)$$

Performing the integrations and simplifying results in

$$A_z = \frac{I_z e^{-jk(r-z'\cos\theta)}}{4\pi rk \sin^2\theta} \{j\cos\theta\sin[k(L_z/2 + z)] - \cos[k(L_z/2 + z)]\} \qquad (4.14)$$

Next, the normalized electric field can be found from A_z

$$E_\theta = \frac{\cos\left(\dfrac{kL_z}{2}\cos\theta\right) - \cos(kL_z/2)}{\sin\theta} \qquad (4.15)$$

When $L_z = \lambda/2$, then the directivity is 1.64 or 2.1 dB.

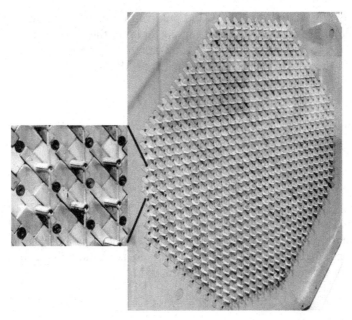

Figure 4.9. Hexagonal array with dipole elements in a square lattice. (Courtesy of the National Electronics Museum.)

Figure 4.9 is a hexagonal array of dipoles in a square lattice. This array was part of the Molecular Electronics for Radar Applications (MERA) project started by the U.S. Air Force (USAF) in 1964 as an exploratory development effort to examine the concept of implementing an X band all solid-state radar [5].

Example. Find the normalized electric field of a half-wavelength dipole.
Substituting $L_z = \lambda/2$ into (4.15) results in

$$E_\theta = \frac{\cos(\pi \cos\theta) + 1}{\sin\theta} \qquad (4.16)$$

The input impedance of a dipole is a function of its length and diameter and is derived in Chapter 6. For a thin, resonant dipole that is slightly less than $\lambda/2$ long, the input impedance is about $73\,\Omega$.

Crossed dipoles are frequently used to create an element with polarization diversity. One dipole controls the electric field parallel to it, and the orthogonal dipoles control the electric fields parallel to them. Varying the amplitude and phase of the signal at each dipole modifies the electric field amplitude and phase in orthogonal directions resulting in any polarization from linear through elliptical to circular [6]. The relative amplitude and phase of the two dipoles also determines its gain pattern.

In order to find the directivity and polarization of the crossed dipoles, the electric fields are calculated from the currents on the dipoles. If the dipoles are assumed to be short (L_x and $L_y \ll \lambda$), then the crossed dipole current is the sum of the constant currents on each short dipole.

$$I(r') = I_x\hat{x} + I_y\hat{y} \tag{4.17}$$

Substituting this current into the equation for the magnetic vector potential for a short dipole yields

$$A = \frac{\mu e^{-jkr}}{4\pi r}(I_x L_x\hat{x} + I_y L_y\hat{y}) \tag{4.18}$$

where r is the distance from the origin to the field point at (x, y, z), $L_{x,y}$ is the dipole length in the x and y directions, and $I_{x,y}$ is the constant current in x and y directions. In the far field, the electric field in rectangular coordinates is found from the magnetic vector potential by

$$E = -j\omega A \tag{4.19}$$

The transmitted electric field is given by

$$E = -j\frac{\omega\mu e^{-jkr}}{4\pi r}(I_x L_x\hat{x} + I_y L_y\hat{y})$$

Converting this rectangular form of the electric field into spherical coordinates produces the relative far-field components [7]

$$E_\theta = I_x L_x \cos\theta\cos\varphi + I_y L_y \cos\theta\sin\varphi \tag{4.20}$$

$$E_\varphi = -I_x L_x \sin\varphi + I_y L_y \cos\varphi \tag{4.21}$$

The directivity is given by

$$D(\theta, \varphi) = 4\pi \frac{|E_\theta(\theta, \varphi)|^2 + |E_\varphi(\theta, \varphi)|^2}{\int_0^{2\pi}\int_0^\pi \left[|E_\theta(\theta, \varphi)|^2 + |E_\varphi(\theta, \varphi)|^2\right]\sin\theta\, d\theta\, d\varphi} \tag{4.22}$$

If the polarization of the incident wave is written in spherical coordinates, then the polarization loss factor is [7]

$$\text{PLF} = \frac{E_{\theta t}}{\sqrt{E_{\theta t}^2 + E_{\varphi t}^2}}\frac{E_{\theta r}}{\sqrt{E_{\theta r}^2 + E_{\varphi r}^2}} + \frac{E_{\varphi t}}{\sqrt{E_{\theta t}^2 + E_{\varphi t}^2}}\frac{E_{\varphi r}}{\sqrt{E_{\theta r}^2 + E_{\varphi r}^2}} \tag{4.23}$$

where $0 \le \text{PLF} \le 1$ with $\text{PLF} = 1$ a perfect match. The t and r subscripts represent transmit and receive, respectively.

Example. Calculate the currents needed to produce linear and circular polarization for crossed dipoles.

The results are shown in the following table:

Polarization	I_x	I_y
Linear x	1	0
Linear y	0	1
Linear 45°	1	1
Circular RH	1	j
Circular LH	1	−j

The High-Frequency Active Auroral Research Program (HAARP) conducts upper atmospheric research in Alaska [8]. The HF transmitter delivers 3.6 MW to a rectangular planar array of 180 crossed dipole elements pointing at the sky. The transmitted signal is partially absorbed in a small volume a few hundred meters thick and a few tens of kilometers in diameter at an altitude between 100 and 350 km (depending on operating frequency). Observations provide new information about the dynamics of plasmas in the ionosphere and new insight into solar–terrestrial interactions. Figure 4.10 is a picture of the HAARP transmit array laid out in a square grid. A tower holds each pair of crossed dipoles above the earth. The crossed dipoles permits control over the polarization of the transmitted signal.

A monopole is one arm of a dipole antenna placed orthogonal to a ground plane. The ground plane acts like a mirror and forms an image of the arm, thus creating a complete dipole. If the ground plane is infinite, the element pattern only exists when $0 \le \theta \le 90°$. A monopole above an infinite, perfectly conducting ground has a normalized element pattern given by

$$e(\theta) = \sin\theta, \qquad 0 \le \theta \le 90° \tag{4.24}$$

The quarter-wavelength monopole above an infinite ground plane has a directivity of $D = 3.28$ at $\theta = 90°$. The input impedance of a monopole is about half that of a dipole [9]. Decreasing the size of the ground plane causes the peak of the directivity to decrease in angle from $\theta = 90°$ [10]. Monopoles are used in AM and FM broadcast antenna arrays. A three element broadcast array is shown in Figure 4.11.

An alternative to modifying the element weighting or spacing to obtain low sidelobes is to individually rotate the elements [11]. Rotating the elements changes the gain and polarization of the element pattern in a given cut of the antenna pattern. The change in gain can be accounted for in the element pattern, but the polarization must be taken into account by multiplying by the element polarization. For instance, a uniform array of dipoles in the x–z plane that are tilted from the z axis have an array far-field pattern given by

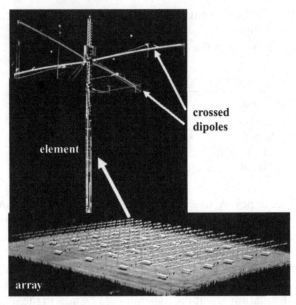

Figure 4.10. Photograph of the HAARP array of crossed dipoles. (Courtesy of HAARP.)

Figure 4.11. Photograph of a 3-element broadcast array in Altoona, PA.

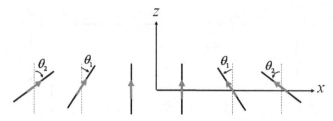

Figure 4.12. Orientation of the dipoles in a 3-element array.

$$FF(\theta, \phi) = \sum_{n=1}^{N_z} \sum_{m=1}^{N_x} \left(\hat{\theta} \cos\theta_{m,n} + \hat{\phi} \sin\theta_{m,n} \right) e_{m,n}(\theta, \phi) w_{m,n} e^{jk[(m-1)d_x \sin\theta\cos\phi + (n-1)d_z\cos\theta]}$$

(4.25)

The rotation of the elements is demonstrated with the four center elements of a linear dipole array in Figure 4.12. The two center elements are oriented to receive θ polarization. Elements at symmetric locations about the center of the array are mirror images of each other. Currents for the co-polarization all flow in the same direction and add in phase. The cross-polarized currents, on the other hand, flow in opposite directions in order to place a null in the cross-polarization pattern at boresight.

Assume that a linear array of dipoles lies along the x axis and in the x–z plane. The primary cut for the array antenna pattern is in the x–y plane. If the dipoles are parallel to the z axis, then the element pattern in the x–y plane is isotropic (ignoring mutual coupling), and the polarization is $\hat{\theta}$. If the dipoles are parallel to the x axis, then the element pattern in the x–y plane is cosine-shaped and the polarization is $\hat{\phi}$. Tilting the dipole from the z-axis transitions its element pattern in the x–y plane from isotropic to cosine. The variations of the gain patterns due to element tilt alone provides some control over the amplitude of the signal received by an element from an off-boresight angle. The polarization of the rotated element plays an important role in determining the strength of the received signal. The power received by an element is multiplied by a polarization efficiency to account for the polarization mismatch between an incident wave and an antenna's receive polarization. Figure 4.13 shows the θ-polarization gain patterns for a dipole rotated 0°, 45°, 60°, and 75° from the z axis. The overall θ-gain decreases with tilt angle. Also, the element pattern becomes less isotropic as θ increases. The change in gain due to element rotation should be enough to create a low-sidelobe taper for a given cut of the antenna pattern.

Figure 4.14a is a diagram of a 10×10 dipole planar array in the x–z plane with the dipoles parallel to the z axis. The array is θ-polarized with no cross-polarized pattern. The element spacing is $d_x = \lambda/2$ and $d_z = \lambda/2$. It has a gain of 22 dB with a relative sidelobe level of 13.1 dB.

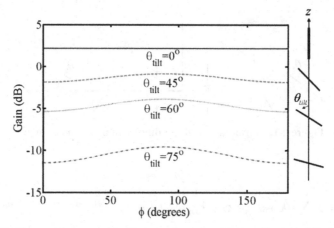

Figure 4.13. Theta gain pattern in the x–y plane for dipoles tilted 0°, 45°, 60°, and 75° from the z axis.

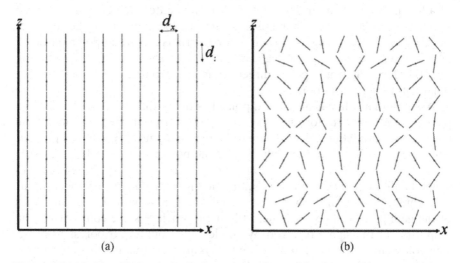

Figure 4.14. A 10×10 θ-polarized planar array located in the x–z plane. (a) Before optimization. (b) After optimization.

Through the use of a genetic algorithm, the dipole tilt angles were found to minimize the maximum sidelobe level. The cost function calculates the co-polarization and cross-polarization antenna patterns over the range of $\phi = 0°$ to $\phi = 90°$ and $\theta = 0°$ to $\theta = 90°$. First, the co-polarization maximum sidelobe level is found. Second, the maximum of the cross-polarization pattern is found. The cost is then the larger of the two values. This results in the co-polarization maximum sidelobe level equaling to the cross-polarization maximum gain with the main beam oriented at $\phi = \theta = 90°$.

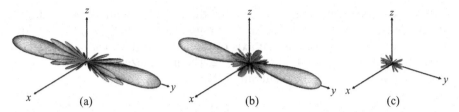

Figure 4.15. Array gain. **(a)** No optimization. **(b)** Optimized θ-gain. **(c)** Optimized ϕ-gain.

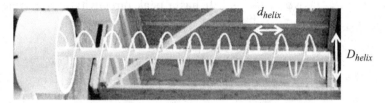

Figure 4.16. Diagram of a helical antenna. (Courtesy of ARA Inc. / Seavey Engineering.)

The optimized element rotations in a 10×10-element array are shown in Figure 4.14b. The gain is 19.9 dB with a maximum relative sidelobe level of 19.4 dB. Figure 4.15a is the θ-polarized gain before optimization and there is no cross-polarization. The optimized θ-polarized gain is shown in Figure 4.15b with the cross-polarization in Figure 4.15c. Since the currents on the dipoles have an x component as well as a z component, the pattern has both θ and ϕ components. The gain of the optimized array decreased by 2.1 dB, and the maximum sidelobe level decreased by 6.3 dB.

4.2.2. Helical Antenna

A helical antenna looks like a corkscrew (Figure 4.16). The loops are considered very small and closely spaced when [12]

$$N_{\text{helix}} \sqrt{d_{\text{helix}}^2 + (\pi D_{\text{helix}})^2} \ll \lambda \tag{4.26}$$

where D_{helix} is the diameter of loop, d_{helix} is the spacing between loops, and N_{helix} is the number of loops. If this condition is met, then the helical antenna acts like a monopole with a doughnut-shaped radiation pattern and is said to operate in the normal mode [12], where *normal* means that the maximum gain is perpendicular to the axis of the helix. The advantage of a normal mode helix is that it behaves like a monopole but is shorter at the same frequency. Its radiation resistance is given by

$$Z = (25.3h/\lambda)^2 \tag{4.27}$$

When the loops and spacing get larger, the helical antenna looks and acts like an end-fire array of loop antennas. In an array, the helical antenna radiates in the axial mode (along the axis of the helix) with a peak at $\theta = 0°$. It works best when $d_{helix} \approx \lambda/4$ and $kD_{helix} \approx 2$. The normalized electric field radiated by an axial mode helical antenna is [13]

$$E(\theta) = \sin\left(\frac{\pi}{2N}\right)\frac{\sin(N_{helix}\psi/2)}{\sin(\psi/2)}\cos\theta \tag{4.28}$$

where $\psi = kd_{helix}\left(\cos\phi - 1 - \dfrac{\lambda}{d_{helix}}\right)$. It has a peak directivity at a wavelength of λ_0 given by [14]

$$D = 8.3\left(\frac{\pi D_{helix}}{\lambda_0}\right)^{\sqrt{N_{helix}+2}-1}\left(\frac{N_{helix}d_{helix}}{\lambda_0}\right)^{0.8}\left(\frac{0.2217\pi D_{helix}}{d_{helix}}\right)^{\sqrt{N_{helix}}/2} \tag{4.29}$$

Axial mode helical antennas are elliptically polarized with an axial ratio of [12]

$$AR = \frac{2N_{helix}+1}{2N_{helix}} \tag{4.30}$$

The helical antenna is right-hand polarized when the helix increases in the counter clockwise direction when looking from the end toward the feed. The helix in Figure 4.16 is left-hand circularly polarized. Figure 4.17 shows a plot of the total electric field radiated by a helical antenna with $N_{helix} = 6$, $d_{helix} = 0.25\lambda$, and $D_{helix} = \lambda/\pi$. The directivity is 9.9 dB using (4.29) and 9.8 dB

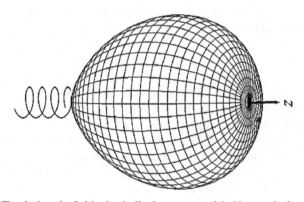

Figure 4.17. Total electric field of a helical antenna with $N_{helix} = 6$, $d_{helix} = 0.25\lambda$, and $D_{helix} = \lambda/\pi$.

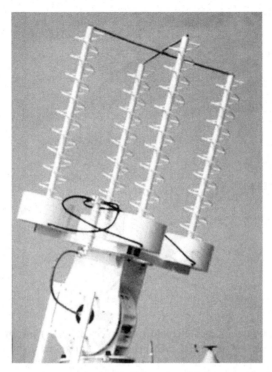

Figure 4.18. Photograph of a quad helix array. (Courtesy of ARA Inc./Seavey Engineering.)

using numerical integration of (4.28). The input impedance of an axial fed helix is [12]

$$Z = 140\pi D_{\text{helix}}/\lambda \qquad (4.31)$$

Helical antennas are frequently used for space communications systems where circular polarization is required. Figure 4.18 is a quad helix array (ARA/Seavey Model Number 9106-800) that operates from 405 to 450 MHz. It has a gain of 18 dBic (dB above an isotropic point source with the same circular polarization). The array is 157.5 cm² and 185.5 cm tall.

Example. Plot the three-dimensional antenna pattern for for an 8-element array of z-directed helical antennas with element spacing $\lambda/2$ along the x axis. The helical antenna has $N_{\text{helix}} = 6$, $d_{\text{helix}} = 0.25\lambda$, and $D_{\text{helix}} = \lambda/\pi$.

Figure 4.19 is a plot of the antenna pattern in decibels. The directivity of 16.3 dB is found by substituting (4.28) into (4.5) and numerically integrating. The helical antenna has the advantage of high directivity while still fitting into the $\lambda/2$ element spacing grid.

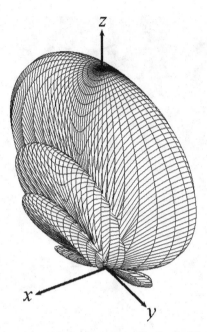

Figure 4.19. Array factor times element pattern (in decibels) for an 8-element array with element spacing $\lambda/2$ along the x axis. The element pattern is shown in Figure 4.17.

4.3. APERTURE ANTENNAS

An aperture antenna is typically a hole in a waveguide. The hole may be a slot in the side or the open end. A horn extends the size of the open end of the waveguide in order to increase the gain of an open ended waveguide. The input impedance is normally 50 Ω. Reflector and microstrip antennas are frequently categorized as aperture antennas as well. The gain of the horn or open-ended waveguide can be significantly increased by placing it at the focal point of a reflector to create another common type of aperture antenna.

4.3.1. Apertures

An $L \times W$ rectangular slot lying in the x–y plane (Figure 4.20a) has normalized far-field components given by [15]

$$E_\theta = \sin\phi \frac{\sin X}{X} \frac{\sin Y}{Y} \qquad (4.32)$$

$$E_\phi = \cos\theta \cos\phi \frac{\sin X}{X} \frac{\sin Y}{Y} \qquad (4.33)$$

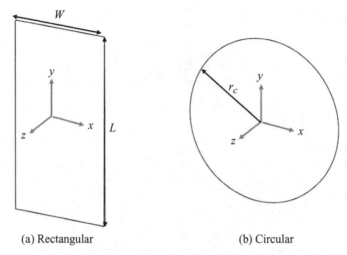

(a) Rectangular   (b) Circular

Figure 4.20. Slots in the *x–y* plane centered at the origin. **(a)** Rectangular. **(b)** Circular.

where

$$X = \frac{kW}{2}\sin\theta\cos\phi$$
$$Y = \frac{kL}{2}\sin\theta\sin\phi$$ (4.34)

The directivity is proportional to the area of the slot in wavelengths squared.

$$D = \frac{4\pi WL}{\lambda^2}$$ (4.35)

A plot of the total field, $E_T = \sqrt{E_\theta^2 + E_\phi^2}$, radiated by a slot with $L = 0.5\lambda$ and $W = 0.1\lambda$ is shown in Figure 4.21. This small slot has a directivity of –2 dB.

Slot elements can be circular as well. A circular slot of radius r_c that is lying in the *x–y* plane (Figure 4.20b) with a uniform aperture field has far-field components given by

$$E_\theta = \sin\phi\,\frac{J_1(kr_c\sin\theta)}{kr_c\sin\theta}$$ (4.36)

$$E_\phi = \cos\theta\cos\phi\,\frac{J_1(kr_c\sin\theta)}{kr_c\sin\theta}$$ (4.37)

The directivity is based on the aperture area

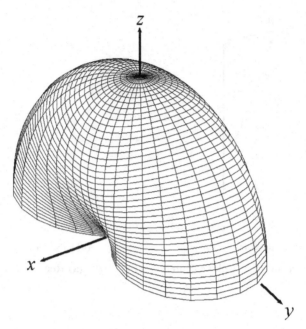

Figure 4.21. Total electric field of a rectangular slot with $a = 0.5\lambda$ and $b = 0.1\lambda$.

$$D = \frac{4\pi\left(\pi r_c^2\right)}{\lambda^2} \tag{4.38}$$

A plot of the total field radiated by a circular slot with $r_c = 5\lambda$ is shown in Figure 4.22. Its directivity is 30 dB.

Example. Find the antenna pattern of a 4-element linear array of reflector antennas that have a diameter of 10λ. Assume the edges of the reflectors touch.

First approximate the reflector apertures by a uniform circular slot. A plot of the three-dimensional directivity pattern appears in Figure 4.22. An individual element in the array has a directivity of 30 dB as calculated from (4.38). A 4-element array has a directivity of 36 dB. In this case, the directivity of the element times the directivity of the array factor equals the directivity of the array of elements. Figure 4.23 are the *E*- and *H*-plane antenna pattern cuts for this array. Since the phase centers are separated by 10λ, grating lobes in the array factor occur at $\theta_g = \pm5.7°, \pm11.5°, \pm17.5°, \pm23.6°, \pm30°, \pm36.9°, \pm44.4°, \pm53.1°, \pm64.1°,$ and $\pm90°$. These angles correspond to every third sidelobe rising above its surrounding sidelobes. The element pattern nulls occur close to the peak of the grating lobes, thus lowering the peak of the grating lobes.

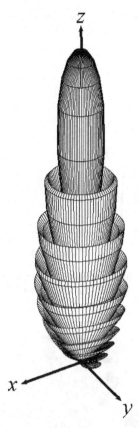

Figure 4.22. Total electric field (in decibels) radiated by a 10λ-diameter circular aperture.

The Very Large Array (VLA) is a radio telescope in New Mexico built from 27 dish antennas that are 25 m in diameter (Figure 4.24) [16]. Elements are placed in a Y-shape as shown in Figure 4.25. The size of the array is regularly changed by moving the elements along a railroad track. There are four configurations with maximum width of 36 km, 10 km, 3.6 km, or 1 km. This array operates from 73 MHz up to 50 GHz. Table 4.1 lists the frequencies of operation and the resolutions associated with various configurations of the VLA.

The impedance of a thin slot antenna can be found from the impedance of a complementary dipole using Booker's relation given by [17]

$$Z_{\text{slot}} = \frac{Z_0^2}{4Z_{\text{dipole}}} = \frac{Z_0^2}{4\left(R_{\text{dipole}}^2 + X_{\text{dipole}}^2\right)}\left(R_{\text{dipole}} - jX_{\text{dipole}}\right) \tag{4.39}$$

where $Z_{\text{dipole}} = R_{\text{dipole}} + jX_{\text{dipole}}$. Thus, when the dipole is capacitive, the slot is inductive, and when the dipole is inductive, the slot is capacitive. Lengthening

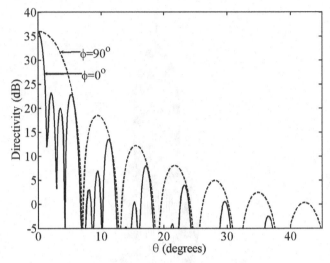

Figure 4.23. Principal cuts of the array factor times element pattern (in decibels) for a linear array with 4 elements spaced 10λ apart along the x axis. The element pattern is shown in Figure 4.22.

Figure 4.24. One reflector element of the VLA. (Courtesy of NRAO / AUI / NSF.)

Figure 4.25. Y-shaped VLA configuration. (Courtesy of NRAO / AUI / NSF.)

TABLE 4.1. Operating Characteristics of the VLA [16]

Frequency (GHz):	.075	.333	1.5	5	8.3	15	23	43
Primary beam (arcmin):	600	150	30	9	5.4	3	2	1
Highest resolution (arcsec):	24.0	6.0	1.4	0.4	0.24	0.14	0.08	0.05

a dipole makes it more capacitive, whereas lengthening a slot makes it more inductive.

4.3.2. Open-Ended Waveguide Antennas

The metallic rectangular waveguide in Figure 4.26 has a wave propagating in the z direction and standing waves in the x and y directions. A propagating wave is either TE (transverse electric) with $E_z = 0$ or TM (transverse magnetic) with $H_z = 0$. Propagating waves are represented by complex exponential functions, whereas standing waves are represented by trigonometric functions. The wave vector has x, y, and z components and they are related by

$$k^2 = k_x^2 + k_y^2 + k_z^2 \tag{4.40}$$

The wavenumber in the x direction is given by

$$k_x = \frac{m\pi}{a} \tag{4.41}$$

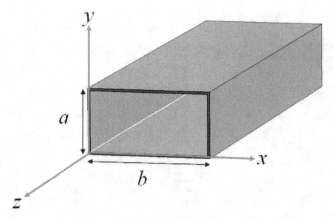

Figure 4.26. Rectangular waveguide.

and the wavenumber in the y direction is given by

$$k_y = \frac{n\pi}{b} \tag{4.42}$$

The propagating wave constant is found from (4.40)

$$k_z = \sqrt{\left(\frac{2\pi}{\lambda}\right)^2 - \left(\frac{m\pi}{a}\right)^2 - \left(\frac{n\pi}{b}\right)^2} \tag{4.43}$$

In order for a wave to propagate, k_z must be real. If it is imaginary, then the wave is evanescent. In other words, it has an exponential decay as it travels down the waveguide. The apparent wavelength of the propagating wave in the z direction is λ_z or more commonly referred to as the waveguide wavelength, λ_g. Waves will propagate whenever (4.43) is greater than zero.

$$k > \sqrt{\left(\frac{m\pi}{a}\right)^2 + \left(\frac{n\pi}{b}\right)^2} \tag{4.44}$$

This occurs when the frequency exceeds the cutoff frequency found from (4.44).

$$f_{cmn} = \frac{1}{2\pi\sqrt{\mu\varepsilon}} \sqrt{\left(\frac{m\pi}{a}\right)^2 + \left(\frac{n\pi}{b}\right)^2} \tag{4.45}$$

When $f > f_{cmn}$, then the wave propagates in the z direction; otherwise the wave attenuates in the z direction.

The normalized electric fields in the waveguide for a TE wave have a two-dimensional Fourier series representation given by

$$E_x(x, y, z) = \cos k_x x \sin k_y y e^{-jk_z z}$$
$$E_y(x, y, z) = -\sin k_x x \cos k_y y e^{-jk_z z}$$

(4.46)

and for the TM wave we have

$$E_z(x, y, z) = \sin k_x x \sin k_y y e^{-jk_z z}$$
$$E_x(x, y, z) = -j \cos k_x x \sin k_y y e^{-jk_z z}$$
$$E_y(x, y, z) = -j \sin k_x x \cos k_y y e^{-jk_z z}$$

(4.47)

Example. Consider a length of air-filled copper X-band waveguide, with dimensions $a = 2.286 \, \text{cm}$, $b = 1.016 \, \text{cm}$. Find the cutoff frequencies of the first few propagating modes.

The cutoff frequencies are found from (4.45).

Mode	m	n	$f_{cmn}(\text{GHz})$
TE	1	0	6.562
TE	2	0	13.123
TE	0	1	14.764
TE, TM	1	1	16.156
TE, TM	1	2	30.248
TE, TM	2	1	19.753

The TE_{10} mode is the dominant mode in the rectangular waveguide, since it has the lowest cutoff frequency. It has a normalized electric field at the opening given by

$$E_y = \sin\left(\frac{\pi x}{a}\right) e^{-jk_z z}$$

(4.48)

The expressions in (4.32) and (4.33) only apply when the field in the aperture is uniform. For the case when the aperture field is given by (4.48), the normalized far field expressions for the open-ended waveguide are [15]

$$E_\theta = \sin\phi \frac{\cos X}{(\pi/2)^2 - X^2} \frac{\sin Y}{Y}$$

(4.49)

$$E_\phi = \cos\theta \cos\phi \frac{\cos X}{(\pi/2)^2 - X^2} \frac{\sin Y}{Y}$$

(4.50)

where X and Y are given by (4.34). The directivity of the open-ended waveguide is [15]

$$D = \frac{3.24\pi(ab)}{\lambda^2} \qquad (4.51)$$

Figure 4.27 is a plot of the antenna pattern for an 8×8 element planar array of open-ended waveguides operating in the TE_{10} mode. The directivity is 20.5 dB which is less than the directivity of the element pattern times the array factor (4.1 dB + 18.6 dB = 22.7 dB). The beamwidth is smaller and there are more sidelobes in the x direction compared to the y direction, because the waveguides in the x direction are 0.763λ wide while in the y direction they are 0.34λ wide. Thus, the element spacing in the x and y directions equals the width of the waveguide in those directions. A uniform aperture the same size as this array has a directivity of 23.2 dB.

Example. Plot the antenna pattern in Figure 4.27 with the beam steered to $(\theta_s, \phi_s) = (30°, 0°)$.

Figure 4.28 shows the pattern. The directivity is reduced to 20.5 dB due to the emergence of the grating lobe at $(\theta, \phi) = (53°, 180°)$.

Open-ended waveguides are almost never used in an array except for a small array feeding a reflector antenna. As an example, Figure 4.29 shows a

Figure 4.27. Array factor times the element pattern (in decibels) of an 8×8 element planar array of open-ended waveguides.

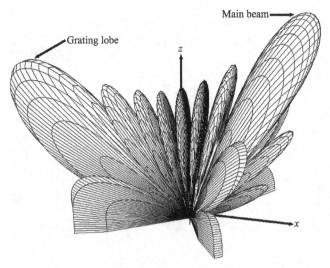

Figure 4.28. Antenna pattern in Figure 4.27 when steered to $(\theta_s, \phi_s) = (30°, 0°)$.

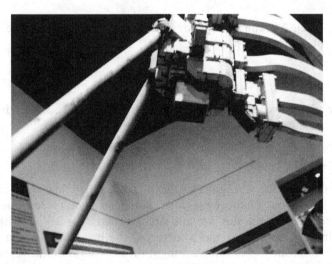

Figure 4.29. Waveguide feed for the RARF. (Courtesy of the National Electronics Museum.)

4-element waveguide feed for the RARF (Reflected Array Radio Frequency) used in AN/APQ-140 Radar [18].

Circular waveguides are also used as elements in arrays. The Electronically Agile Radar (EAR) had 1818 circular waveguide elements [19]. A ceramic

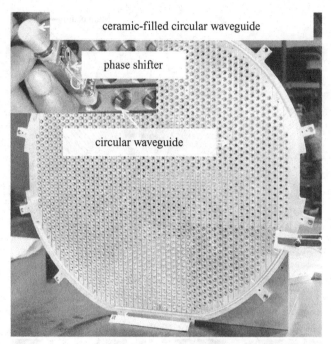

Figure 4.30. The EAR with a blowup of the element/phase shifter module. (Courtesy of the National Electronics Museum.)

loaded circular waveguide element connected to a phase shifter is inserted into each circular waveguide hole in the array backplane (Figure 4.30).

4.3.3. Slots in Waveguides

Slots are placed in the waveguide walls in order to couple a small portion of the fields in and out of the waveguide. Slot resonance (susceptance equals zero) is a function of the slot size and occurs when $2L + 2W \approx \lambda$[22]. As with a dipole, increasing the slot width increases the bandwidth of the slot. Slot admittance is a function of the placement of the slot on the waveguide. Slots can be on either the broad or narrow walls and are positioned to have the desired admittance which in turn causes the field radiated by the slot to have the desired amplitude and phase. The admittance or impedance of a slot depends upon its displacement from the centerline (Δ) and its tilt (ζ) as shown in Figure 4.31. The narrow wall of a rectangular waveguide is too short for a rectangular resonant slot, so either a different slot shape that fits on the wall must be used (like the barbell shape in Figure 4.31), or the slot extends into the broad wall of the waveguide. Examples of the two most commonly used slots in array antennas are shown in Figure 4.32. The field at the slots in the

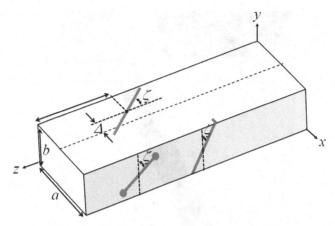

Figure 4.31. Slots in a rectangular waveguide.

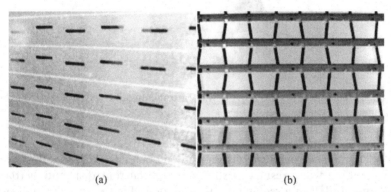

(a)   (b)

Figure 4.32. Examples of slots in a rectangular waveguide. **(a)** Offset longitudinal slots in the broad wall (AN/APG-68 radar antenna). **(b)** Inclined slots in the narrow wall (AWACS antenna). (Courtesy of the National Electronics Museum.)

broad wall of the waveguide in Figure 4.32a are controlled by the element offset, Δ, from the centerline. The field at the slots in the narrow wall of the waveguide in Figure 4.32b are controlled by the element tilt, ζ. Note that the tilted slot in the narrow wall wraps around the edge so that there is a small notch in the top and bottom broadside walls as seen in Figure 4.33. A long slot at the center of the broad wall or a vertical slot in the side wall does not radiate. Longitudinal slots in the broad wall of the waveguide have negligible cross-polarization. Slots have a cross-polarization component when $\zeta \neq 0$. Longitudinal slots are not used in the side wall, because the admittance and power radiated cannot be controlled by changing their position in the y direction. Crossed slots radiate circular polarization if placed properly.

Figure 4.33. Tilted slot in the narrow wall of a rectangular waveguide. (Courtesy of the National Electronics Museum.)

A resonant waveguide array is shorted on one end and fed on the other so that a standing wave results as shown by the current distribution (dark is higher amplitude) on a rectangular waveguide in Figure 4.34. The null (white spot) in the center of the broad wall occurs $\lambda/4$ behind the short. Slots should be placed such that they disturb the current distribution in such a way as to cause radiation. The current flow on both sides of the slot should be in the same direction. The first slot is placed at either $\lambda_g/4$ or $\lambda_g/2$ behind the waveguide termination.

At $(x, y, z) = (a/2, b, \lambda_g/4)$ a null occurs in the standing wave (first white spot on top of the waveguide in Figure 4.34). Placing the center of a resonant slot at $x = a/2$ with its long side parallel to the z direction does not disturb the current flowing on the waveguide (Figure 4.35a). Rotating this slot to $\zeta = 30°$ while keeping the slot center at $x = a/2$ does not perturb the standing wave currents, so the slot does not radiate (Figure 4.35b). Moving the slot in the x direction causes a disruption of the current flow and the standing wave resulting in radiation. Figure 4.35c shows the slot at $\Delta = 0.1\lambda$, and Figure 4.35d shows the slot at $\Delta = 0.2\lambda$. Increasing Δ increases the power radiated by the slot. When the slot in the broadside wall that is parallel to the z axis is placed $\lambda_g/4$ behind the short, the power radiated by the slot is a function of Δ. Consequently,

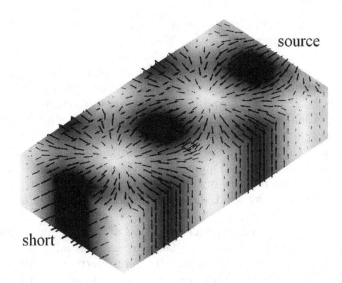

Figure 4.34. Currents on a rectangular waveguide when one end has a short.

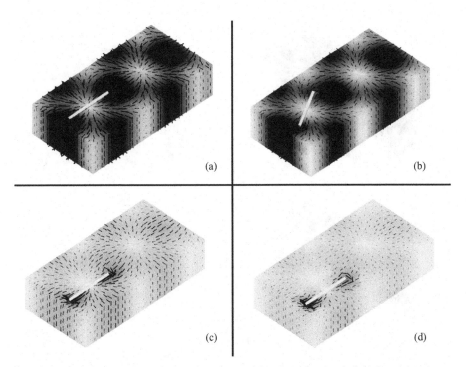

Figure 4.35. Currents on a rectangular waveguide when the slot is $\lambda_g/4$ from the short. **(a)** $\Delta = 0$, $\zeta = 0$. **(b)** $\Delta = 0$, $\zeta = 30°$. **(c)** $\Delta = 0.1\lambda$, $\zeta = 0$. **(d)** $\Delta = 0.2$, $\zeta = 0$.

the elements near the center of a low-sidelobe array have a larger Δ than the elements near the edges.

At $(x, z) = (a/2, \lambda_g/2)$, a peak occurs in the standing wave. Placing the center of a resonant slot at $x = a/2$ with its long side parallel to the z direction does not disturb the current flowing on the waveguide, even though the current is at a peak, because the slot is so thin and the current is flowing in the z direction (Figure 4.36a). Rotating this slot to $\zeta = 30°$ while keeping the slot center at $x = a/2$ disrupts the current flow and causes radiation (Figure 4.36b). Moving the slot in the x direction produces little disturbance to the standing wave currents, thus no radiation (Figure 4.36c). Figure 4.36d shows the slot at $\zeta = 90°$ and blocking the flow of current in the z direction. When the slot is placed at $\lambda_g/2$ behind the short, the power radiated by the slot is a function of ζ.

Shunt slots interrupt the transverse currents (J_x and J_y) and can be represented by a two-terminal shunt admittance. Series slots interrupt J_z and can be represented by a series impedance. If a slot interrupts all three current components, then a Pi or T impedance network represents it. The following formulas were derived by [21][22][23]:

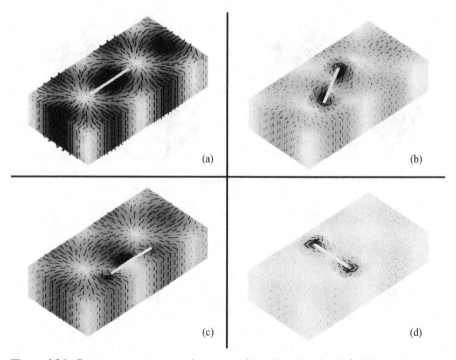

(a)

(b)

(c)

(d)

Figure 4.36. Currents on a rectangular waveguide when the slot is $\lambda_g/2$ from the short. **(a)** $\Delta = 0$, $\zeta = 0$. **(b)** $\Delta = 0$, $\zeta = 30°$. **(c)** $\Delta = 0.2$, $\zeta = 0$. **(d)** $\Delta = 0$, $\zeta = 90°$.

Longitudinal slot in broad face (shunt):

$$g_s = \left[2.09 \frac{\lambda_g a}{\lambda b} \cos^2\left(\frac{\pi \lambda}{2\lambda_g} \right) \right] \sin^2\left(\frac{\pi \Delta}{a} \right)$$

(4.52)

Slot in narrow face (shunt):

$$g_s = \frac{30\lambda^3 \lambda_g}{73\pi a^3 b} \left[\frac{\sin \zeta \cos\left(\frac{\pi\lambda \sin \zeta}{2\lambda_g} \right)}{1 - \left(\frac{\lambda \sin \zeta}{\lambda_g} \right)^2} \right]^2$$

(4.53)

Traverse slot in broad face (series):

$$r_s = \frac{0.131\lambda^3}{ab\lambda_g} \left[I^-(\zeta)\sin \zeta + \frac{\lambda_g I^+(\zeta)\cos \zeta}{2a} \right]^2$$

(4.54)

where g_s is the normalized slot admittance, r_s is the normalized slot resistance, Δ is the offset from waveguide centerline,

$$I^{\pm}(\zeta) = \frac{\cos\left(\frac{\pi \xi^+}{2} \right)}{1 - \left(\xi^+ \right)^2} \pm \frac{\cos\left(\frac{\pi \xi^-}{2} \right)}{1 - \left(\xi^- \right)^2}$$

$$\xi^{\pm} = \frac{\lambda}{\lambda_g}\cos \zeta \pm \frac{\lambda}{2a}\sin \zeta$$

and ζ is the slot tilt angle (Figure 4.31).

The radiation pattern of a thin slot is the same as a dipole except it has the orthogonal polarization. A horizontal slot has vertical polarization, while a vertical slot has horizontal polarization.

Figure 4.37 is the 548-element planar array of slots used for the APG-68 radar system [24]. The elements are longitudinal slots in the broad side of the rectangular waveguide. The AN/APG-68 radar is a solid-state medium range (up to 150 km) pulse-Doppler radar designed for the USAF F-16.

4.3.4. Horn Antennas

A horn antenna increases the aperture size, hence the directivity, of an open-ended waveguide and provides a better match to free space. Figure 4.38 is a diagram of a rectangular horn antenna where the aperture is flared in two orthogonal directions. The flare in Figure 4.38 is linear, but it could also be exponential. If the flare of the horn is only in the direction of the electric field in a TE_{10} waveguide ($A = a$ and $B > b$), then the horn is known as an *E*-plane

Figure 4.37. Photograph of the APG-68. (Courtesy of the National Electronics Museum.)

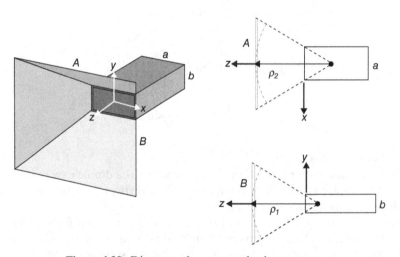

Figure 4.38. Diagram of a rectangular horn antenna.

sectoral horn. If the flare of the horn is only orthogonal to the direction of the electric field in a TE_{10} waveguide ($A > a$ and $B = b$), then the horn is known as an *H*-plane sectoral horn. Flaring a horn in both directions ($A > a$ and $B > b$) produces a pyramidal horn. The sectoral horns can be stacked along the narrow dimension to form a linear array. Since the horns are flared in the plane orthogonal to the array, the beam is also narrow in that plane. For analysis purposes, the *E*-plane is $\phi = \pi/2$, and the *H*-plane is $\phi = 0$ in Figure 4.38.

Normalized expressions for the electric field of an E-plane sectoral horn antenna are [15]

$$E_\theta = \sin\phi(1+\cos\theta)\frac{\cos(X)}{X^2 - \left(\frac{\pi}{2}\right)^2}F(t_1, t_2) \qquad (4.55)$$

$$E_\phi = \cos\phi(1+\cos\theta)\frac{\cos(X)}{X^2 - \left(\frac{\pi}{2}\right)^2}F(t_1, t_2) \qquad (4.56)$$

The corresponding directivity is

$$D_E = \frac{64a\rho_1}{\pi b}\left|F\left(\frac{B}{\sqrt{2\rho_1}}, 0\right)\right|^2 \qquad (4.57)$$

Figure 4.39 is a plot of the electric field of an E-plane sectoral horn with $a = 0.763\lambda$, $b = 0.34\lambda$, $A = 0.763\lambda$, $B = 3.6\lambda$, and $\rho_2 = 7.6\lambda$. It has a narrow beam in the E plane due to the flair of the horn. The normalized electric fields and directivity of an H-plane sectoral horn antenna are [15]

$$E_\theta = \sin\phi(1+\cos\theta)\frac{\sin Y}{Y}\left[e^{j\psi_1}F(t_1', t_2') + e^{j\psi_2}F(t_1'', t_2''')\right] \qquad (4.58)$$

$$E_\phi = \cos\phi(1+\cos\theta)\frac{\sin Y}{Y}\left[e^{j\psi_1}F(t_1', t_2') + e^{j\psi_2}F(t_1'', t_2''')\right] \qquad (4.59)$$

$$D_H = \frac{4\pi b\rho_2}{a}\left|F\left(\frac{\rho_2 - A^2}{A\sqrt{2\rho_2}}, \frac{\rho_2 + A^2}{A\sqrt{2\rho_2}}\right)\right|^2 \qquad (4.60)$$

Figure 4.40 is a plot of the electric field of an E-plane sectoral horn with $a = 0.763\lambda$, $b = 0.34\lambda$, $A = 4.6\lambda$, $B = 0.34\lambda$, and $\rho_1 = 6.9\lambda$. It has a narrow beam in the H plane due to the flair of the horn. The normalized electric fields and directivity of a pyramidal horn antenna are [15]

$$E_\theta = \sin\phi(1+\cos\theta)\left[e^{j\psi_1}F(t_1', t_2') + e^{j\psi_2}F(t_1'', t_2''')\right]F(t_1, t_2) \qquad (4.61)$$

$$E_\phi = \cos\phi(1+\cos\theta)\left[e^{j\psi_1}F(t_1', t_2') + e^{j\psi_2}F(t_1'', t_2''')\right]F(t_1, t_2) \qquad (4.62)$$

$$D_p = \frac{\lambda^2\pi}{32ab}D_E D_H \qquad (4.63)$$

Figure 4.41 is a plot of the electric field of an E-plane sectoral horn with $a = 0.763\lambda$, $b = 0.34\lambda$, $A = 4.6\lambda$, $B = 3.6\lambda$, $\rho_1 = 6.9\lambda$, and $\rho_2 = 7.6\lambda$. Since the

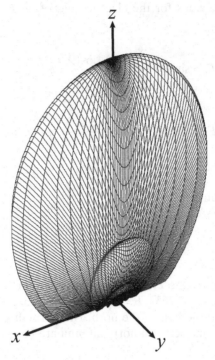

Figure 4.39. Total electric field radiated by a $0.76\lambda \times 3.6\lambda$ E-plane sectoral horn.

aperture is large in both directions, the beamwidth is narrow for all ϕ. Definitions for the variables in the field and directivity equations for the horns are listed below:

$$F(t_1,t_2) = [C(t_2) - C(t_1)] - j[S(t_2) - S(t_1)]$$

$$\psi_1 = \frac{\rho_2}{2k}\left(k\sin\theta\cos\phi + \frac{\pi}{A}\right)^2$$

$$\psi_2 = \frac{\rho_2}{2k}\left(k\sin\theta\cos\phi - \frac{\pi}{A}\right)^2$$

$$t_1 = -\sqrt{\frac{1}{\pi k\rho_1}}\left[\frac{kB}{2} + (k\sin\theta\sin\phi)\rho_1\right]$$

$$t_2 = \sqrt{\frac{1}{\pi k\rho_1}}\left[\frac{kB}{2} - (k\sin\theta\sin\phi)\rho_1\right]$$

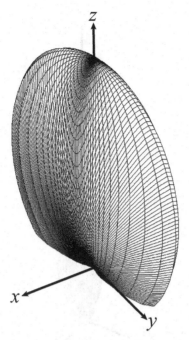

Figure 4.40. Total electric field radiated by a $4.6\lambda \times 0.34\lambda$ H-plane sectoral horn.

$$t_1' = -\sqrt{\frac{1}{\pi k \rho_2}}\left[\frac{kA}{2} + \left(k\sin\theta\cos\phi + \frac{\pi}{A}\right)\rho_2\right]$$

$$t_2' = \sqrt{\frac{1}{\pi k \rho_2}}\left[\frac{kA}{2} - \left(k\sin\theta\cos\phi + \frac{\pi}{A}\right)\rho_2\right]$$

$$t_1'' = -\sqrt{\frac{1}{\pi k \rho_2}}\left[\frac{kA}{2} + \left(k\sin\theta\cos\phi - \frac{\pi}{A}\right)\rho_2\right]$$

$$t_2'' = \sqrt{\frac{1}{\pi k \rho_2}}\left[\frac{kA}{2} - \left(k\sin\theta\cos\phi - \frac{\pi}{A}\right)\rho_2\right]$$

$$X = \frac{ka}{2}\sin\theta\cos\phi$$

$$Y = \frac{kb}{2}\sin\theta\sin\phi$$

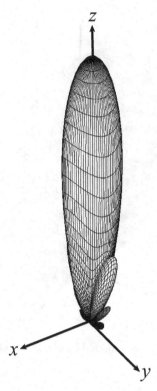

Figure 4.41. Total electric field radiated by a $4.6\lambda \times 3.6\lambda$ pyramidal horn.

$$C(x) = \int_0^x \cos\left(\frac{\pi}{2}t^2\right)dt$$

$$S(x) = \int_0^x \sin\left(\frac{\pi}{2}t^2\right)dt$$

Pyramidal horns are not commonly used in arrays, because the horn flare in both directions means that the elements would be spaced far apart in the array and grating lobes would appear.

Example. Plot the antenna patterns for 8-element linear arrays of E-plane sectoral horns, H-plane sectoral horns, and pyramidal horns using the horn patterns in Figures 4.39 to 4.41.

A linear array of E-plane sectoral horns would be along the x axis. Assuming that the elements are placed side by side, then the element spacing would be $d_x = 0.763\lambda$. Figure 4.42 shows the array pattern in decibels. A linear array of

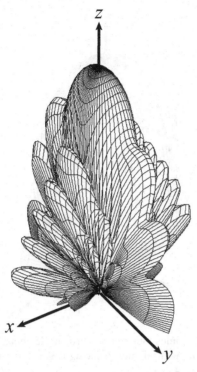

Figure 4.42. Array factor times element pattern (in decibels) for an 8-element array of *E*-plane sectoral horns with element spacing 0.76λ along the x axis. The element pattern is shown in Figure 4.39.

H-plane sectoral horns would be along the y axis. Assuming that the elements are placed side by side, then the element spacing would be $d_y = 0.34\lambda$. Figure 4.43 shows the array pattern in decibels. A linear array of pyramidal horns along the x axis has an element spacing of $d_x = 4.6\lambda$. This large spacing produces grating lobes that are obvious in the *H*-plane cut of the array pattern in Figure 4.44.

Figure 4.45 is a picture of a low sidelobe 80-element array of *H*-plane sectoral horns. The large air conditioning unit on the back of the array keeps the feed network and phase shifters at a constant temperature in order to minimize errors and keep sidelobes low. Figure 4.46 is a plot of the far-field sum pattern with a maximum relative sidelobe level of 31 dB. Figure 4.47 is a plot of the far-field difference pattern with a maximum relative sidelobe level of 28.5 dB. Each element had a 6-bit digitally controlled ferrite phase shifter.

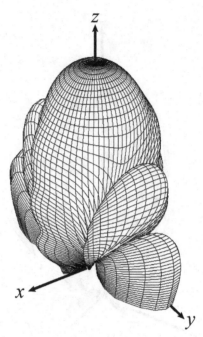

Figure 4.43. Array factor times element pattern (in decibels) for an 8-element array of H-plane sectoral horns with element spacing 0.34λ along the y axis. The element pattern is shown in Figure 4.40.

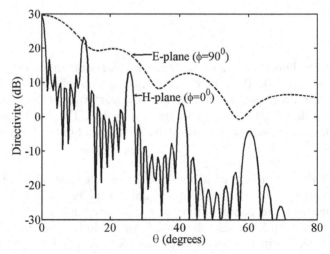

Figure 4.44. Principal cuts of the array factor times element pattern (in decibels) for a linear array with 8 pyramidal horn elements spaced 4.6λ along the x axis. The element pattern is shown in Figure 4.41.

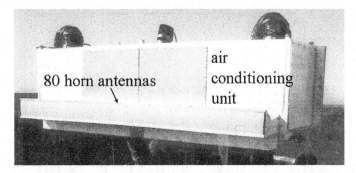

Figure 4.45. Linear array of 80 *H*-plane sectoral horns.

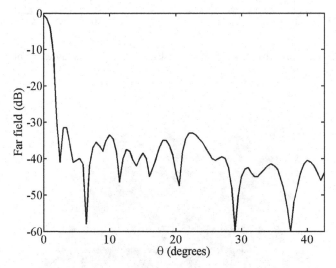

Figure 4.46. Far-field sum pattern of the 80-element array of *H*-plane sectoral horns.

4.4. PATCH ANTENNAS

A simple patch antenna consists of a flat, thin piece of metal in some geometric shape on top of a dielectric substrate with a metal ground plane on the bottom. Patches are very popular array elements because they

1. Do not protrude from the surface.
2. Can be cheaply made with the same technology as printed circuits.
3. Integrate well with printed circuit boards.
4. Are easy to conform to a nonplanar surface.
5. Are rugged.

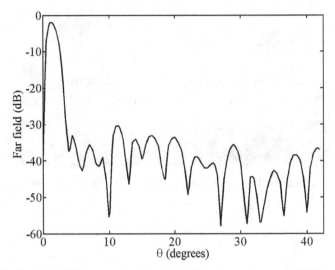

Figure 4.47. Far-field difference pattern of the 80-element array of *H*-plane sectoral horns.

Figure 4.48. Rectangular patch made from copper tape and polycarbonate. (Courtesy of Amy Haupt.)

The most common patch shape is a rectangle with a disk being second. Many other shapes, including fractal, have been explored as well. An example of a simple patch antenna is shown in Figure 4.48 [25]. The patch is made from a 10.3- by 10.5-cm piece of copper tape placed in the center of a 10.28- by 10.45-cm piece of thick polycarbonate ($\varepsilon_r = 2.7$) with a copper tape ground plane. Such a patch is very easy to build and has a very narrow impedance bandwidth of 54 MHz or 2.8% at 2.075 GHz (Figure 4.49).

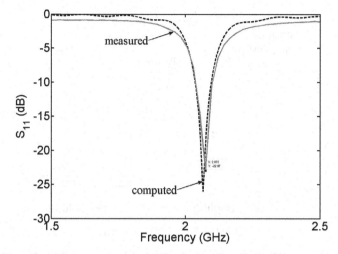

Figure 4.49. S_{11} of the patch in Figure 4.48. (Courtesy of Amy Haupt.)

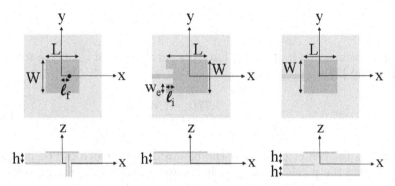

Figure 4.50. Some methods of feeding a patch antenna.

Patches can be fed in a number of ways. The left patch in Figure 4.50 is fed by a coaxial cable from beneath the ground plane. A hole is drilled through the ground, substrate and patch. Then the inner conductor of the coaxial cable is place in the hole and soldered to the top of the patch. The outer conductor is soldered to the ground plane. The hole in the ground plane must be big enough so that the inner conductor does not touch the ground plane. The center patch in Figure 4.50 is fed via a microstrip transmission line. Part of the patch is cut out to match the impedance of the patch to that of the transmission line. The patch on the right in Figure 4.50 is fed from below by an open-ended microstrip transmission line. This approach as well as aperture-fed patches require multilayered substrate boards.

Patches tend to be narrowband, so a variety of techniques have been developed to increase their bandwidth. The bandwidth of a $L \times W$ rectangular patch

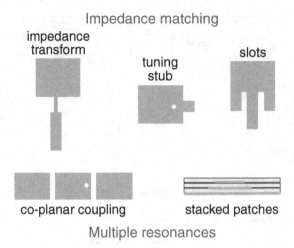

Figure 4.51. Techniques for increasing the bandwidth of a rectangular patch antenna.

on top of a substrate h thick and having a permittivity of ε_r has the following relationship [26]:

$$\text{BW} \propto \frac{3.77(\varepsilon_r - 1)Wh}{\varepsilon_r^2 L \lambda_0} \qquad (4.64)$$

where λ_0 is the center frequency. The bandwidth of a patch can be increased by matching the patch to the feed line, adding extensions, placing pins connecting the patch to the ground plane in strategic locations, making the substrate thicker, decreasing the dielectric constant of the substrate, stacking patches, and adding parasitic patches (Figure 4.51) [27].

Polarization of the patch is controlled through the location of the feed, shape of the patch, slots in the patch, or shorting pins. There are two approaches to creating elliptically polarized patch antennas [28]. First, feed the patch at a single point and change the shape of the patch or place slots inside the patch. Second, feed the patch at two points so the amplitude and phase difference creates the desired polarization. In both cases, two modes are induced in the patch.

The normalized fields radiated by a rectangular microstrip patch are

$$E_\theta = \cos\phi \cos\left(\frac{k_0 L}{2}\sin\theta\cos\phi\right)\frac{\sin\left(\frac{k_0 W}{2}\sin\theta\sin\phi\right)}{\frac{k_0 W}{2}\sin\theta\sin\phi} \qquad (4.65)$$

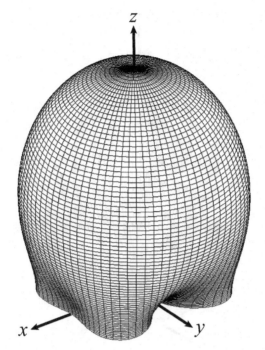

Figure 4.52. Total electric field of a microstrip patch.

$$E_\phi = -\cos\theta\sin\phi\cos\left(\frac{k_0 L}{2}\sin\theta\cos\phi\right)\frac{\sin\left(\frac{k_0 W}{2}\sin\theta\sin\phi\right)}{\frac{k_0 W}{2}\sin\theta\sin\phi} \qquad (4.66)$$

where k_0 is the wavenumber in free space. These formulas assume the patch consists of two radiating slots that are W wide and separated by a distance L. The directivity of a rectangular patch is estimated to be [15]

$$D = \frac{2(k_0 W)^2}{\cos(k_0 W) - 2 + k_0 W \int_0^{k_0 W} \sin x/x\,dx + \sin(k_0 W)/(k_0 W)} \qquad (4.67)$$

Figure 4.52 is a plot of the total electric field ($\sqrt{E_\theta^2 + E_\phi^2}$) a $0.4\lambda \times 0.4\lambda$ square patch.

Example. Plot the antenna pattern for an 8- × 4-element planar array of patches ($0.4\lambda \times 0.4\lambda$) with $d_x = d_y = 0.5\lambda$.
   Figure 4.53 shows the array pattern.

An estimate of the dimensions of a rectangular patch are given by [28]

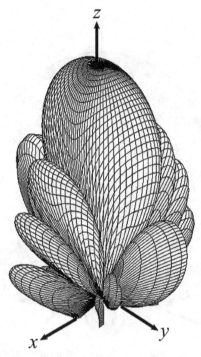

Figure 4.53. Array factor times the element pattern (in decibels) of an 8- × 4-element planar array of rectangular patches.

$$W = \sqrt{\frac{h\lambda_0}{\varepsilon_r}} \left[\ln\left(\frac{\lambda_0}{h\varepsilon_r}\right) - 1 \right] \qquad (4.68)$$

$$L = \frac{\lambda_0}{2\sqrt{\varepsilon_{\text{eff}}}} - \frac{0.824(\varepsilon_{\text{eff}} + 0.3)(W + 0.264h)}{(\varepsilon_{\text{eff}} - 0.258)(W + 0.8h)} \qquad (4.69)$$

$$\varepsilon_{\text{eff}} = \frac{\varepsilon_r + 1}{2} + \frac{\varepsilon_r - 1}{2\sqrt{1 + 10h/W}} \qquad (4.70)$$

The input resistance at resonance is [15]

$$R_{in}(y = y_0) = \frac{\cos^2(\pi y_0/L)}{2(G_1 \pm G_{12})} \qquad (4.71)$$

where

$$G_1 = \frac{-2 + \cos(k_0 W) + k_0 W S_i(k_0 W) + \sin(k_0 W)/(k_0 W)}{120\pi^2}$$

$$G_{12} = \frac{1}{120\pi^2} \int_0^\pi \left[\frac{\sin\left(\frac{k_0 W \cos\theta}{2}\right)}{\cos\theta} \right]^2 J_0(k_0 L \sin\theta) \sin^3\theta\, d\theta$$

$$f_r = \frac{1.8412c}{2\pi a_e \sqrt{\varepsilon_r}} \tag{4.72}$$

$$R_{in}(\rho' = \rho_0) = \frac{J_1^2(k\rho_0)}{(G_{rad} + G_{cd}) J_1^2(ka_e)} \tag{4.73}$$

where

$$a_e = a\sqrt{1 + \frac{2h}{\pi a \varepsilon_r}\left[\ln\left(\frac{\pi a}{2h}\right) + 1.7726\right]}$$

$$G_{rad} = \frac{(k_0 a_e)^2}{480} \int_0^{\pi/2} \left(J_{02}'^2 + J_{02}^2 \cos^2\theta\right)\sin\theta\, d\theta$$

$$G_{cd} = \left[\frac{\pi(\pi\mu_0 f_r)^{-3/2}}{h\sqrt{\sigma}} + \frac{\tan\delta}{\mu_0 f_r}\right]\left[(ka_e)^2 - 1\right]\frac{\varepsilon}{4h}$$

$$J_{02}' = J_0(k_0 a_e \sin\theta) - J_2(k_0 a_e \sin\theta)$$

$$J_{02} = J_0(k_0 a_e \sin\theta) + J_2(k_0 a_e \sin\theta)$$

The microstrip array design in Figure 4.54a has a conical beam with a null at $\theta = 0°$ and a peak between $\theta = 0°$ and $\theta = 90°$ at 5.2 GHz [29]. The conical beam results when the structure is not completely circularly symmetric but is

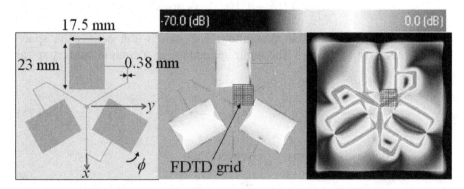

Figure 4.54. Conical beam antenna made from three patches. **(a)** Diagram. **(b)** Currents. **(c)** Electric field in the plane of the antenna. (Courtesy of Remcom.)

self-congruent for rotations of $\phi = 360°/N$, where N is the number of elements in the circular array. Both the currents on the microstrip patches and the electric field in the plane of the array are calculated using the finite difference time domain (FDTD) approach. The currents and electric field are shown in Figures 4.54b and 4.54c.

4.5. BROADBAND ANTENNAS

The bandwidth of a phased array is determined by a number of factors. As mentioned in Chapter 2, time delay units are needed to extend the array bandwidth when the main beam moves more than a half-beamwidth in any direction over the instantaneous bandwidth when steered to the maximum scan angle. Grating lobes due to large element spacing also limit the band-width at the highest frequency. Finally, the elements, feed network, and associated hardware must function appropriately over the bandwidth of the array. This section describes four types of broad band elements. Many other examples can be found in the literature [30].

4.5.1. Spiral Antennas

Spiral antennas have two or more arms that wrap around each other [31]. The arms of an Archimedian spiral in the x–y plane progressively move farther from the center with increasing ϕ. Arm n of an N_{arm} arm Archimedean spiral is described by [32]

$$r = r_{\mathrm{in}} + r_a\phi \tag{4.74}$$

where r_{in} is the distance from the center to the start of an arm and

$$r_a = \begin{cases} (\delta_s + \delta_w)/\pi & (2\,\mathrm{arms}) \\ 2(\delta_s + \delta_w)/\pi & (4\,\mathrm{arms}) \end{cases}$$

Usually, this antenna is complementary: The strip width equals the space between the strips ($\delta_s = \delta_w$). The input impedance of a complementary spiral is [33]

$$Z_{\mathrm{strip}}Z_{\mathrm{space}} = \frac{Z^2}{4} \tag{4.75}$$

If the area between the arms is free space, then

$$Z_{in} = \frac{Z_0^2}{2} = 188.5\,\Omega \tag{4.76}$$

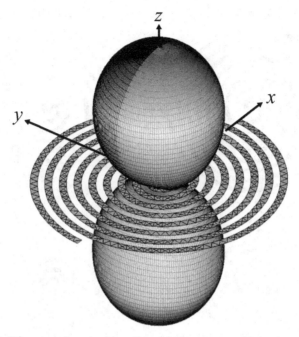

Figure 4.55. Diagram of an Archimedian spiral antenna with its antenna pattern.

For a spiral antenna, the circumference of the circle that just contains the spiral is approximately equal to the maximum wavelength radiated by the spiral [34]. As such, the low frequency of operation is determined by the outer radius while the high frequency is determined by the inner radius or smallest dimension.

$$\frac{c}{2\pi r_{\text{out}}} \le f \le \frac{c}{2\pi r_{\text{in}}} \tag{4.77}$$

Figure 4.55 shows the far-field pattern of a 2-arm self-complementary Archimedian spiral at 1 GHz with $N = 4$ turns. This element pattern was calculated numerically. The Archimedean spiral antenna radiates from a region where the circumference of the spiral equals one wavelength. This is called the active region of the spiral. Each arm of the spiral is fed 180° out of phase; so when the circumference of the spiral is one wavelength, the currents at complementary or opposite points on each arm of the spiral add in phase in the far field. Reflections from the end of the arm increase the lower end of the bandwidth predicted by (4.77). Placing loads at the end of the arms can reduce these reflections [35]. Figure 4.56 shows the extent of the current on the arms of the spiral as a function of frequency. At 250 MHz, the current (darker means higher amplitude) extends the length of the arms.

250 MHz

500 MHz

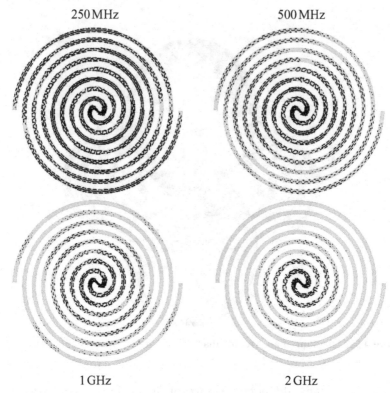

1 GHz

2 GHz

Figure 4.56. At 250 MHz, the Archimedian spiral antenna currents extend to the end of the arms. As the frequency increases, the current limits move closer to the center.

As the frequency increases, the current extends less out the arms. The spiral has a very broad bandwidth as shown by the plot of the computed S_{11} in Figure 4.57.

Spiral antennas radiate out both the front and back of the element. The back radiation is undesirable and may be reduced by placing the spiral above a ground plane, using a lossy cavity behind the spiral, or using a conical spiral. A ground plane is generally a bad idea, because it significantly reduces the spiral's bandwidth. The lossy cavity is the most popular option, even though it adds depth to the spiral and induces loss in the element. A conical spiral is not conformal to the surface and is more difficult to manufacture [36].

Placing this antenna in an array limits the diameter of the spiral to less than the element spacing. For a linear array or planar array with a rectangular lattice, grating lobes limit the maximum spacing between elements based on the steering angle to $d = 0.5\lambda_{min}/\sin\theta_s$. The lower limit is based upon the diameter of the spiral which has to be less than the element spacing ($d = \lambda_{max}/\pi$). As a result, the bandwidth of the array has an upper limit of

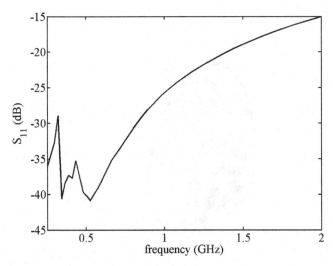

Figure 4.57. S_{11} of the Archimedian spiral antenna assuming a 250-Ω feed.

$$BW = \frac{\lambda_{max}}{\lambda_{min}} = \begin{cases} \dfrac{\pi}{2\sin\theta_s} & \text{for linear and rectangular lattice planar arrays} \\[2mm] \dfrac{1.15\pi}{2\sin\theta_s} & \text{for hexagonal lattice planar arrays} \end{cases} \tag{4.78}$$

Example. Plot the antenna pattern for an 8-element array of spirals shown in Figure 4.55.

Figure 4.58 shows the array pattern.

Figure 4.59 is a photograph of one of the spiral antennas that was in a 4-element interferometer array on the Gemini spacecraft. This array was part of the rendezvous radar in the recovery section of the spacecraft. It located and tracked the target vehicle during rendezvous maneuvers. The array/inter-ferometer consisted of four spirals: one each for transmit, reference, elevation, and azimuth.

Two identical linear arrays of Archimedean spiral elements operating at the same frequency but having orthogonal polarizations can be interleaved such that they have low sidelobes [37]. Each self-complementary spiral has five turns and is 34 mm in diameter. They are center-fed with a 1-V sources. The design specifications include ±30° scan with an axial ratio (AR) less than 3 dB, a VSWR less than 2, and a relative sidelobe level (RSLL) under −10 dB from 3 GHz to 6 GHz. The 80-element array has 40 RHCP spirals and 40 LHCP spirals. The spirals have a center-to-center separation of 38 mm, so the grating lobes start around 3.94 GHz.

Figure 4.60 shows a plot of the axial ratio of the array for three steering angles. The AR shows little variation as a function of steering angle and is under 3 dB above 4 GHz for all steering angles.

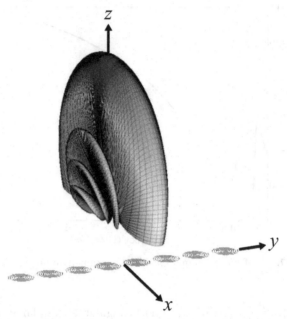

Figure 4.58. Computed far-field pattern of an 8-element linear array of Archimedian spirals along the x axis with $d = 0.55\lambda$.

Figure 4.59. Picture of a spiral element for the Gemini rendezvous radar. (Courtesy of the National Electronics Museum.)

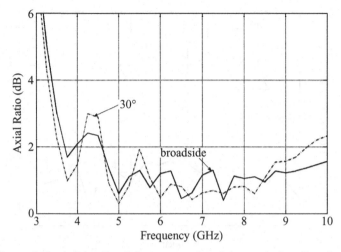

Figure 4.60. Axial ratio of the array at four beam pointing directions.

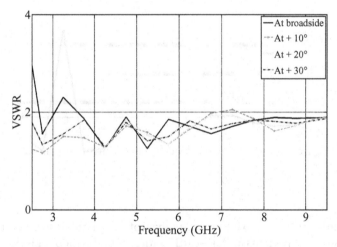

Figure 4.61. Plot of the VSWR at four beam pointing directions.

Element 20 has the worst VSWR when connected to a 250-Ω feed line. Figure 4.61 shows that the VSWR rapidly decreases until 2.5 GHz. There is a small resonance at 3.25 GHz when the beam is steered to 20° or 30°. The VSWR bandwidth extends from 2.75 GHz to greater than 10 GHz (except around 3.25 GHz, but this exception only occurs for element 20).

Figure 4.62 shows the maximum RSLL for an optimized 80-element array, with 40 RHCP spirals and 40 LHCP spirals. This maximum RSLL is shown for various steering angles, up to 30°. At broadside, the RSLL stays under

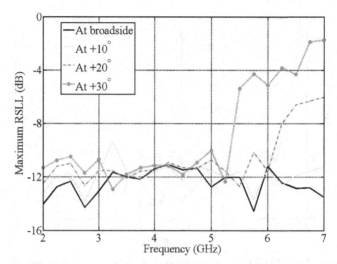

Figure 4.62. Plot of the maximum RSLL at four beam pointing directions.

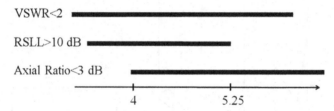

Figure 4.63. The array bandwidth defined three different ways.

−11 dB until 7 GHz (maximum frequency of the simulation). For the other steering angles, it remains under −10 dB until 5.25 GHz (theory gives 5.26 GHz), the frequency at which the grating lobe appears at a 30° steering angle. Whatever the steering angle, the RSLL is quite flat. This is obviously due to the choice of the cost function.

If the bandwidth is defined simultaneously for a RSLL less than −10 dB, an AR less than 3 dB and a VSWR less than 2, all of this for steering angles from −30° to +30°, this array has an approximate bandwidth of 4–5.25 GHz, or 27%. Figure 4.63 shows the AR, VSWR, and RSLL bandwidths. This figure clearly shows that the AR establishes the lowest frequency in the bandwidth, while the RSLL sets the highest frequency. Array bandwidth may be a function of more than just impedance bandwidth of the element and the formation of grating lobes.

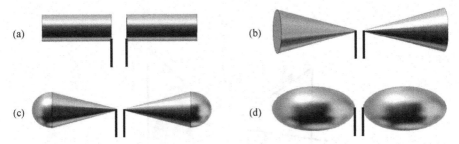

Figure 4.64. Some broadband dipole-like antennas. **(a)** Fat dipole. **(b)** Bicone. **(c)** Bicone with spherical caps. **(d)** Ellipsoidal.

4.5.2. Dipole-Like Antennas

A dipole antenna has a very narrow bandwidth because the current only resonates along the length of the dipole. Making the dipole fat (Figure 4.64a), encourages the currents to pursue other paths like circling the dipole, thus adding another dimension to the current flow and thereby broadening the bandwidth. Balanis notes that a dipole with a length-to-diameter ratio of 5000 has a 3% bandwidth, while one with a length-to-diameter ratio of 260 has a 30% bandwidth [15]. Another method of increasing the bandwidth of a dipole is to make each wire into a cone with the apex of the cones at the feed point (Figure 4.64a). This type of an antenna is called a biconical antenna. The bandwidth of a biconical antenna is further extended by adding spherical caps on the ends of the cones as in Figure 4.64c. Sharp edges cause radiation, so the spherical caps reduce the amount of radiation from the cone edge. Further smoothing creates ellipsoidal or spherical arms on the dipole (Figure 4.64d) [38].

Two-dimensional versions of these dipole-like antennas are broadband as well. For instance, the bow-tie antenna is a two-dimensional version of the bicone antenna [39]. Circular caps are often placed at the end of the bow-tie antennas as well [39]. These antennas also come in monopole flavors too [40].

Example. Plot the antenna pattern for an 8- × 4-element planar array of y-oriented bow-tie elements ($0.4\lambda \times 0.4\lambda$) spaced $\lambda/2$ apart in the x and y directions.

Figure 4.65 shows the element pattern and Figure 4.66 shows the array pattern. The array pattern is compressed in the y–z plane, because the element pattern is narrow in that direction and there are more elements in the y direction.

The broadband-printed bent dipole shown in Figure 4.67 has the S_{11} plot shown in Figure 4.68. This element has an S_{11} below −10 dB from 7.9 to 12.9 GHz which is a 48% bandwidth. The three-dimensional element pattern in Figure 4.69 has a directivity of 4.8 dB. If this element is placed in a 3 × 5 array with element spacing $\lambda/2$ in the x and y directions, then the product of

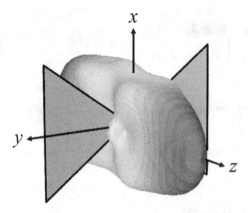

Figure 4.65. Diagram of a bow-tie antenna antenna with its antenna pattern.

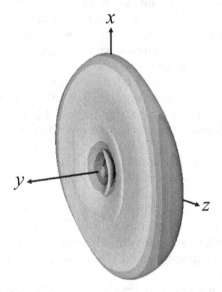

Figure 4.66. Computed far-field pattern of pattern of an 8-element linear array along the x axis with $d = 0.55\lambda$.

the element pattern times the array factor is shown in Figure 4.70. This array pattern has a directivity of 16.6 dB at 10 GHz.

4.5.3. Tapered Slot Antennas

A tapered slot antenna (TSA) or Vivaldi antenna [41] is a flared slot line just like a horn antenna is a flared waveguide. The TSA is an ideal wideband array element because [42]

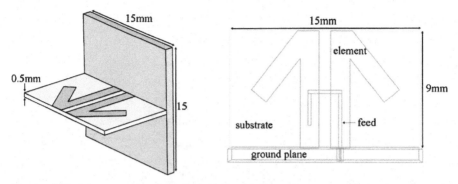

Figure 4.67. Diagram of a broadband-printed bent dipole. (Courtesy of James R. Willhite, Sonnet Software, Inc.)

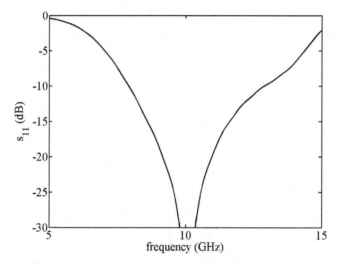

Figure 4.68. Plot of S_{11} for the element in Figure 4.67.

1. The bandwidth with VSWR <2 can be up to 10:1.
2. The $\cos\theta$ element pattern exists over all scan planes out to $\theta_s = 50°$–$60°$.
3. No need to use lossy materials to obtain bandwidth.
4. A balun can be easily incorporated into the design.
5. Beamwidth remains constant over the operating bandwidth [43].

These advantages have led to the widespread use of TSAs in the design of new phased arrays.

Figure 4.71 is a picture of the AN/APG-81 phased array with over 1000 elements and a blowup of the slotted tab element that has a rectangular

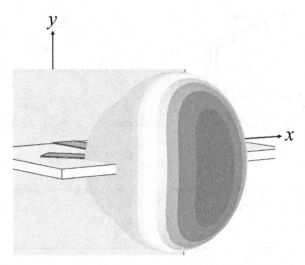

Figure 4.69. Element pattern for the broadband-printed bent dipole at 10 GHz.

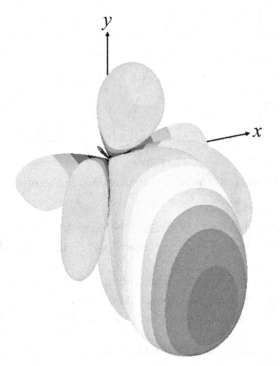

Figure 4.70. Array pattern for a 3- × 5-element array.

Figure 4.71. Slotted tab element in the AN/APG-81 antenna array. (Courtesy of Northrop Grumman and available at the National Electronics Museum.)

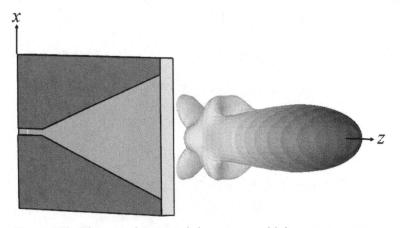

Figure 4.72. Diagram of a tapered slot antenna with its antenna pattern.

opening. Oftentimes the opening is a linear taper instead of a rectangular tab. Figure 4.72 is a diagram of a 10-GHz 50-Ω TSA with its antenna pattern. The slot line is 5 mm long and 1.39 mm wide on a 13.27–mm × 15.34–mm dielectric substrate (ε_r = 3) 2 mm thick. The aperture width is 5.67 mm. The reflection coefficient as a function of frequency is shown in Figure 4.73. It is below −10 dB from 9.2 to 11.4 GHz for a 22% bandwidth. The far-field pattern of an 8-element linear array along the x axis with d = 0.55λ is shown in Figure 4.74.

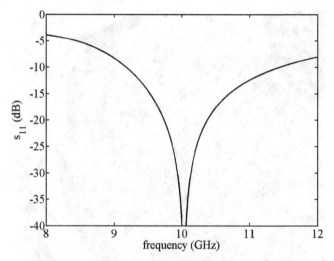

Figure 4.73. Plot of S_{11} versus frequency for the TSA.

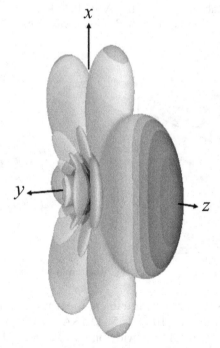

Figure 4.74. Far-field pattern pattern of an 8-element linear array along the x axis with $d = 0.55\lambda$.

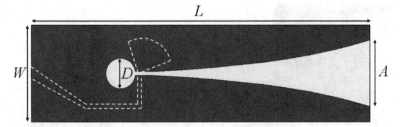

Figure 4.75. Diagram of a Vivaldi antenna.

A curved opening provides a smoother transition to free space than the linear opening which increases the bandwidth. Optimizing the shape of the opening yields better performance [44]. A typical Vivaldi design appears in Figure 4.75 [45]. Optimizing the exponential taper, length, and width of the slot along with the feeding mechanism produces excellent broadband elements. The feed design limits the high frequency of the bandwidth, whereas the aperture size limits the low frequency of the bandwidth. Some general guidelines for the design of a Vivaldi antenna array include [42]:

1. $d_x, d_y \approx 0.45\lambda$.
2. Increasing the element length, lowers the minimum frequency in the bandwidth.
3. Substrate:
   a. Its thickness should be ≈ 0.1 times the element width.
   b. Increasing ε_r lowers the minimum frequency in the bandwidth. It may also lower the maximum frequency as well.
4. Increasing R_a lowers the minimum frequency in the bandwidth but increases variations in the impedance over the bandwidth.
5. Increasing D increases the antenna resistance at lower frequencies.

Surrounding the TSA with shorting pins or vias that connect the top and bottom metal surfaces on either side of the substrate eliminates unwanted resonances that limit the bandwidth [46].

The Vivaldi antenna in Figure 4.76 is a Schaubert design [47] with $L = 97$ mm, $W = 27.02$ mm, $A = 18.6$ mm, $D = 8.47$ mm, and $R = 0.03$. The element spacing is $d_x = d_y = 0.09\lambda$ at the lowest frequency of 1.0 GHz. The substrate is 3.15 mm thick and has a relative dielectric constant of 2.2. The slot line and strip line feed are 1.1 mm wide. The strip line feed has an input impedance of 50 Ω and narrows to 82 Ω. A picture of the element and the array are shown in Figure 4.76. This 144 element ($9 \times 8 \times 2$ pol) array operates from 1 to 5 GHz. The planar array consists of parallel linear arrays with each linear array printed on a single circuit board. An identical array is placed orthogonal to the first planar array in order to get the two orthogonal polarizations. The

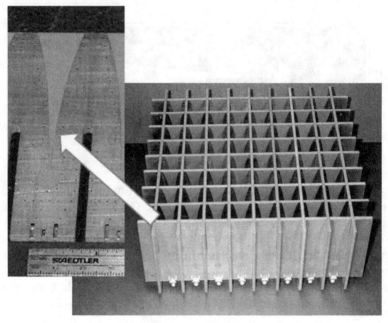

Figure 4.76. Vivaldi planar antenna array with close-up picture of element. (Courtesy of Daniel Schaubert, University of Massachusetts.)

circuit boards have notches cut in them between each element in order to place them in an "egg crate assembly" as shown in Figure 4.76.

Figure 4.77 is a picture of a Vivaldi element for the direction finding array in Figure 4.78 [48]. The element has a receive bandwidth of 30 MHz to 3 GHz and a transmit bandwidth at 1 kW of 100 MHz to 4 GHz. It is 88.9 cm by 42.2 cm and weighs only 4.7 pounds. The gain variation over the bandwidth is shown in Figure 4.79, and its beamwidth in the vertical and horizontal planes are shown in Figure 4.80.

4.5.4. Dielectric Rod Antennas

A dielectric rod antenna is formed from a piece of dielectric that extends from the open end of a waveguide. If the dielectric is polystyrene (a common type of plastic), then the antenna is called a polyrod. The radiation pattern of the polyrod is a function of its cross section, length, dielectric constant, and shape. A high dielectric constant results in a thinner and lighter rod, but a narrower bandwidth. Dielectric rod antennas commonly use a cylindrical waveguide with a shaped dielectric rod protruding from the waveguide as shown in Figure 4.81. To maximize directivity, the diameter of the dielectric rod should be [12]

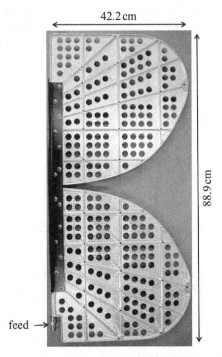

Figure 4.77. Vivaldi antenna array for DF array. (Courtesy of US Navy SPAWAR.)

Figure 4.78. DF array of Vivaldi elements. (Courtesy of US Navy SPAWAR.)

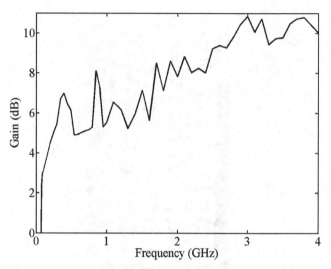

Figure 4.79. Measured gain of the Vivaldi element in Figure 4.77. (Data for plot courtesy of US Navy SPAWAR.)

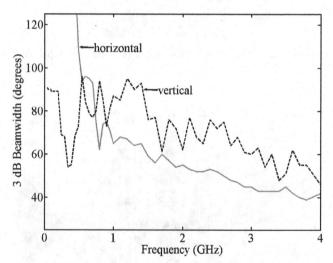

Figure 4.80. Measured beamwidths of the Vivaldi element in Figure 4.77. (Data for plot courtesy of US Navy SPAWAR.)

$$\frac{D}{\lambda} = 0.2 + \frac{3}{\varepsilon_r^{1.5}\sqrt{1 + 2L/\lambda}} \tag{4.79}$$

for $L > 2\lambda$ and $2 < \varepsilon_r < 5$. Cylindrical rods can be tapered to get a better match and reduce sidelobes. An approximate expression for the far field of a polyrod is [12]

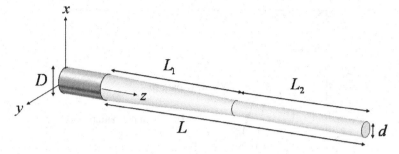

Figure 4.81. Polyrod antenna.

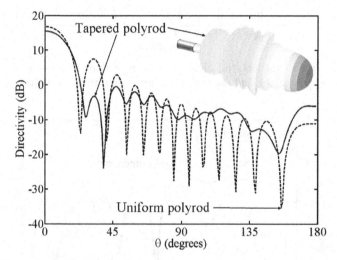

Figure 4.82. Antenna pattern cuts for the tapered and uniform polyrod antennas. A three-dimensional view of the tapered polyrod antenna is in the upper right-hand corner.

$$E(\theta) = \frac{\sin(\zeta)}{\zeta}$$

(4.80)

where $\zeta = \pi \left[\frac{L}{\lambda}(\cos\theta - 1) - 0.5 \right]$.

Example. Calculate the far field radiated by the polyrod antenna shown in Figure 4.81 at 10 GHz with $D = 0.5\lambda$, $d = 0.3\lambda$, $L = 6\lambda$, $L_1 = 3\lambda$, $L_2 = 3\lambda$, and $\varepsilon_r = 3$ and for a uniform polyrod ($d = 0.5\lambda$).

Figure 4.82 shows the antenna pattern for a uniform polyrod antenna (dashed line) calculated using (4.80). No simple formula exists for calculating the antenna pattern of the tapered polyrod, so the pattern was numerically

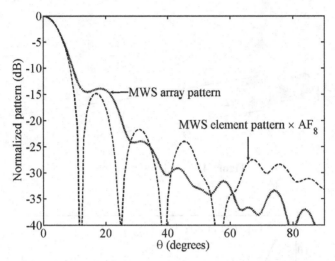

Figure 4.83. Plot of the polyrod element pattern, 8-element array factor, element pattern times array factor, and the 8-element array with mutual coupling.

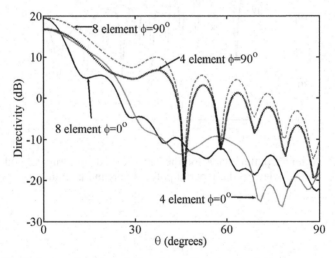

Figure 4.84. Orthogonal cuts of the 4- × 8-element planar arrays of polyrod elements.

calculated using Microwave Studio. The tapered polyrod has lower maximum sidelobe levels at the expense of a decrease in directivity.

Figure 4.83 shows the far field of an 8-element array of the polyrod elements with $d = 0.6\lambda$. The dotted line is the array factor times the element pattern in Figure 4.82. The solid line is the array pattern of the 8-element array taking into account mutual coupling. Nulls and low sidelobes are not approxi-

Figure 4.85. Mark 8 planar phased array with polyrod elements. (Courtesy of the National Electronics Museum.)

mated well using the simple formula given by the element pattern times the array factor. Figure 4.84 compares the orthogonal antenna pattern cuts of 4-element and 8-element linear arrays of tapered polyrods ($d = 0.6\lambda$) when mutual coupling is included in the calculations. The arrays lie along the x axis, so the patterns at $\phi = 90°$ are nearly identical except that the 8-element array is about 3 dB higher, because it has twice the elements of the 4-element array.

Bell Labs built the X-band surface fire control radar, the Mark 8 using an array of 42 polyrods, configured as a planar array having 14 × 3 elements as shown in Figure 4.85. This was the first use of the polyrod antenna in an array, and the first microwave phased array. Each column of three elements received the same phase shift. Rotary phase shifters adjusted the phase of each column of elements to steer the beam. The array had 2° azimuth and 6.5° elevation beamwidths [49].

REFERENCES

1. P. Hannan, The element-gain paradox for a phased-array antenna, *IEEE Trans. Antennas Propagat.*, Vol. 12, No. 4, 1964, pp. 423–433.

2. D. Carter, Phase centers of microwave antennas, *IRE Trans. Antennas Propagat.*, Vol. 4, No. 4, 1956, pp. 597–600.

3. J. Huang, A technique for an array to generate circular polarization with linearly polarized elements, *IEEE AP-S Trans.*, Vol. 34, No. 9, September 1986, pp. 1113–1124.

4. R. C. Hansen, *Phased Array Antennas*, New York: John Wiley & Sons, 1998.

5. D. N. McQuiddy, Jr., R. L. Gassner, P. Hull et al., Transmit/receive module technology for X-band active array radar, *Proc. IEEE*, Vol. 79, No. 3, 1991, pp. 308–341.

6. R. T. Compton, On the performance of a polarization sensitive adaptive array, *IEEE AP-S Trans.*, Vol. AP-29, September 1981, pp. 718–725.

7. R. L. Haupt, Adaptive crossed dipole antennas using a genetic algorithm, *IEEE AP-S Trans.*, Vol. 52, No. 8, August 2004, pp. 1976–1982.

8. http://www.haarp.alaska.edu/haarp/index.html

9. R. E. Collin, *Antennas and Radiowave Propagation*, New York, McGraw-Hill, 1985.

10. R. G. Fitzgerrell, Monopole impedance and gain measurements on finite ground planes, *IEEE Trans. Antennas Propagat.*, Vol. 36, No. 3, 1988, pp. 431–439.

11. R. L. Haupt and D. W. Aten, Low sidelobe arrays via dipole rotation, *IEEE AP-S Trans.*, Vol. 57, No. 5, May 2009, pp. 1574–1577.

12. J. D. Kraus and R. J. Marhefka, *Antennas for All Applications*, 3rd ed., New York: McGraw-Hill, 2002.

13. W. L. Stutzman and G. A. Thiele, *Antenna Theory and Design*, 2nd ed., New York: John Wiley & Sons, 1998.

14. H. King and J. Wong, Characteristics of 1 to 8 wavelength uniform helical antennas, *IEEE Trans. Antennas Propagat.*, Vol. 28, No. 2, 1980, pp. 291–296.

15. C. A. Balanis, *Antenna Theory: Analysis and Design*, 3rd ed., Hoboken, NJ: Wiley-Interscience, 2005.

16. National RadioAstronomy Observatory, An overview of the very large array, http://www.vla.nrao.edu/genpub/overview/, June 17, 2009.

17. R. S. Elliott, *Antenna Theory and Design*, revised edition, Hoboken, NJ: John Wiley & Sons, 2003.

18. National Electronics Museum, RARF Antenna, 2009, Display Information.

19. National Electronics Museum, EAR Antenna, 2009, Display Information.

20. R. S. Elliott, *Antenna Theory and Design*, revised edition, Hoboken, NJ: John Wiley & Sons, 2003.

21. A. J. Stevenson, Theory of slots in rectangular waveguides, *J. Appl. Phys.*, Vol. 19, January 1948, pp. 24–38.

22. A. Oliner, The impedance properties of narrow radiating slots in the broad face of rectangular waveguide: Part I—Theory, *IRE Trans. Antennas Propagat.*, Vol. 5, No. 1, 1957, pp. 4–11.

23. A. Oliner, The impedance propeties of narrow radiating slots in the broad face of rectangular waveguide: Part II—Comparison with measurement, *IRE Trans. Antennas and Propagat.*, Vol. 5, No. 1, 1957, pp. 12–20.

24. AN/APG-68(V) airborne fire-control radar, Jane's Radar and Electronic Warfare Systems, 21-Dec-2007.

25. A. J. Haupt, Antennas, IEEE Aerospace Junior Engineering Conference, March 2008.

26. D. R. Jackson and N. G. Alexopoulos, Simple approximate formulas for input resistance, bandwidth, and efficiency of a resonant rectangular patch, *IEEE Trans. Antennas Propagat.*, Vol. 39, No. 3, 1991, pp. 407–410.

27. R. Garg, P. Bhartia, I. Bahl, and A. Ittipiboon, *Microstrip Antenna Design Handbook*, Norwood, MA: Artech House, 2001.

28. M. Kara, Formulas for the computation of the physical properties of rectangular microstrip antenna elements with various substrate thicknesses, *Microwave Opt. Technol. Lett.*, Vol. 12, No. 4, 1996, pp. 234–239.

29. N. J. McEwan, R. A. Abd-Alhameed, E. M. Ibrahim et al., A new design of horizontally polarized and dual-polarized uniplanar conical beam antennas for HIPERLAN, *IEEE Trans. Antennas Propagat.*, Vol. 51, No. 2, 2003, pp. 229–237.

30. Z. N. Chen, *Broadband Planar Antennas*, Hoboken, NJ: John Wiley & Sons, 2006.

31. J. Dyson, The equiangular spiral antenna, *IRE Trans. Antennas Propagat.*, Vol. 7, No. 2, 1959, pp. 181–187.

32. W. Curtis, Spiral antennas, *IRE Trans. Antennas Propagat.*, Vol. 8, No. 3, 1960, pp. 298–306.

33. M. Gustafsson, Broadband Self-Complementary Antenna Arrays, C. E. Baum, A. P. Stone, and J. S. Tyo, ed., *Ultra-Wideband Short-Pulse Electromagnetics 8*, New York, Springer, 2007.

34. J. Kaiser, The Archimedean two-wire spiral antenna, *IRE Trans. Antennas Propagat.*, Vol. 8, No. 3, 1960, pp. 312–323.

35. M. Lee, B. A. Kramer, C. Chi-Chih et al., Distributed lumped loads and lossy Transmission Line Model for Wideband Spiral Antenna Miniaturization and Characterization, *IEEE Trans. Antennas Propagat.*, Vol. 55, No. 10, 2007, pp. 2671–2678.

36. J. Dyson, The unidirectional equiangular spiral antenna, *IRE Trans. Antennas Propagat.*, Vol. 7, No. 4, 1959, pp. 329–334.

37. R. Guinvarc'h and R. L. Haupt, Dual polarization interleaved spiral antenna phased array with an octave bandwidth, *IEEE Trans. Antennas Propagat.*, Vol. 58, No. 2, 2010, pp. 1–7.

38. B. Allen, *Ultra-Wideband: Antennas and Propagation for Communications, Radar and Imaging*, Chichester: John Wiley & Sons, 2007.

39. G. H. Brown and J. O. M. Woodward, Experimentally determined radiation characteristics of conical and triangular antennas, *RCA Rev.*, Vol. 13, No. 4, December 1952, pp. 425–452.

40. Z. N. Chen, *Broadband Planar Antennas*, Hoboken, NJ: John Wiley & Sons, 2006.

41. P. J. Gibson, The Vivaldi aerial, 9th European Microwave Conference Proceedings, 1979, pp. 103–105.

42. W. F. Croswell, T. Durham, M. Jones, D. Schaubert, P. Friederich, and J. G. Maloney, Wideband arrays, in *Modern Antenna Handbook*, C. A. Balanis, ed., Hoboken, NJ: John Wiley & Sons, 2008, pp. 581–630.

43. J. S. Mandeep and Nicholas, Design an X band Vivaldi antenna, *Microwaves & RF*, July 2008, pp. 59–65.

44. H. Oraizi and S. Jam, Optimum design of tapered slot antenna profile, *IEEE AP-S Trans.*, Vol. 51, No. 8, 2003, pp. 1987–1995.

45. T.-H. Chio and D. H. Schaubert, Parameter study and design of wide-band widescan dual-polarized tapered slot antenna arrays, *IEEE Trans. Antennas Propagat.*, Vol. 48, No. 6, 2000, pp. 879–886.

46. H. Holter, C. Tan-Huat, and D. H. Schaubert, Elimination of impedance anomalies in single- and dual-polarized endfire tapered slot phased arrays, *IEEE Trans. Antennas Propagat.*, Vol. 48, No. 1, 2000, pp. 122–124.

47. H. Holter, C. Tan-Huat, and D. H. Schaubert, Experimental results of 144-element dual-polarized endfire tapered-slot phased arrays, *IEEE Trans. Antennas Propagat.*, Vol. 48, No. 11, 2000, pp. 1707–1718.

48. R. Horner, B. Calder, R. Mangra et al., Tapered slot antenna cylindrical array, US7518565, April 14, 2009.

49. L. Brown, *A Radar History of World War II: Technical and Military Imperatives*, Philadelphia: Institute of Physics Publishing, 1999.

5 Nonplanar Arrays

Array elements usually lie in a plane, because the feed network for nonplanar array is more complicated. In addition, placing array elements on a curved surface complicates design, construction, and testing. A nonplanar array, like the one in Figure 5.1, with elements distributed in three-dimensional space has an array factor given by

$$\text{AF} = \sum_{n=1}^{N} w_n e^{jk[x_n(u-u_s)+y_n(v-v_s)+z_n\cos\theta]} \tag{5.1}$$

where

$$u = \sin\theta\cos\phi$$

$$u_s = \sin\theta_s\cos\phi_s$$

$$v = \sin\theta\sin\phi$$

$$v_s = \sin\theta_s\sin\phi_s$$

$$u^2 + v^2 = \sin^2\theta(\sin^2\phi + \cos^2\phi) \le 1$$

$$u_s^2 + v_s^2 \le 1$$

$$\theta_s = \text{elevation steering angle}$$

$$\phi_s = \text{azimuth steering angle}$$

A conformal array has the (x_n, y_n, z_n) lying on a curved surface. The maximum gain of a conformal array equals the sum of the array element gain pattern values in the main beam direction when they are phase compensated. This chapter presents a number of approaches to distributed and conformal array configurations.

5.1. ARRAYS WITH MULTIPLE PLANAR FACES

Linear and planar arrays have limited scan regions due to element patterns, mutual coupling, and grating lobes. One way to increase the array scanning

Antenna Arrays: A Computational Approach, by Randy L. Haupt
Copyright © 2010 John Wiley & Sons, Inc.

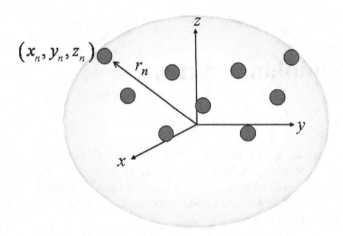

Figure 5.1. Arbitrary distribution of elements in the *x–y* plane.

Shape	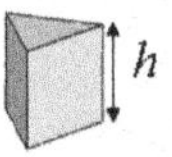	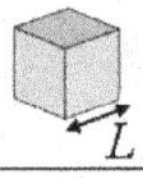	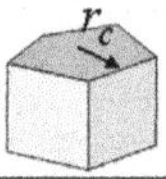	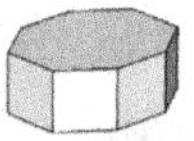	
Vertex angle	60°	90°	108°	140°	∞
Face scan	$\pm 60^o$	$\pm 45^o$	$\pm 36^o$	$\pm 30^o$	$\pm \dfrac{180^o}{N_r}$

Figure 5.2. Three-dimensional polygonal surfaces for full azimuth scanning.

region uses multiple linear/planar arrays that face in different directions. Each planar array lies on a side of a three-dimensional polygon surfaces [1]. Figure 5.2 shows some polygons with faces that can accommodate arrays for complete azimuth scanning. A 360° azimuth scan requires that each array face equally divides the azimuth for scanning. As a result, one face has an azimuth scan defined by

$$\theta_{s_{max}} = \pm \frac{360°}{2N_p} \tag{5.2}$$

where N_p are the number of sides to the polygon. The size of one face of the polygon is determined by the required gain and beam width.

PAVE (Precision Acquisition Vehicle Entry) PAWS (Phased Array Warning System) (AN/FPS-115) is a UHF array built by Raytheon in 1978 to detect and track sea-launched ballistic missiles (Figure 5.3) [2]. It has two faces

Figure 5.3. PAVE PAWS (AN/FPS-115) array. (Courtesy of the National Electronics Museum.)

Figure 5.4. MESA array with installation on a Boeing 737 in the lower right corner. (Courtesy of Northrop Grumman and available at the National Electronics Museum.)

inclined 120° to each other with 1792 T/R (transmit/receive) modules (per face) connected to dipole radiating elements that operate from 420 to 450 MHz. Each element has a 4-bit phase shifter. The modules are grouped into 56 subarrays of 32 elements. The receive antenna gain is 34 dB.

The multirole electronically scanned array (MESA) is a different multiple face array concept that provides 360° coverage on a Boeing 737 without rotating the antenna (Figure 5.4). The MESA antenna (called the Wedgetail) has three electronically scanning arrays that share 288 L-band T/R modules [3]. The port and starboard arrays scan ±60°, while the "Top Hat" endfire array provides ±30° front and back scanning as shown in Figure 5.5. The array shape is very aerodynamic, with the "Top Hat" providing additional lift for the airplane.

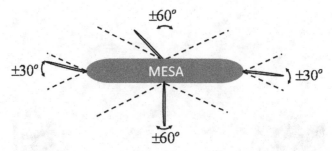

Figure 5.5. MESA array azimuth scanning.

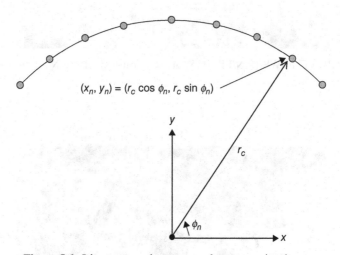

Figure 5.6. Linear array bent to conform to a circular arc.

5.2. ARRAYS ON SINGLY CURVED SURFACES

Figure 5.6 shows a linear array bent around a circular arc. If the element spacing remains constant along the arc, then the spacing between the elements is approximately equal to

$$d = r_c(\phi_{n+1} - \phi_n), \qquad n = 1, 2, \ldots, N-1 \tag{5.3}$$

where r_c is the radius of curvature. If the radius of the circular arc is large, then the linear array is only slightly curved; and when $r_c = \infty$, the arc becomes a straight line. End elements of the conformal arc array are at $(r_c \cos \phi_1, r_c \sin \phi_1)$ and $(r_c \cos \phi_N, r_c \sin \phi_N)$ when the midpoint of the arc is at $(0, r_c)$. The main beam radiates away from the convex side of the arc or in the $+y$ direction.

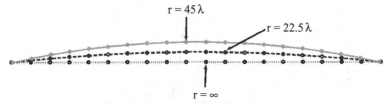

Figure 5.7. Conformal linear array with $r_c = \infty$, 45λ, and 22.5λ.

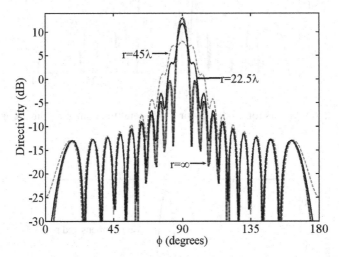

Figure 5.8. Array factors for uniformly weighted conformal linear arrays with $r_c = \infty$, 45λ, and 22.5λ.

As r_c decreases, the array curvature increases, and the changes to the array factor become more dramatic. For instance, r_c values of ∞, 45λ, and 22.5λ produce the array bends shown in Figure 5.7. If the array has uniform weighting, then the resulting array factors appear in Figure 5.8. The directivity decreases from 13.0 to 11.8 to 8.0 dB as the curvature decreases from ∞ to 45λ to 22.5λ. Relative sidelobe levels near the main beam rise from 13.3 to 8.3 to 0.8 dB, but the sidelobes far from the main beam exhibit little change. Figure 5.9 shows the array factors associated with Figure 5.7 when the weighting is a 25 dB Taylor $\bar{n} = 5$ amplitude taper. In this case, the directivity decreases from 12.6 to 11.7 to 9.2 dB. This decrease in directivity is less than that of the corresponding uniform array, because the end elements that are most displaced from the center have the lowest amplitude. Again, the pattern within $\pm 45°$ of the main beam is distorted, while the sidelobes beyond that range are not. The distortion in this case is in the form of the first sidelobe ($r_c = 45\lambda$) and first two sidelobes ($r_c = 22.5\lambda$) being absorbed into the main beam.

Curvature compensation restores the array factor to the desired array factor by applying a phase shift, δ_n, that offsets the y translation.

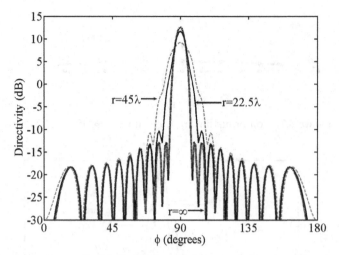

Figure 5.9. Array factors for 25-dB Taylor $\bar{n} = 5$ weighted conformal linear arrays with $r_c = \infty$, 45λ, and 22.5λ.

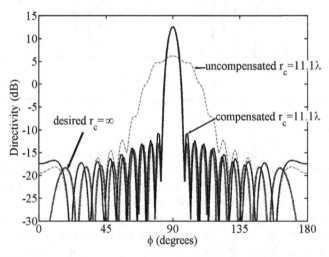

Figure 5.10. Compensated array factor for a 25-dB Taylor $\bar{n} = 5$ weighted conformal linear arrays with $r_c = 11.1\lambda$.

$$\delta_n = -k\left(x_n \cos\phi_s + y_n \sin\phi_s\right) \tag{5.4}$$

where the main beam is steered to ϕ_s. This equation implies that the phase compensation changes at each scan angle and is independent of the amplitude taper. Applying (5.4) to the array factor of a 20-element $d = 0.5\lambda$ spaced array having $r_c = 11.1\lambda$ with a 25-dB Taylor $\bar{n} = 5$ results in the array factor shown

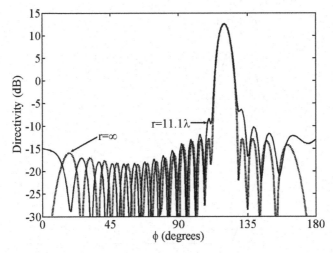

Figure 5.11. Compensated array factor for a 25-dB Taylor $\bar{n} = 5$ weighted conformal linear arrays with $r_c = 11.1\lambda$. when the main beam points at $\phi_s = 120°$.

in Figure 5.10. The compensation restores the main beam and sidelobes to the desired levels. The directivity increases from 6.2 to 12.5 dB (which is the same as the corresponding linear array with $r_c = \infty$), and the maximum sidelobe level is 23.3 dB below the peak of the main beam. Since the array has a smaller extent in the x direction, the array factor for the curved array array factor has the number of sidelobes of a linear array factor that is $2x_c$ long.

Example. Graph the array factor for a 25-dB Taylor $\bar{n} = 5$ weighted conformal linear arrays with $r_c = 11.1\lambda$ when the main beam points at $\phi_s = 120°$.

Figure 5.11 shows the compensated array factor superimposed on the $r_c = \infty$ array factor. The sidelobe peaks at angles greater than about 100° increased between 2 and 7 dB, while the rest of the sidelobes look like those of the desired pattern.

The phase correction in (5.4) works well to restore the pattern of a linear or planar array conforming to the surface of a curved surface like a cylinder. Finding the weights that produce a more complicated pattern, such as a flat-top main beam, requires numerical methods to adjust either the amplitude and phase [4–6] or phase alone [7].

In almost all conformal antennas, the element pattern peak is normal to the surface, so the peaks will not all point in the same direction as with linear and planar arrays. Figure 5.12 is a diagram of a conformal array displaying both isotropic and directional element patterns. The isotropic element patterns all contribute the same amplitude in a given direction, while the

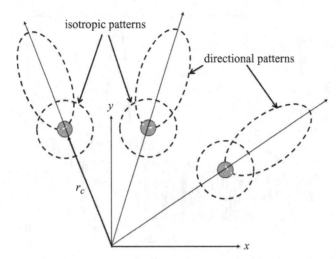

Figure 5.12. Comparison of a conformal array with directional elements to one with isotropic elements.

directional element patterns do not. Since the element pattern is a different function of ϕ for each element, the element pattern remains inside the sum sign of the array factor.

$$\text{AF} = \sum_{n=1}^{N} e(\phi - \phi_n) w_n e^{jk[x_n \sin\theta \cos\phi + y_n \sin\theta \sin\phi + z_n \cos\theta]} \tag{5.5}$$

where $e(\phi - \phi_n)$ is the element pattern of element n having a peak at $\phi = 0°$. Now, the antenna pattern is not the product of a single element pattern times an array factor.

As with a linear array, the directivity and sidelobe level changes when including the element patterns. Typically, the element pattern maximum points normal to the surface where the element lies. The antenna pattern changes when the peaks of all the element patterns point toward 90°. In this case, the antenna pattern is the product of the element pattern and the array factor. Consider a 20-element array with elements spaced $\lambda/2$ over an arc of radius 11.1λ. If the peak of the element patterns point normal to the arc, then the element patterns are represented by

$$e(\phi - \phi_n) = \begin{cases} \cos(\phi - \phi_n), & -90° \le \phi \le 90° \\ 0, & \text{otherwise} \end{cases} \tag{5.6}$$

When all the element pattern peaks point toward 90°, then $\phi_n = 90°$. Figure 5.13 shows little difference between the antenna patterns of the array with the elements pointing normal and the elements pointing 90° except for angles far

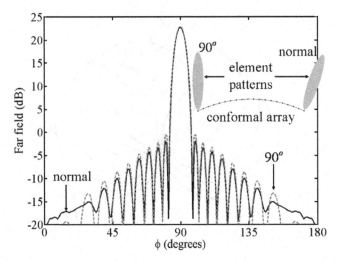

Figure 5.13. Comparing the antenna patterns of two conformal arrays. One has the maximum element patterns normal to the surface while the other has the element pattern peaks pointing to $\phi = 90°$.

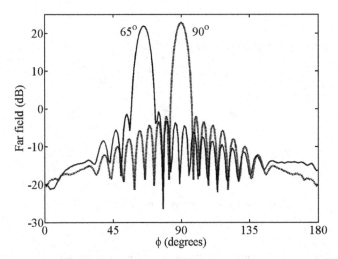

Figure 5.14. Main beam of the conformal array steered to $\phi = 65°$.

from the main beam. Figure 5.14 shows the beam steered to 25° off boresight for the array with $\phi_n = 90°$. The directivity decreases from 22.7 to 21.9 dB, while the relative sidelobe level increases about 2 dB.

A polar plot of all the element patterns for a 10-element array appears in Figure 5.15. Dotted lines indicate the directions of the element pattern peaks. It is easy to see that the element pattern of element 1 contributes little field in the direction of the maximum of element pattern 10.

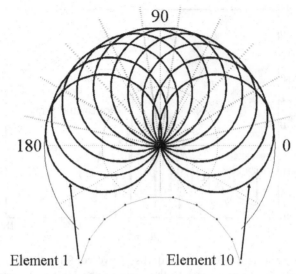

Figure 5.15. Polar plot of all the element patterns for a 10-element conformal array when their peaks point normal to the surface.

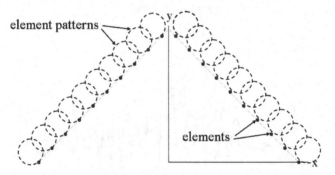

Figure 5.16. Linear array bent at an angle of 45°.

Example. A uniformly weighted 20-element linear array with element spacing of 0.5λ is bent at an angle of 90° as shown in Figure 5.16. Plot the uncompensated array factor, the compensated array factor, the antenna pattern with element patterns pointing in the direction of the normal to the surface, and the antenna pattern steered to 67.5° and 45°.

The uncompensated pattern has a directivity of –15.1 dB at $\phi = 90°$ with large 13-dB lobes at 45.5° and 135°. Phase compensation restores the directivity at $\phi = 90°$ to 16 dB. The large lobes are reduced to 12.6 dB and move to 28° and 152° These patterns are shown in Figure 5.17. The compensated antenna pattern when all the element pattern peaks point toward $\phi = 90°$ is

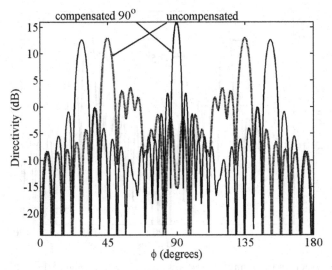

Figure 5.17. Compensated and uncompensated antenna patterns for the array in Figure 5.16.

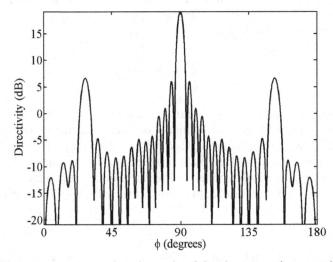

Figure 5.18. Antenna pattern when the peaks of the elements point toward $\phi = 90°$ for the array in Figure 5.16.

shown in Figure 5.18. Now, the main beam has a directivity of 19.2 dB. Large lobes at 28° and 152° have a directivity of only 6.6 dB. Figure 5.19 shows the compensated antenna pattern steered to 67.5° and 45°. The main beam directivity is 15.4 dB at 67.5° and 13 dB at 45°.

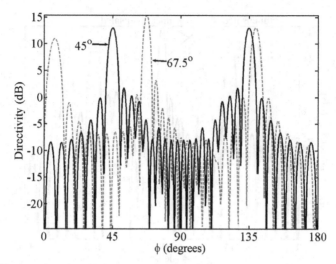

Figure 5.19. Steering the compensated antenna pattern in Figure 17 to 67.5° and 45° when $\phi_n = 90°$.

A Wullenweber antenna [8] is a large direction finding circular array with a narrow beam that scans 360° in azimuth. Only a small group (N_{wul}) of adjacent elements are active at a time with the array factor

$$\text{AF}_{wul} = \sum_{n=1}^{N_{wul}} w_n e^{jkr_c \sin\theta \cos(\phi-\phi_n)} \tag{5.7}$$

where $N_{wul} < N$. The main beam points normal to the center of that group of elements. A commutating feed connects the active elements to the output (Figure 5.20). As the commutating feed rotates, the group of active elements changes by one element, and the main beam points in a new azimuth direction. Phase compensation is only applied to the active elements. To form a coherent beam in the desired direction ϕ_s, the element weights in (5.7) are given by

$$w_n = e^{-kr_c \cos(\phi_s-\phi_n)} \tag{5.8}$$

A 60-element Wullenweber array with 0.5λ spacing between elements is shown in Figure 5.20 when the indicated 15 elements (large dark dots on the center circle) are turned on. Its corresponding array factor is shown in Figure 5.21.

The Wullenwever array was invented during World War II by Dr. Hans Rindfleisch [9]. The name Wullenwever was a World War II German cover name for the antenna and not the name of a person. The first one, which was built in Skisby, Denmark, used 40 vertical radiator elements, placed on a 120-m-diameter circle with 40 reflecting elements installed behind the radiator

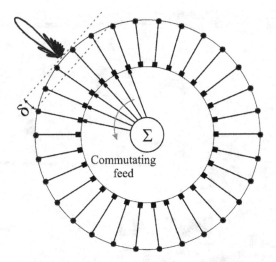

Figure 5.20. Diagram of a Wullenweber array.

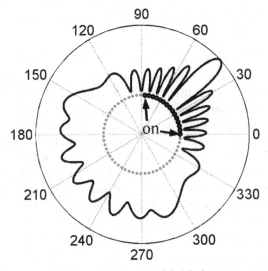

Figure 5.21. A 60-element Wullenweber array with 15 elements turned on (large dark dots on center circle) and its array factor.

elements around a circle having a diameter of 112.5 m. After the war, the Soviet Union imported the technology and built many Wullenwevers for HF direction finding, which included tracking Sputnik. The United States became interested in the technology in the late 1950s and early 1960s. At some point, the Americans changed the name from Wullenwever to Wullenweber. Its immense size and huge circular reflecting screen inspired such names as

Figure 5.22. AN/FLR-9 Wullenweber antenna array near Augsburg, Germany.

"elephant cage" and "dinosaur cage." The University of Illinois built a large Wullenweber array with 120 monopoles over 2 to 20 MHz. Tall wooden poles supported a 1000-ft-diameter circular screen of vertical wires located within the ring of monopoles. In 1959 the Navy began construction of the AN/FRD-10 HF/DF arrays which were based on the experimental investigations at the University of Illinois. The FRD-10 had two concentric rings: The high-frequency array was 260 m in diameter with 120 sleeve monopoles, while the low-frequency array was 230 m in diameter with 40 folded dipoles. Inside each ring there was also a large wire screen, supported by 80 towers. AN/FLR-9 Wullenweber antenna array near Augsburg, Germany is shown in Figure 5.22.

5.2.1. Circular Adcock Array

An Adcock array [10] with many elements arranged in a circle [11] was introduced in Chapter 4. Table 5.1 shows how the signals from adjacent elements combine to form four beams. Opposite beams form difference patterns as with the standard 4-element Adcock array. Subtract beam 3 from beam 1 to get

$$e^{jkr\cos\phi} + e^{j.707kr(\cos\phi+\sin\phi)} - e^{-jkr\cos\phi} - e^{-j.707kr(\cos\phi+\sin\phi)}$$
$$= 2j\sin(kr\cos\phi) + 2j\sin[0.707kr(\cos\phi+\sin\phi)] \tag{5.9}$$

Subtract beam 4 from beam 2 to get

TABLE 5.1. Element Signals and Beams Formed by the 8-Element Adcock Array

Element	Signal	Sum Adjacent Pairs	Beam
1	$e^{jkr\cos\phi}$	$e^{jkr\cos\phi} + e^{j.707kr(\cos\phi+\sin\phi)}$	1
2	$e^{j.707kr(\cos\phi+\sin\phi)}$		
3	$e^{jkr\sin\phi}$	$e^{jkr\sin\phi} + e^{j.707kr(-\cos\phi+\sin\phi)}$	2
4	$e^{j.707kr(-\cos\phi+\sin\phi)}$		
5	$e^{-jkr\cos\phi}$	$e^{-jkr\cos\phi} + e^{-j.707kr(\cos\phi+\sin\phi)}$	3
6	$e^{-j.707kr(\cos\phi+\sin\phi)}$		
7	$e^{-jkr\sin\phi}$	$e^{-jkr\sin\phi} + e^{j.707kr(\cos\phi-\sin\phi)}$	4
8	$e^{j.707kr(\cos\phi-\sin\phi)}$		

$$e^{jkr\sin\phi} + e^{j.707kr(-\cos\phi+\sin\phi)} - e^{-jkr\sin\phi} - e^{-j.707kr(-\cos\phi+\sin\phi)}$$
$$= 2j\sin(kr\sin\phi) + 2j\sin[0.707kr(-\cos\phi+\sin\phi)] \quad (5.10)$$

The phase at element n is given by

$$e^{jk(x_n\cos\phi+y_n\sin\phi)} = e^{jk(r\cos\phi_n\cos\phi+r\sin\phi_n\sin\phi)} \quad (5.11)$$

An omnidirectional pattern results from the sum of the signals from all the elements

$$AF_{sum} = \sum_{n=1}^{N} e^{jkr(\cos\phi_n\cos\phi+\sin\phi_n\sin\phi)} \quad (5.12)$$

which for this array is

$$e^{jkr\cos\phi} + e^{j.707kr(\cos\phi+\sin\phi)} + e^{jkr\sin\phi} + e^{j.707kr(-\cos\phi+\sin\phi)} + e^{-jkr\cos\phi} + e^{-j.707kr(\cos\phi+\sin\phi)}$$
$$+ e^{-jkr\sin\phi} + e^{j.707kr(\cos\phi-\sin\phi)} = 2\cos(kr\cos\phi) + 2\cos(kr\sin\phi)$$
$$+ 2e^{j.707kr\sin\phi}\cos(0.707kr\cos\phi) + 2e^{-j.707kr\sin\phi}\cos(0.707kr\cos\phi)$$
$$= 2\cos(kr\cos\phi) + 2\cos(kr\sin\phi) + 4\cos(0.707kr\sin\phi)\cos(0.707kr\cos\phi)$$

$$(5.13)$$

Example. Design an 8-element Adcock array that works at 1 GHz (Figure 5.23). Show the effects of increasing frequency.

Start with a circular array that has a radius of 7.5 cm. The sum, difference, and omni patterns at 1 GHz are shown in Figure 5.24. Increasing the frequency to 1.5-GHz results in the patterns shown in Figure 5.25. At this point, the diameter of the circle is 0.75λ and sidelobes are beginning to form. The effect of frequency on the direction capability is demonstrated in Figure 5.26. At the design frequency of 1 GHz, the calculated and actual directions are very close.

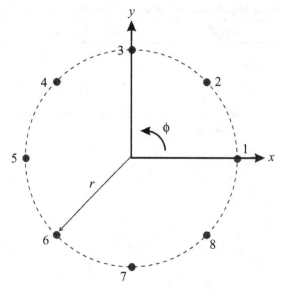

Figure 5.23. Diagram of an 8-element Adcock array.

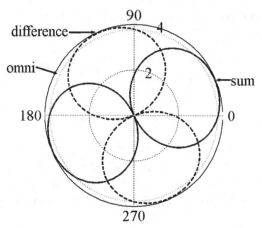

Figure 5.24. Sum (solid), difference (dashed), and omni- (dotted) array factors for the 8-element Adcock array at 1 GHz.

There are small deviations at 1.5 GHz. At 2.0 GHz the error is so large that the array can only tell if the signal comes from 0°, 90°, 180°, or 270°.

Figure 5.27 is an example of a linear array of z-directed dipoles wrapped around a cylinder that is a perfect electrical conductor (PEC). The element field of view extends $\pm(90° + \gamma)$, where $\gamma = \cos^{-1}(r_c/(r_c + h))$. Consider an array

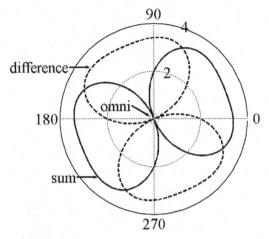

Figure 5.25. Sum (solid), difference (dashed), and omni- (dotted) array factors for the 8-element Adcock array at 1.5 GHz.

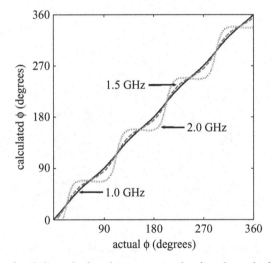

Figure 5.26. Graph of the calculated versus actual azimuth angle for the 8-element Adcock array at three different frequencies.

located in the x–y plane with half wavelength dipoles placed $h = \lambda/4$ away from the perfectly conducting cylinder that is 2λ in diameter and 2λ tall (Figure 5.28). Elements are spaced $\lambda/2$ apart in arc length. The field of view for an element in this array is 234.5°. An element can see up to four other elements to its left or right, while the rest are not in its line of sight. Figure 5.29 are the directivity plots for phase compensated arrays with 1, 3, and 5 elements. A single element has a directivity of 7.1 dB, while 3 and 5 elements are 10.9 dB

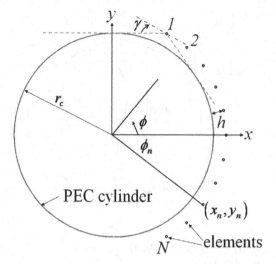

Figure 5.27. Top view of dipole array with cylindrical ground plane.

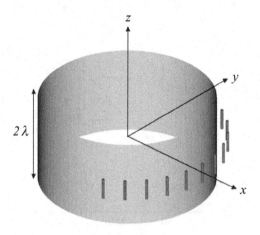

Figure 5.28. Three-dimensional diagram of dipole array conformal to a perfectly conducting cylinder.

and 13.1 dB, respectively. Blockage is not an issue for these small arrays. Increasing the number of elements to 11 does block the elements on one end of the array from "seeing" the elements on the other end. If the 11-element array is not phase compensated, then the directivity is 7.0 dB. Compensation improved the directivity to 16.2 dB as shown in Figure 5.30. Letting $\phi_s = 60°$ and applying the phase shift from (5.4) results in the array pattern in Figure 5.31. The actual peak of the beam is at $\phi = 58°$, where the directivity is 14.4 dB. The directivity at $\phi = 60°$ is 14.16 dB.

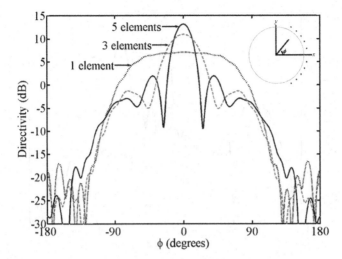

Figure 5.29. Directivity of an array of dipoles on a perfectly conducting cylinder with $N = 1, 3$, and 5 elements.

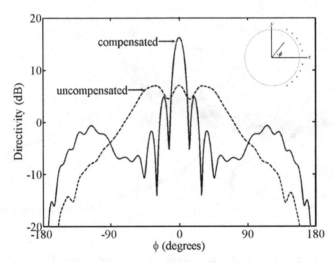

Figure 5.30. Directivity of an 11-element array of dipoles on a perfectly conducting cylinder when phase is compensated and uncompensated.

Figure 5.32 shows an array of spirals wrapped around the center of an ellipsoid-shaped airship [12]. This array operates from 800 to 1200 MHz. The elements are two-arm, self-complementary Archimedean spirals, 12.5 cm in diameter. The microstrip spiral element was etched on a 0.8-mm-thick, $\varepsilon_r = 3.38$ substrate. The spiral is unbalanced and has an input impedance of 188 Ω, so a surface mount transformer is used as a balun and a match to the 50-Ω coax.

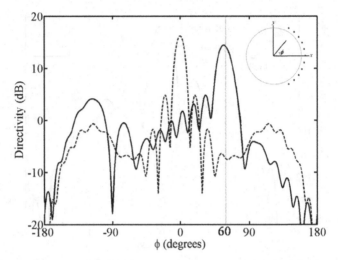

Figure 5.31. Compensated main beam steered to $\phi_s = 0°$ (dashed line) and $\phi_s = 60°$ (solid line).

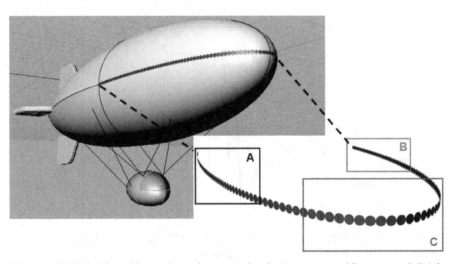

Figure 5.32. Airship with conformal array of spiral antennas. (Courtesy of Régis Guinvarc'h, SONDRA, Supelec.)

In this case, the curvature is not uniform but is a function of the location on the ellipsoid surface. Figure 5.33 shows the measured and computed uncompensated array patterns of an 8-element array on the front of the airship with a 20° radius of curvature. Figure 5.34 shows the measured and computed uncompensated array patterns of an 8-element array when the side of the airship has a 5° radius of curvature. The pattern of the side array has a distinct main beam, while the front array does not.

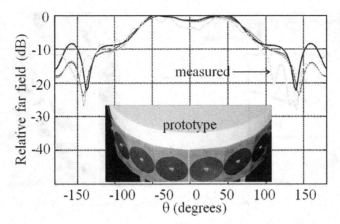

Figure 5.33. Calculated and measured antenna patterns for a 20° radius of curvature (front). (Courtesy of Régis Guinvarc'h, SONDRA, Supelec.)

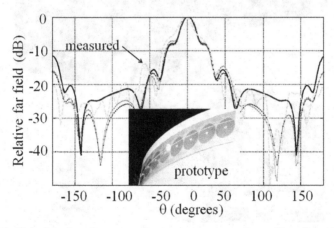

Figure 5.34. Calculated and measured antenna patterns for a 5° radius of curvature (side). (Courtesy of Régis Guinvarc'h, SONDRA, Supelec.)

AIRLINK® is an airborne antenna system for in-flight telephone, fax, and data transmission that uses the INMARSAT (International Maritime Satellite) system of geostationary satellites. The antenna array is very thin and curved (Figure 5.35) to conform to the outside of an airplane. Figure 5.36 shows the array being installed on the outside of an airplane. The array operates at 1530 to 1559 MHz on receive and 1626.5 to 1660.5 MHz on transmit. It has a gain of greater than 12 dB with a VSWR of less than 1.5. Figure 5.37 is a picture of the Airlink array with rectangular microstrip elements that are spaced half

Figure 5.35. Front and side views of the Airlink antenna. (Courtesy of Ball Aerospace & Technologies Corp.)

Figure 5.36. Airlink antenna being installed on an airplane (Courtesy of Ball Aerospace & Technologies Corp.)

a wavelength (9.27 cm) apart on a triangular grid. The array is capable of scanning ±60° in azimuth and elevation.

Figure 5.38 is an artist concept of a multilayer conformal array antenna showing installation on curved surface. A thin dielectric covering protects the crossed bow-tie elements from the environment. The output from the element is fed to a phase shifter mounted below the surface. Low-dielectric-constant foam supports the elements above the ground plane. Elements are placed in subarrays as shown by the small rectangular portions of the array.

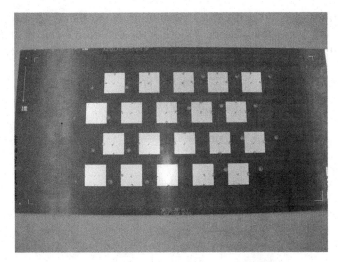

Figure 5.37. Airlink array elements. (Courtesy of Ball Aerospace & Technologies Corp.)

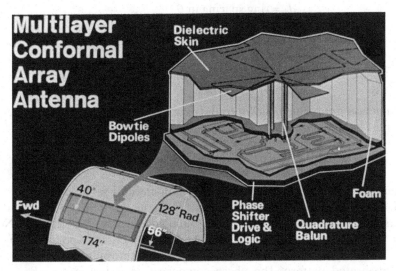

Figure 5.38. Multilayer conformal array antenna. (Courtesy of the National Electronics Museum.)

5.3. ARRAYS CONFORMAL TO DOUBLY CURVED SURFACES

Arrays that conform to doubly curved surfaces are relatively rare, because they are difficult to build. A doubly curved surface has curvature in two orthogonal directions. The spherical array is the most common type of conformal array on a doubly curved surface. A spherical array of point sources has an array factor given by (5.1). Assume that the elements are placed in rings that are conformal to the sphere with a center element at (0, 0, 0). Element m in ring n on a sphere of radius r are located at (x_{mn}, y_{mn}, z_{mn}), where

$$x_{mn} = r_n \cos\left(\frac{2\pi(m-1)}{N_n}\right)$$

$$y_{mn} = r_n \sin\left(\frac{2\pi(m-1)}{N_n}\right)$$

$$z_{mn} = r_s \cos\left(\frac{nd_\theta}{r_s}\right)$$

$$r_n = r_s \sin\left(\frac{nd_\theta}{r_s}\right)$$

$$N_n = floor\left(\frac{2\pi r_n}{d_{\phi\text{desired}}}\right)$$

$$d_\theta = \text{ring spacing in } \theta$$

$$d_{\phi\text{desired}} = \text{desired ring spacing in } \phi$$

$$d_{\phi n} = \frac{2\pi}{N_n} \geq d_{\phi\text{desired}}$$

The function *floor* rounds down to the nearest integer. The phase compensation must be applied to the elements in the form

$$\delta_{mn} = -k\left(x_{mn}\sin\theta_s\cos\phi_s + y_{mn}\sin\theta_s\sin\phi_s + z_{mn}\cos\theta_s\right) \quad (5.14)$$

Example. Plot the uncompensated and compensated far-field patterns of a 5-ring array placed on a sphere of radius 4λ. Assume $d_\theta = 0.5\lambda$ and $d_{\phi\text{desired}} = 0.5\lambda$.

First, the values of d_ϕ, N_n, and r_n are found and displayed in Table 5.2. Next, the element locations are calculated and shown on the sphere in Figure 5.39. The array factor is calculated using (5.1) and shown in Figure 5.39 as the uncompensated pattern. The phase compensation is just $\delta_{mn} = -kz_{mn}$. The compensated pattern is shown in Figure 5.39.

TABLE 5.2. List of Spherical Array Variables

Ring	1	2	3	4	5
d_ϕ	0.52λ	0.52λ	0.51λ	0.50λ	0.51λ
N_n	6	12	18	24	29
r_n	0.5λ	0.99λ	1.47λ	1.92λ	2.34λ

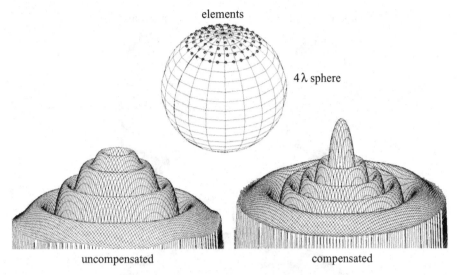

Figure 5.39. Uncompensated and compensated array factors for the 5-ring array paced on a sphere.

The previous example used isotropic point sources, with no mutual coupling, no ground plane, no element patterns, and no blockage. A more realistic example includes all these factors. Assume a spherical array of dipoles that are 0.48λ long and 0.25λ above a perfectly conducting partial sphere that has a radius of 4λ. Figure 5.40 shows a 3-ring array above a $30°$ sphere section with its uncompensated and compensated far-field patterns. The partial sphere angle is defined as the angle between the radius to the center of the partial sphere and the radius to the edge of the partial sphere. The uncompensated gain of 17.7 dB increases to 20 dB with phase compensation. Increasing the size of the partial sphere to $45°$ produces little change to the compensated pattern that now has a gain of 20.3 dB, but decreases the uncompensated pattern gain to 14.7 dB (Figure 5.41). Increasing the number of rings on the $45°$ sphere to 5 results in the array with uncompensated and compensated patterns shown in Figure 5.42. The gain of the uncompensated pattern is 12.8 dB, and the gain of the compensated pattern is 23.4 dB. As more rings are added, the main beam of the uncompensated pattern gets worse because the z distance of each additional ring gets larger. Steering the beam from $0°$ to $60°$ in $15°$ increments is shown in Figure 5.43 for $\phi = 0$. The gain decreases from 23.4 dB at $0°$ to 20.9 dB at $60°$.

As with a circular array, usually only a subsection of a spherical array is active at a time. The feed network dynamically connects to the appropriate elements to form an aperture on a portion of the sphere that faces the direction of the desired main beam location. The active elements on the sphere are contained within a cone having an apex at the origin, a cone half-angle of α_{max}, and the main beam pointing perpendicular to its base. Figure 5.44 shows the

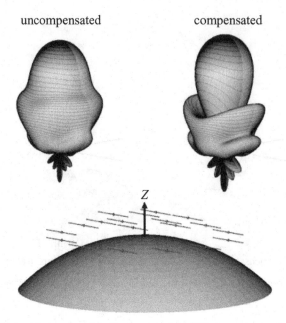

Figure 5.40. Three-ring array of dipoles placed $\lambda/4$ above a perfectly conducting portion of a 30° sphere with uncompensated and compensated far-field patterns.

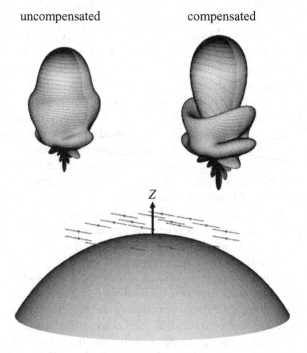

Figure 5.41. Three-ring array of dipoles placed $\lambda/4$ above a perfectly conducting portion of a 45° sphere with uncompensated and compensated far-field patterns.

uncompensated   compensated

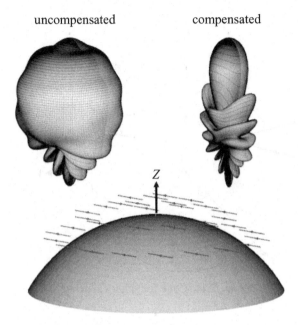

Figure 5.42. Five-ring array of dipoles placed $\lambda/4$ above a perfectly conducting portion of a 45° sphere with uncompensated and compensated far-field patterns.

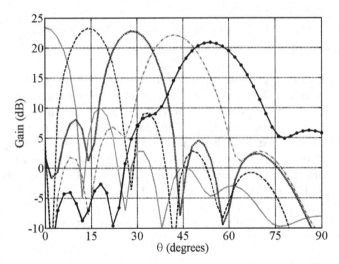

Figure 5.43. Array pattern in Figure 5.42 steered to $\phi = 0°$ and $\theta = 0°$ to 60° in 15° increments.

circular boundary of active elements on the sphere. The arrow from the origin through the center of the base points in the direction of the main beam. Thus, the array has a circular boundary. This implementation has many advantages over a multifaced array [13]:

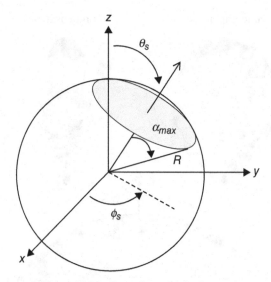

Figure 5.44. Cone for spherical array.

1. Lower polarization losses.
2. Lower mismatch losses.
3. No beam shift versus frequency (implies wider bandwidth).
4. No gain loss with beam steering.

Planar arrays usually cannot scan beyond about 60°, giving spherical arrays a significant advantage. The radiating elements in the active sector are assumed to have symmetry in the plane perpendicular to the axis of the antenna beam, and consequently a spherical array can be used to cover the hemisphere with practically identical beams. Thus, in contrast to other array configurations which suffer from beam degradation as the beam is steered over wide angular regions, spherical arrays provide uniform patterns and gain over the entire field of view. Fabrication and assembly of the curved radiators and feed network is challenging.

The gain of the spherical array is based on its projected aperture shown in Figure 5.45.

$$G = \frac{4\pi A'}{\lambda^2} \qquad (5.15)$$

where the projected area of active sector is

$$A' = \pi R^2 \sin^2 \alpha_{max} \qquad (5.16)$$

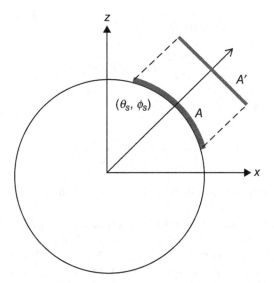

Figure 5.45. Projected aperture of spherical array.

Substituting (5.16) into (5.15) results in [14]

$$G = (kR)^2 \sin^2 \alpha_{max} \tag{5.17}$$

The polarization efficiency of a spherical array is a function of the size of the active area on the sphere [14].

$$p = \frac{2\left[2^{2/3} - \left(2 - \sin^2 \alpha_{max} \right)^{3/2} \right]^2}{9 \sin^4 \alpha_{max}} \tag{5.18}$$

while the planar array polarization efficiency is a function of the scan angle [14].

$$p = \frac{1 + \cos^2 \theta_s}{2} \tag{5.19}$$

Example. Compare the polarization efficiencies of a planar array scanned to 60° and a spherical array with $\alpha_{max} = 60°$.
The planar array has $p = 0.625$ and the spherical array has p = 0.809.

The spherical array bandwidth criterion requires the phase difference between the center element and an edge element is less than $\pi/2$ as the frequency changes from f_0 to f [14].

$$\left| k_0 R (1 - \cos \alpha_{max}) - k R (1 - \cos \alpha_{max}) \right| = \frac{\pi}{2}$$

$$\frac{2\pi R}{c} |f_0 - f| (1 - \cos \alpha_{max}) = \frac{\pi}{2}$$

(5.20)

If $\Delta f = 2 |f_0 - f|$, then (5.20) is written as

$$\frac{\Delta f}{f_0} = \frac{\pi}{k_0 R (1 - \cos \alpha_{max})}$$

(5.21)

This criterion gives about 1-dB loss at the edge of the band.

Example. Compare the bandwidth of a planar array scanned to 60° and a spherical array with $\alpha_{max} = 60°$, $R = 5\,\mathrm{m}$, and $f = 2\,\mathrm{GHz}$.

Substituting into (5.21) produces $\Delta f / f_0 = 3\%$ or $\Delta f = 60\,\mathrm{MHz}$. If the circular planar array has a radius of $R \sin \alpha_{max} = 4.33\,\mathrm{m}$, then its beamwidth is 1° with a bandwidth of 1%.

In order to overcome the disadvantages of a spherical array, the sphere approximates a polyhedron (a solid figure with four or more faces). Cheaper planar array technology is then used for panels on the flat faces while the array still approximates a spherical shape. A geodesic sphere is a skeleton of a sphere built using only straight supports that form polygons (most commonly triangles). Figure 5.46 shows two examples of geodesic spheres. If only part of a geodesic sphere exists, then it is known as a geodesic dome. The one on the left consists of pentagons and hexagons, while the one on the right is made from triangles. A geodesic dome is mechanically very stable and supports itself without needing internal columns or interior load-bearing walls.

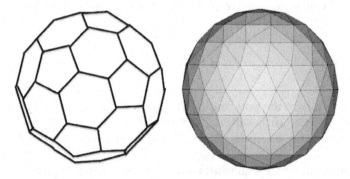

Figure 5.46. Geodesic spheres are made from polygons. The one on the left is made from hexagons and pentagons while the one on the right is made from triangles.

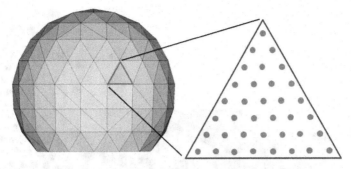

Figure 5.47. Geodesic dome array made from triangular planar arrays with 36 elements in a triangular lattice.

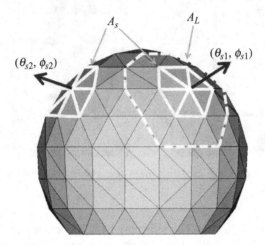

Figure 5.48. Active sector of the array depends upon the beam pointing direction.

An example of a geodesic dome array built from triangles is shown in Figure 5.47. Each triangular section is a panel containing 36 elements on a triangular grid. This array scans horizon to horizon by activating the appropriate subarrays to form an active aperture that points in the desired direction [15]. Each array element or subarray must have an RF on/off switch. When the subarray is active, then the switch is on and the element phase shifter is adjusted to form a coherent beam in the desired direction. Figure 5.48 demonstrates the concept of an active region that is a function of the beam pointing direction. The size of the active section depends upon α_{max}. For instance, the number of triangles in the active region may be 6 for a small α_{max} or the region enclosed by the dashed line in Figure 5.48 for a large α_{max} when the beam pointing direction is (θ_{s1}, ϕ_{s1}). Moving the beam to (θ_{s2}, ϕ_{s2}) moves the active region to the other side of the sphere (Figure 5.48). Beam scanning is accomplished by activating different sectors of the dome. A representative triangular

Figure 5.49. Subarray panel in a near-field test range. (Courtesy of Boris Tomasic, USAF AFRL [15].)

section was built and tested. The subarray has 78 circular patch elements in a triangular grid (Figure 5.49). The antenna pattern was measured in the near field and compared well with the predicted pattern (Figure 5.50).

In order to make the manufacture and testing of the geodesic dome array as cheap as possible, all the polygons composing the surface should be of identical size. It turns out that there are only five types of regular polyhedron. Their characteristics are shown in Table 5.3. The more faces there are on a polyhedron, the more closely it approximates a sphere. As such, the icosahedron (composed of 20 equilateral triangles) seems the most likely candidate.

The 10-m-diameter Geodesic Dome Phased Array Antenna (GDPAA) Advanced Technology Demonstration (ATD) antenna shown in Figure 5.51 was built by Ball Aerospace & Technologies Corp. This array was developed to demonstrate multi-beam phased array capability for next generation Air Force Satellite Control Network stations [17]. Hexagonal and pentagonal panels in the geodesic dome contain many hexagonal subarrays. The hexagon subarrays have 37 elements (Figure 5.52) in a triangular grid. Only 36 of the elements are active, because the center element is devoted to calibration. The GDPAA tracks multiple satellites using multiple beams as shown in Figure

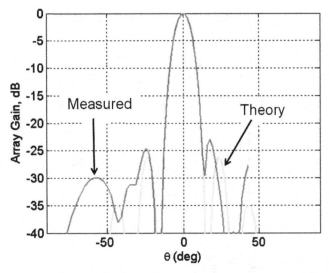

Figure 5.50. Measured and computed subarray patterns. (Courtesy of Boris Tomasic, USAF AFRL [15].)

5.53. A very sophisticated beam management algorithm was developed to allow active section to overlap and cross while multiple beams simultaneously track multiple targets [15].

5.4. DISTRIBUTED ARRAY BEAMFORMING

A distributed array collects signals from antennas at various locations and coherently combines them to form a beam. The signal combination occurs in real time or much later as is often done with radio telescopes. Data collected by each element is precisely time stamped at the element which has a precisely known location. The strict timing and location accuracy requires constant calibration, otherwise the phase errors would be too large to form a desirable coherent main beam. Distributed arrays are typically very sparse, so grating lobes are a problem that must be addressed through signal processing, element placement, and limited scanning.

Consider the fictional Very Long Baseline Interferometry (VLBI) array shown on the map of the United States in Figure 5.54. Such an antenna might be used in radio astronomy to form a huge aperture. The large distances between the reflector antenna elements make it impossible to use transmission lines to coherently combine signals to form a beam at the data collection site. Small element location errors produce major antenna pattern distortions described in Chapter 2. In order to determine the element signal phases, the data assembled by each element are tagged with the time from atomic clocks

TABLE 5.3. Characteristics of Regular Polyhedron [16]

	Tetrahedron	Hexahedron	Octahedron	Dodecahedron	Icosahedron
Composition:	(4 Equilateral Triangles)	(6 Squares)	(8 Equilateral Triangles)	(12 rectangular Pentagons)	(20 Equilateral Triangles)
Total Area:	$1.73e^2$	$6e^2$	$3.46e^2$	$20.65e^2$	$8.66e^2$
Volume:	$0.12e^2$	e^2	$0.47e^2$	$7.66e^2$	$2.18e^2$
Number of Verticies:	4	8	6	20	12
Number of Edges:	6	12	12	30	30
Number of Faces:	4	6	8	12	20
Edge Length, e:	$1.63R$	$1.16R$	$1.41R$	$0.71R$	$1.05R$
Radius of Inscribed Sphere:	$0.33R$	$0.58R$	$0.58R$	$0.80R$	$0.80R$
Face Area:	$0.43e^2$	e^2	$0.43e^2$	$1.72e^2$	$0.43e^2$
Radius of Circumscribed Sphere, R:	$0.61e$	$0.87e$	$0.71e$	$e\sqrt{15+\sqrt{3}}/4$	$e\sqrt{10+2\sqrt{5}}/4$

Figure 5.51. GDPAA ATD antenna. (Courtesy of Ball Aerospace & Technologies Corp.)

Figure 5.52. Hexagonal subarray for the GDPAA. (Courtesy of Ball Aerospace & Technologies Corp.)

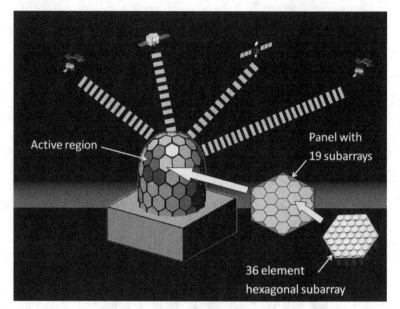

Figure 5.53. Operational concept of the GDPAA. (Courtesy of USAF AFRL.)

Figure 5.54. Fictional VLBI array in the United States.

along with a GPS time stamp. The data are collected at a central location (shown by a star on the map). The timing of the data at each element is adjusted according to the atomic clock time stamp, and the estimated times of arrival of the signal at the elements. Each antenna is a different distance

Figure 5.55. ALMA concept [(copyright © ALMA (ESO/NAOJ/NRAO)].

from the radio source, so an artificial time delay is added to the appropriate elements. If the position of the antennas is not known to sufficient accuracy or atmospheric effects are significant, fine adjustments to the delays in the signal processing must be made until a coherent beam is formed.

The VLA presented in Chapter 4 is one example of a VLBI array. Another example is the new VLBI array called the Atacama Large Millimeter/submillimeter Array, or ALMA [18]. It is an international collaboration between Europe, North America, and East Asia, with the Republic of Chile, to build and operate a radio telescope composed of 66 12-m and 7-m parabolic dishes in Chile's Andes Mountains (Figure 5.55), at 5000 m above sea level. The telescope operates at frequencies from 31.25 GHz to 950 GHz. Array configurations from 250 m to 15 km will be possible. The ALMA antennas will be movable between flat concrete slabs built on the site. Figure 5.56 is a picture of one of the vehicles moving an antenna. Coherent beams can be formed in real time through the use of optical fiber feeds to a local data collection center.

5.5. TIME-VARYING ARRAYS

The properties of an array may quickly change with time through purposely time modulating array parameters or through the array element positions changing with time. In effect, the three-dimensional spatial problem becomes a four-dimensional space–time problem [19].

5.5.1. Synthetic Apertures

A synthetic aperture is a large aperture built over time by moving a smaller antenna and taking snapshots along the path. Radio telescopes frequently use

Figure 5.56. Vehicle used to move an ALMA prototype element [(copyright © ALMA (ESO/NAOJ/NRAO)].

the rotation of the earth to fill in an aperture. For instance, the radio telescopes at A1 and A2 in Figure 5.57 move to A1′ and A2′ then to A1″ and A2″ as the earth turns. Continually adding the snapshots together using proper phasing forms an image with much higher resolution than is only A1 and A2 were used alone.

A synthetic aperture radar SAR is usually mounted on an airplane or satellite that passes over relatively immobile targets. The length of the antenna collection determines the resolution in the azimuth (along-track) direction of the image. If the antennas collect data over time, T, when the platform is moving at velocity, v, then the antenna is vT long, and the virtual beamwidth (resolution) of the antenna is approximately λ/vT. Figure 5.58 shows a cartoon of the space shuttle with a SAR. The footprint on the ground is one element pattern in the synthetic array. Pulses are transmitted as the radar moves relative to the ground. The returned echoes are Doppler-shifted (negatively as the radar approaches a target; positively as it moves away). Comparing the Doppler-shifted frequencies to a reference frequency allows many returned signals to be "focused" on a single point, effectively increasing the length of the virtual antenna [20].

Radar signals received by the SAR depend on the polarization and look angle of the antenna as well as the variations in shape and materials in the landscape [21]. SAR arrays frequently have dual polarized elements, so the array can transmit vertical and receive vertical or horizontal, or transmit

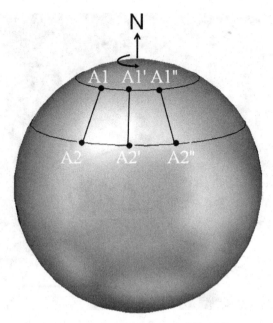

Figure 5.57. Synthetic aperture built using two radio telescopes at A1 and A2.

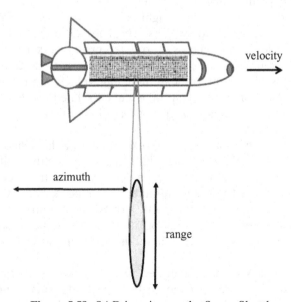

Figure 5.58. SAR imaging on the Space Shuttle.

Figure 5.59. SIR-B antenna before unfolding during deployment in space. (Courtesy of Ball Aerospace & Technologies Corp.)

horizontal and receive vertical or horizontal. Using these four polarization combinations produces different "views" of the terrain, because scattering is a function of polarization and look angle.

SAR can detect changes (that might result from motion) in sequential images. When the pixels of two images of the same scene, but taken at different times, are aligned and then cross-correlated pixel by pixel, then the pixels that change between images are uncorrelated and easy to detect. Alterations in the landscape, such as subterranean tunneling, vehicle motion, glacier flow, earthquakes, and volcanic activity, produce uncorrelated pixels in the before and after images. The motion of a ground-based moving target such as a car is also detectable.

The Shuttle Imaging Radar B (SIR-B) experiment was launched on October 5, 1984, onboard the shuttle Challenger [22]. SIR-B collected 15 million square kilometers of ocean and land surface data during a 16-hour period. The radar was mechanically tilted to obtain multiple radar images of a given target at different angles during successive shuttle orbits. The antenna array was put together in subarray panels that fold together for storage in the space shuttle as shown in Figure 5.59. Figure 5.60 shows the array unfolding to its full aperture size outside the space shuttle.

Figure 5.61 is a SAR image centered on Salt Lake City, Utah [22]. This image was acquired by the Spaceborne Imaging Radar-C/X-Band Synthetic Aperture Radar (SIR-C/X-SAR) aboard the space shuttle Endeavour on

Figure 5.60. SIR-B antenna array unfolding for deployment on the space shuttle. (Courtesy of Ball Aerospace & Technologies Corp.)

April 10, 1994. The dark area is the Great Salt Lake. It is dark, because the surface is relatively smooth and the reflected signal obeys Snell's law, so it bounces away from the radar. Vegetation, buildings, and mountains are rough surfaces and scatter in all directions. Consequently, the SAR receives signals from these areas and produce bright spots on the image for specular reflections.

The Spaceborne Imaging Radar-C/X-band Synthetic Aperture Radar (SIR-C/X-SAR) fit inside the Space Shuttle's cargo bay as shown in Figure 5.62. Its mission was to make measurements of vegetation type, extent, and deforestation; soil moisture content; ocean dynamics, wave and surface wind speeds and directions; volcanism and tectonic activity; and soil erosion and desertification [22]. The SIR-C/X-SAR had antennas operating at 1.3 GHz, 5.2 GHz, and 10 GHz (Figure 5.63). The L-band and C-band phased arrays have both horizontal and vertical polarizations. The X-SAR array is a 12-m × 0.4-m X-band vertically polarized slotted waveguide array, which uses a mechanical tilt to change the beam-pointing direction between 15° and 60°. The beam electronically scans in the range direction ±23° from the nominal 40° off nadir position. Both SIR-C and X-SAR operate as either stand-alone radars or together. Roll and yaw maneuvers of the shuttle allow data to be acquired on either side of the shuttle nadir (ground) track. The width of the imaged swath

Figure 5.61. SAR image of Salt Lake City, UT. (Courtesy of NASA JPL.)

on the ground varies from 15 to 90 km, depending on the orientation of the antenna beams and the operational mode.

5.5.2. Time-Modulated Arrays

In the time domain, the array factor for a linear array can be written as

$$\text{AF} = \sum_{n=1}^{N} w_n e^{j[\omega t + k(n-1)du]} \tag{5.22}$$

Time-modulated arrays open and close switches in a prescribed sequence that causes the average array pattern to simulate a low sidelobe array amplitude taper. When the switch is on, the element connects to the feed network. If the switch is off, the element contributes no signal to the output. The switch on time, τ_n, determines the average element weights. Each switch rapidly turns

Figure 5.62. SIR-C concept. (Courtesy of Ball Aerospace & Technologies Corp.)

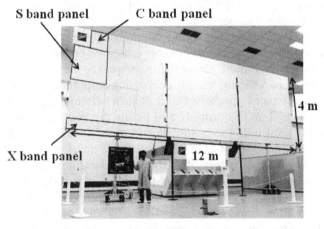

Figure 5.63. Photograph of the SIR-C antenna. (Courtesy of Ball Aerospace & Technologies Corp.)

on and off in synchronization to an external periodic signal having a period of T. An element whose switch is always on has a weight of 1, while an element whose switch is always off has a weight of 0. The amplitude weight on element n is the fraction of the period that a switch is on.

$$w_n = \frac{\tau_n}{T} \tag{5.23}$$

The switches cannot control the phase of the signal.

The new time-varying weights result in the following time dependent array factor:

$$AF = \sum_{m=-\infty}^{\infty} e^{j[(\omega - m\omega_0)t]} \sum_{n=1}^{N} w_{mn} e^{jk(n-1)du} \tag{5.24}$$

where

$$w_{mn} = \frac{1}{T} \int_0^T w_n(t) e^{-jm\omega_0 t} dt$$

$$\omega_0 = \frac{2\pi}{T}$$

This equation implies that the array factor not only exists at the fundamental frequency, ω_0, but array factors also exist at all the harmonics of the switching frequency.

Time modulation was demonstrated using an X-band experimental corporate-fed array of eight collinear slots that had three diode switches [23]. The center two elements were always on while each of the three switches controlled a pair of elements located symmetrically about the array center. Isolators on either end of the switch prevented impedance changes due to turning the switch on and off. The switching sequence begins with all the array elements turned on. A monotonically decreasing amplitude taper results when the outer pair (element numbers 1 and 8) turn off, then the next outer pair (element numbers 2 and 7) turn off, and finally the next (element numbers 3 and 6) turn off until only the center pair (4 and 5) remain on. The switches operated at a frequency of 10 kHz. The various on times were selected to give the correct low-sidelobe pattern after the signal was filtered.

The time-modulated low-sidelobe pattern gain is less than the gain of an equivalent static (no switching) amplitude tapered array due to the power lost through the additional array factors at the harmonics of the modulation frequency [24]. Expressions for the power loss are found in reference 24. A different array pattern exists at each harmonic. As an example, an array with a time-modulated 40-dB sidelobe taper has 3.53 dB less gain than a 40-dB sidelobe static amplitude tapered array. If the static array has an initial low-sidelobe amplitude taper, then the time modulation will be more efficient. For instance, if an array with a static 30-dB amplitude taper is lowered to 40 dB through time modulation, then it has a gain that is 0.51 dB less than that of a 40-dB sidelobe static amplitude tapered array.

5.5.3. Time-Varying Array Element Positions

Some communications and sensing applications need high-gain arrays in space. Element positions in these arrays vary when the surface of the array flexes or individual elements on independent platforms change position. The support structure for a large array on the ground is made with steel and concrete. In space, however, the lightweight materials used in construction cannot provide the same solid support. Even though strong structural supports can be made viable for space, vibrations induced by antenna deployment, temperature differences, orbit changes, and so on, cause the array aperture to undulate. Small vibrations at high antenna frequencies are disastrous for the main-beam gain. Element displacements add unwanted phase to the signals received at each element. Without proper compensation, these signals do not add coherently to form a main beam, resulting in antenna performance degradation.

Example. A low-sidelobe 50-element linear array with element spacing of 0.5λ has a sinusoidal deflection with an amplitude of $\lambda/2$ as shown in Figure 5.64. Plot the array factors.

If the array is not phase compensated, then the main beam is severely distorted as shown in Figure 5.65. Phase compensation restores the array factor to near perfection if the element locations are known.

Elements of an array that are on different moving platforms are another type of time-varying aperture. For instance, antenna elements may be on individual unmanned aerial vehicles flying in formation. Beamforming consists of a wireless network that collects and coherently adds the signals from all the elements.

The Techsat 21 space-based radar [25] is a proposed cluster of microsatellites that cooperate to form one large virtual satellite (Figure 5.66). Each satellite has roughly a 4-m² phased array, operating at 10 GHz. There are 4 to 20 satellites per cluster with cluster diameters of 100–1000 m. Orbits vary from 700 to 1000 km above earth. Satellites in the cluster communicate with each other and share the processing, communications, and mission functions. The benefits of this approach include [26]:

1. The ability to create extremely large apertures.
2. Increased system reliability, because the failure of a single satellite does not result in complete system failure.

Figure 5.64. Small $\lambda/2$ deflection in a linear array of 50 point sources.

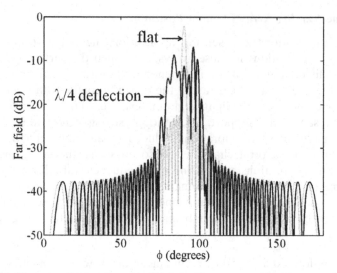

Figure 5.65. The array factor due a to $\lambda/2$ deflection compared to an array factor with no deflection.

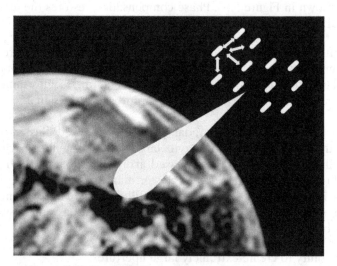

Figure 5.66. Proposed Techsat 21 configuration in orbit.

3. Easier upgrades to the satellites, since the satellites can be replaced one at a time instead of all at once.
4. Low costs due to satellite mass production and small launch payloads.
5. Improved survivability to natural and man-made threats due to the separation between satellites.

6. The ability to reconfigure and optimize the satellite formation for a given mission, enhanced survivability, and increased reliability.

Each satellite transmits a different frequency or other orthogonal signal, but receives all the signals from the other satellites in the formation. Although the absolute position of a satellite must be accurately known, the relative positions between satellites have the stringent requirement of a fraction of a wavelength in order to coherently sum the signals to form a beam. As shown in Chapter 2, the random element position errors determine the gain of the array.

GPS provides absolute timing to 100 ns. Differential GPS can provide relative position knowledge of 10 cm and timing to 20 ns, and an ultra stable oscillator provides local time precision of 5 ps over the maximum signal integration time of 5 s. Intersatellite communications regularly update position and timing measurements. The actual satellite position with respect to other satellites in the array is measured using laser ranging, to provide relative positions with an error significantly less than a wavelength at the X-band. Thus, positional errors of the satellites are modeled as deviations from the nominal positions, having (a) a random but known component on the order of meter and (b) a random but unknown component on the order of 3 mm. Phase compensation corrects the known errors and the unknown errors raise the sidelobe level and lower the gain.

Example. An array with 64 elements that are randomly distributed in a $3.5\lambda \times 3.5\lambda \times 3.5\lambda$ volume. Plot the incoherent array factor and the array factor with a coherent beam steered to $(\theta_s, \phi_s) = (45°, 0°)$.

The randomly distributed volume array is shown in Figure 5.67. The uniformly weighted array factor appears in Figure 5.68. It consists of many very small lobes of about the same height. If the elements are phased to form a coherent beam at $(\theta, \phi) = (45°, 0°)$, then the array factor in Figure 5.69 results with a desired main beam.

The High-Frequency Surface Wave Radar (HFSWR) is deployed along a coast to detect boats far from shore [27]. Finding an area along the water to build a long array is often difficult, so it has been proposed to place the array elements on buoys that are tied to the sea bottom. Ocean waves cause movement of the elements that are a function of the length of the cable and the sea state or height of the waves. A computer model of the ocean calculated the element locations as a function of x, y, and z [28]. As long as the location of the element is known, then appropriate phase corrections can be applied to compensate the pattern.

As an example, consider the case of a 10-element linear array with a desire spacing of $\lambda/2$ on top of an ocean with 0.9- to 1.5-m waves (sea state 3). The array configuration changes with time as shown by the time varying azimuth antenna pattern in Figure 5.70. The sidelobes increase from −13 dB to as much

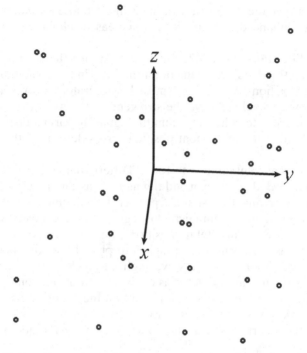

Figure 5.67. Random array of 64 point sources in a $3.5\lambda \times 3.5\lambda \times 3.5\lambda$ volume.

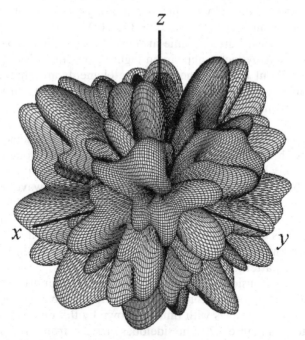

Figure 5.68. Incoherent array factor for the random array.

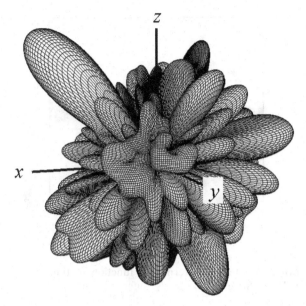

Figure 5.69. Array factor with a coherent beam pointing at $(\theta, \phi) = (45°, 0°)$.

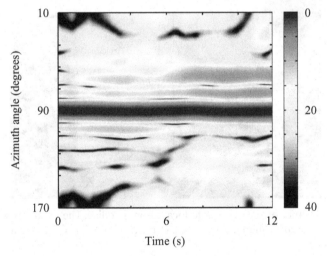

Figure 5.70. Uncorrected antenna pattern as a function of time. (Courtesy of Régis Guinvarc'h, SONDRA, Supelec.)

as $-6.6\,dB$ at $t = 10\,s$. These waves do not cause a change in the beam-pointing direction. Phase corrections result in the pattern shown in Figure 5.71. The array pattern is restored except in the far-out sidelobe region. This concept will be demonstrated by building a small array of monopoles located on as

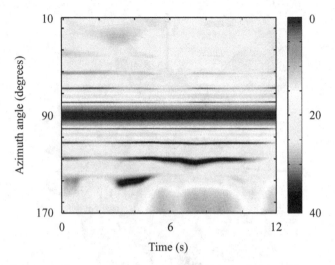

Figure 5.71. Corrected antenna pattern as a function of time. (Courtesy of Régis Guinvarc'h, SONDRA, Supelec.)

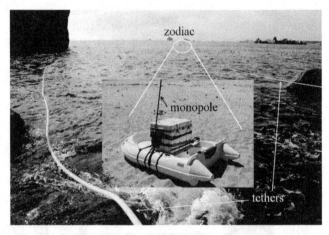

Figure 5.72. Picture of the monopole located on a zodiac. The elements were on the ocean and tethered to the land by ropes. (Courtesy of Régis Guinvarc'h, SONDRA, Supelec.)

shown in Figure 5.72. One zodiac with the HF radar equipment and monopole antennas are tethered to shore during testing. Additional zodiacs with monopoles will be added in the future.

Platform vibrations also result in time-varying elements in an array. The element positions can be estimated by mounting some near field reference sources on rigid frames that do not vibrate. The beamformer uses the estimated position information to correct the distorted pattern of the antenna array. A 5.8-GHz 8-element array with artificially displaced element positions was built to emulate the vibration effect [29]. The proposed calibration approach estimated the position with deviations under 4% of the free space wavelength.

REFERENCES

1. L. Josefsson and P. Persson, Conformal array synthesis including mutual coupling, *Electron. Lett.*, Vol. 35, No. 8, 1999, pp. 625–627.

2. M. T. Borkowski, Solid state transmitters, in *Radar Handbook*, M. Skolnik, ed., New York: McGraw-Hill, 2008, pp. 11.1–11.36.

3. R. Hendrix, Aerospace system improvements enabled by modern phased array radar—2008, IEEE *Radar Conference*, 2008, 26-30 May 2008, pp. 1–6.

4. O. M. Bucci, G. D' Elia, and G. Romito, Power synthesis of conformal arrays by a generalised projection method, *IEE Proc., Microwaves, Antennas Propagat.*, Vol. 142, No. 6, 1995, pp. 467–471.

5. J. A. Ferreira and F. Ares, Pattern synthesis of conformal arrays by the simulated annealing technique, *Electron. Lett.*, Vol. 33, No. 14, 1997, pp. 1187–1189.

6. E. Botha and D. A. McNamara, Conformal array synthesis using alternating projections, with maximal likelihood estimation used in one of the projection operators, *Electron. Lett.*, Vol. 29, No. 20, 1993, pp. 1733–1734.

7. O. M. Bucci and G. D'Elia, Power synthesis of reconfigurable conformal arrays with phase-only control, *IEE Proc. Microwaves Antennas Propagat.*, Vol. 145, No. 1, 1998, pp. 131–136.

8. M. R. Frater and M. J. Ryan, *Electronic Warfare for the Digitized Battlefield*, Boston: Artech House, 2001.

9. M. R. Morris. Wullenweber antenna arrays, May 28, 2009, http://www.navycthistory.com/WullenweberArticle.txt.

10. F. Adcock, *Improvement in Means for Determining the Direction of a Distant Source of Electromagnetic Radiation*, 1304901919, B.P. Office, 1917.

11. E. J. Baghdady, New developments in direction-of-arrival measurement based on Adcock antenna clusters, *Aerospace and Electronics Conference*, 1989, pp. 1873–1879.

12. F. Chauvet, Conformal, wideband, dual polarization phased antenna array for UHF radar carried by airship, The University Pierre et Marie Curie Paris 6, 2008.

13. B. Tomasic, J. Turtle, and L. Shiang, Spherical arrays—Design considerations, *18th International Conference on Applied Electromagnetics and Communications*, 2005, pp. 1–8.

14. B. Tomasic, J. Turtle, and S. Liu, *The Geodesic Sphere Phased Array Antenna for Satellite Communication and Air/Space Surveillance—Part I*, AFRL-SN-HSTR-2004-031, January 2004.

15. L. Shiang, B. Tomasic, and J. Turtle, The geodesic dome phased array antenna for satellite operations support—Antenna resource management, Antennas and Propagation Society International Symposium, June 2007, pp. 3161–3164.

16. Chemical Rubber Company, *CRC Standard Mathematical Tables and Formulae*, Boca Raton, FL: CRC Press, 1991, p. v.

17. B. Tomasic, J. Turtle, L. Shiang et al., The geodesic dome phased array antenna for satellite control and communication—Subarray design, development and demonstration, *IEEE International Symposium on Phased Array Systems and Technology*, 2003, pp. 411–416.

18. About ALMA Technology, May 27, 2009, http://www.almaobservatory.org/.

19. H. E. Shanks and R. W. Bickmore, Four-dimensional electromagnetic radiators, *Can. J. Phys.*, Vol. 37, 1957, pp. 263–275.

20. G. W. Stimson, *Introduction to Airborne Radar*, El Segundo, CA: Hughes Aircraft Co., 1983.

21. R. K. Raney, Space-based remote sensing radars, in *Radar Handbook*, M. Skolnik, ed., New York: McGraw-Hill, pp. 18.1–18.70, 2008.

22. Imaging Radar @ Southport, May 28, 2009; http://southport.jpl.nasa.gov/.

23. W. Kummer, A. Villeneuve, T. Fong et al., Ultra-low sidelobes from time-modulated arrays, *IEEE Trans. Antennas Propagat.*, Vol. 11, No. 6, 1963, pp. 633–639.

24. J. C. Bregains, J. Fondevila-Gomez, G. Franceschetti et al., Signal radiation and power losses of time-modulated arrays, *IEEE Trans. Antennas Propagat.*, Vol. 56, No. 6, 2008, pp. 1799–1804.

25. H. Steyskal, J. K. Schindler, P. Franchi et al., Pattern synthesis for TechSat21—A distributed space-based radar system, *IEEE, Antennas Propagat. Mag.*, Vol. 45, No. 4, 2003, pp. 19–25.

26. R. Burns, C.A. McLaughlin, J. Leitner, and M. Martin, "TechSat 21: formation design, control, and simulation," IEEE Aerospace Conference Proceedings, 2000, pp. 19–25

27. L. Sevgi, A. Ponsford, and H. C. Chan, An integrated maritime surveillance system based on high-frequency surface-wave radars. 1. Theoretical background and numerical simulations, *IEEE, Antennas Propagat. Mag.*, Vol. 43, No. 4, 2001, pp. 28–43.

28. A. Bourges, R. Guinvarc'h, B. Uguen et al., A simple pattern correction approach for high frequency surface wave radar on buoys, First European Conference on Antennas and Propagation, 6–10 Nov. 2006, pp. 1–4.

29. W. Qing, H. Rideout, Z. Fei et al., Millimeter-wave frequency tripling based on four-wave mixing in a semiconductor optical amplifier, *Photonics Technol. Lett., IEEE*, Vol. 18, No. 23, 2006, pp. 2460–2462.

6 Mutual Coupling

An antenna never exists in isolation. Radiation from a transmitting source illuminates the receive antenna as well as everything in its environment. The time-varying fields strike various objects in the environment that absorb and reradiate the fields. As a result, the antenna performs differently in its environment than in free space. Mutual coupling is the interactions between an antenna and its environment is called mutual coupling. It has three components [1]:

1. Radiation coupling between two nearby antennas.
2. Interactions between an antenna and nearby objects, particularly conducting objects.
3. Coupling inside the feed network of an antenna array.

This chapter examines the radiation coupling between elements in an antenna array. This is the final step toward the design of realistic phased array antennas.

Figure 6.1a is a diagram of an isolated antenna that is matched to its transmission line: The antenna impedance (Z_a) equals the transmission line impedance (Z_0). A wave travels from the transmitter to the antenna (V^+) with no reflection ($V^- = 0$), so the reflection coefficient is zero, $\Gamma = 0$. In Figure 6.1b an identical second antenna is placed near the first one. The second antenna radiates a wave that is received by the first antenna. This received wave travels from the antenna back to the transmitter ($|V^-| > 0$). Thus, the signal received from the second antenna looks like a reflected wave in the first antenna. A reflected wave is associated with an impedance mismatch, so it appears as though the first antenna is not matched to the transmission line. Since the reflection coefficient is no longer zero, the antenna impedance differs from the transmission line impedance by

$$Z_L = Z_0 \frac{1+\Gamma}{1-\Gamma} \qquad (6.1)$$

Antenna Arrays: A Computational Approach, by Randy L. Haupt
Copyright © 2010 John Wiley & Sons, Inc.

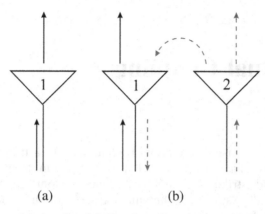

Figure 6.1. Reflected and radiated fields of an array element. **(a)** An antenna matched to its transmission line radiates with no reflections. **(b)** A second antenna radiates fields that are received by the first antenna.

If the excitation on element 2 changes, then the apparent reflected signal in element 1 changes. As a result, the element 1 impedance calculated by (6.1) changes. Thus, changing the array excitation, such as beam steering, causes the active element impedance of all the elements to change. The transmitter/ receiver that is matched to an isolated element is no longer matched to an element in the array. Even if the impedance of the transmitter/receiver changes to match the element at one excitation, then the transmitter/receiver would no longer be matched if the excitation changes. In order to receive the most power from an incident wave, the element impedance should minimize the total scattering from the array, because less scattering means more of the signal was received by the array.

When antennas are placed close together, as in an array, if one antenna transmits, then all of the others receive some of the signal depending upon the distance and orientation of the antennas [2]. Even when no element is transmitting, an incident wave causes scattering from the elements which effectively makes each element a transmitter. The current on an element is the vector sum of all the currents induced by the fields from the other elements. These element interactions must be understood before designing the array.

This chapter starts by introducing the concept of mutual impedance and the analytical expressions for coupling between two dipoles. There are two parts to examining the effects of mutual coupling on large arrays [3]. The first looks at elements that are surrounded on all sides by many elements. This element appears to be in an infinite array. The second examines the edge effects on the elements close to the edge of a large array or the elements of a small array. It is possible to find the self impedance of the element, and then find the mutual impedances due to the presence of the neighboring elements. The mutual impedances change as a function of scan angle and, under certain conditions, even result in a total reflection of the signal which produces scan blindness at that angle.

6.1. MUTUAL IMPEDANCE

Self-impedance is the impedance of the isolated antenna. The voltage at the antenna terminal is given by Ohm's law:

$$V_1 = Z_{11}I_1 \tag{6.2}$$

If there is no current source at element 1, then the voltage at element 1 due to the current on element 2 is

$$V_1 = Z_{12}I_2 \tag{6.3}$$

Z_{12} is the mutual impedance between elements 1 and 2. The total voltage at element 1 is the sum of (6.2) and (6.3).

$$V_1 = Z_{11}I_1 + Z_{12}I_2 \tag{6.4}$$

The impedance at element 1 is no longer Z_{11} when element 2 is present. Instead, the impedance is known as the driving point or active impedance of the element and is given by

$$Z_{d1} = \frac{V_1}{I_1} = Z_{11} + Z_{12}\frac{I_2}{I_1} \tag{6.5}$$

The value of the mutual impedance depends upon the distance between the elements, the polarization of the elements, the orientation of the elements, and the element patterns.

Mutual coupling occurs in receive arrays, when a plane wave induces a current on elements 1 and 2. The current induced on element 2 radiates and is received by element 1. Adding more elements causes more interactions. If there are N elements, then there are N self-impedances and $N(N-1)$ mutual impedances. The mutual impedance between any two elements in the array is found by dividing the open-circuit voltage at one element by the current at the other element.

$$Z_{mn} = \frac{V_m}{I_n} \tag{6.6}$$

After calculating all the self- and mutual impedances, they can be placed in the $N \times N$ impedance matrix.

$$ZI = V \tag{6.7}$$

or

$$
\begin{bmatrix}
Z_{11} & Z_{12} & \cdots & Z_{1N} \\
Z_{21} & Z_{22} & \cdots & Z_{2N} \\
\vdots & \vdots & \ddots & \vdots \\
Z_{N1} & Z_{N2} & \cdots & Z_{NN}
\end{bmatrix}
\begin{bmatrix}
I_1 \\
I_2 \\
\vdots \\
I_N
\end{bmatrix}
=
\begin{bmatrix}
V_1 \\
V_2 \\
\vdots \\
V_N
\end{bmatrix}
\tag{6.8}
$$

The impedance matrix contains all the self- and mutual impedances of an N-element array. Array element m couples to element n through the mutual impedance Z_{mn} in row m and column n of the matrix. Reciprocity dictates that $Z_{mn} = Z_{nm}$. The driving point impedance for an element in an N element array follows from (6.5):

$$Z_{dn} = \frac{I_1}{I_n} Z_{n1} + Z_{n2} \frac{I_2}{I_n} + \cdots + Z_{nN} \frac{I_N}{I_n} \tag{6.9}$$

Alternatively, the coupling can be formulated with admittances instead of impedances. Multiplying both sides of (6.7) by the admittance matrix (inverse of impedance matrix) produces

$$YZI = YV = I \tag{6.10}$$

with the admittance matrix defined as

$$
Y = Z^{-1} =
\begin{bmatrix}
Y_{11} & Y_{12} & \cdots & Y_{1N} \\
Y_{21} & Y_{22} & \cdots & Y_{2N} \\
\vdots & \vdots & \ddots & \vdots \\
Y_{N1} & Y_{N2} & \cdots & Y_{NN}
\end{bmatrix}
\tag{6.11}
$$

The driving point admittance is then defined by

$$Y_{dn} = \frac{I_n}{V_n} = \cdots + Y_{n,n-1} \frac{V_{n-1}}{V_n} + Y_{n.n} + Y_{n,n+1} \frac{V_{n+1}}{V_n} + \cdots \tag{6.12}$$

Admittance is preferred over impedance for arrays of slots.

6.2. COUPLING BETWEEN TWO DIPOLES

This section presents the analytical equations for mutual coupling between two dipoles. Although these calculations are only valid for two dipoles, they may be applied to any pair of elements in a linear or planar array with or without a ground plane. The elements of the $N \times N$ impedance matrix are found by assuming it is composed of $N(N-1)$ two-dipole mutual impedances and N two dipole self-impedances.

The 2×2 impedance matrix for a two-dipole array has elements given by

$$Z_{11} = \frac{V_1}{I_1}\bigg|_{I_2=0} \qquad Z_{12} = \frac{V_1}{I_2}\bigg|_{I_1=0}$$

$$Z_{21} = \frac{V_2}{I_1}\bigg|_{I_2=0} \qquad Z_{22} = \frac{V_2}{I_2}\bigg|_{I_1=0} \tag{6.13}$$

The driving point impedance at the first element is given by (6.5), and the one at the second element is given by

$$Z_{d2} = \frac{V_2}{I_2} = Z_{22} + Z_{21}\left(\frac{I_1}{I_2}\right) \tag{6.14}$$

The Z_d is the impedance seen by the transmission line or waveguide of the feed network. When the antenna element is designed and tested in isolation, then Z_{22} is the impedance. Once the second element is present, then it becomes a determining factor in the impedance of the first antenna element.

The reciprocity theorem is used to find the values of the impedances in (6.6)

$$\iiint_V E_{y1} J_{y2} dv' = \iiint_V E_{y2} J_{y1} dv' \tag{6.15}$$

Since the current only flows in the y direction and the dipole is very thin,

$$I_{y1}(y') = 2\pi a J_{y1}(y') = I_m \sin[k(L_y/2 - |y'|)] \tag{6.16}$$

The y-polarized near field of a dipole oriented along the y axis is

$$E_y = -j\frac{Z_0 I_m}{4\pi}\left[\frac{e^{-jkR_1}}{R_1} + \frac{e^{-jkR_2}}{R_2} - 2\cos\left(\frac{kL_y}{2}\right)\frac{e^{-jkr}}{r}\right] \tag{6.17}$$

where

$$r = \sqrt{x^2 + y^2 + z^2}$$

$$R_1 = \sqrt{x^2 + (y - L_y/2)^2 + z^2}$$

$$R_2 = \sqrt{x^2 + (y + L_y/2)^2 + z^2}$$

For a lone dipole, the power at the current maximum is

$$P = V_m I_m = -\int_{-L_y/2}^{L_y/2} I_{y1}(y') E_{y1}(y') dy' \qquad (6.18)$$

The self-impedance is the voltage at the antenna divided by the current fed to the antenna when the antenna is isolated. Assume that the dipole is very thin and that the current runs only along the y axis. This antenna can only radiate and receive electric fields with a y component. The power at the dipole feed point is the integral of the current and the electric field along the dipole.

$$P = I_m^2 Z_{11} = -\int_{-L_y/2}^{L_y/2} I_{y1}(y') E_z^{\text{surface}}(y') dy' \qquad (6.19)$$

where E_y^{surface} is the electric field at the surface of the dipole. Substituting (6.16) and (6.17) into (6.19) produces

$$
\begin{aligned}
Z_{21} &= R_{21} + jX_{21} \\
&= \frac{30}{I_m} \int_{-L_y/2}^{L_y/2} \sin\left[k\left(\frac{L_y}{2} - |y'|\right)\right]\left[\frac{e^{-jkR_1}}{R_1} + \frac{e^{-jkR_2}}{R_2} - 2\cos\left(\frac{kL_y}{2}\right)\frac{e^{-jkr}}{r}\right]dy' \quad (6.20)
\end{aligned}
$$

where

$$r = \sqrt{a^2 + y'^2}$$

$$R_1 = \sqrt{a^2 + (y' - L_y/2)^2}$$

$$R_2 = \sqrt{x^2 + (y' + L_y/2)^2}$$

Solving the integral in (6.20) leads to [4]

$$
\begin{aligned}
R_{11} &= \frac{\eta}{2\pi}\left\{\gamma + \ln(kL_y) - C_i(kL_y) + \frac{1}{2}\sin(kL_y)[S_i(2kL_y) - 2S_i(kL_y)]\right. \\
&\quad \left. + \frac{1}{2}\cos(kL_y)[\gamma + \ln(kL_y/2) + C_i(2kL_y) - 2C_i(kL_y)]\right\} \\
X_{11} &= \left\{\frac{\eta}{4\pi}2S_i(kL_y) + \cos(kL_y)[2S_i(kL_y) - S_i(2kL_y)]\right. \\
&\quad \left. - \sin(kL_y)\left[2C_i(kL_y) - C_i(2kL_y) - C_i\left(\frac{2ka^2}{L_y}\right)\right]\right\}
\end{aligned}
\qquad (6.21)
$$

where $\gamma = 0.5772$ is Euler's constant. For a half-wavelength dipole, $Z_{11} = 73 + j42.5$.

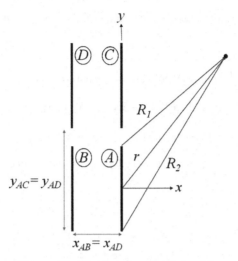

Figure 6.2. Dipoles placed side by side.

The mutual impedance between two dipoles is found in a similar manner, except the electric field comes from the second dipole, so the values of r, R_1 and R_2 are different. Assuming that the two dipoles are of the same length, L_y, then the formulas for the mutual impedances are given by:

1. Parallel dipoles A and B in Figure 6.2. (A line connecting the centers of the dipoles is perpendicular to the dipoles.)

$$R_{21} = \frac{\eta}{4\pi}[2C_i(u_1) - C_i(u_2) - C_i(u_3)]$$
$$X_{21} = -\frac{\eta}{4\pi}[2S_i(w_1) - S_i(v_2) - S_i(v_3)]$$

(6.22)

where

$$u_0 = kx_{AB}$$

$$u_1 = k\left(\sqrt{x_{AB}^2 + L_y^2} + L_y\right)$$

$$u_2 = k\left(\sqrt{x_{AB}^2 + L_y^2} - L_y\right)$$

$$S_i(x) = \int_0^x \frac{\sin \tau}{\tau} d\tau$$

$$C_i(x) = 0.577215665 + \ln x - \int_0^x \frac{1 - \cos \tau}{\tau} d\tau$$

2. Collinear dipoles A and C in Figure 6.2. (Dipoles lie along the same line.)

$$R_{21} = \frac{\eta}{8\pi}\{\cos(v_0)[2C_i(2v_0) - C_i(v_1) - C_i(v_2) + \ln C_i(v_3)]$$
$$+ \sin(v_0)[2S_i(2v_0) - S_i(v_1) - S_i(v_2)]\}$$
$$X_{21} = \frac{\eta}{8\pi}\{\cos(v_0)[-2S_i(2v_0) + S_i(v_1) + S_i(v_2)]$$
$$- \sin(v_0)[-2C_i(2v_0) + C_i(v_1) + C_i(v_2) + \ln C_i(v_3)]\}$$

$$(6.23)$$

where

$$v_0 = ky_{AC}$$
$$v_1 = 2k(y_{AC} + L_y)$$
$$v_2 = 2k(y_{AC} - L_y)$$
$$v_3 = (y_{AC}^2 - L_y^2)/y_{AC}^2$$

3. Offset dipoles A and D in Figure 6.2. (The dipoles are parallel, but a line connecting the centers of the dipoles is not perpendicular to the dipoles.)

$$R_{21} = \frac{\eta}{8\pi}\{\cos(w_0)[2C_i(w_1) + 2C_i(w_2) - C_i(w_3) - C_i(w_4) - C_i(w_5) - C_i(w_6)]$$
$$+ \sin(w_0)[2S_i(w_1) - 2S_i(w_2) - S_i(w_3) + S_i(w_4) - S_i(w_5) + S_i(w_6)]\}$$
$$X_{21} = \frac{\eta}{8\pi}\cos(w_0)\{[-2S_i(w_1) - 2S_i(w_2) + S_i(w_3) + S_i(w_4) - S_i(w_5) + S_i(w_6)]$$
$$+ \sin(w_0)[2C_i(w_1) - 2C_i(w_2) - C_i(w_3) + C_i(w_4) - C_i(w_5) + C_i(w_6)]\}$$

$$(6.24)$$

where

$$w_0 = ky_{AD}$$
$$w_1 = k\left(\sqrt{x_{AD}^2 + y_{AD}^2} + y_{AD}\right)$$
$$w_2 = k\left(\sqrt{x_{AD}^2 + y_{AD}^2} - y_{AD}\right)$$
$$w_3 = k\left(\sqrt{x_{AD}^2 + (y_{AD} - L_y)^2} + y_{AD} - L_y\right)$$
$$w_4 = k\left(\sqrt{x_{AD}^2 + (y_{AD} - L_y)^2} - y_{AD} + L_y\right)$$

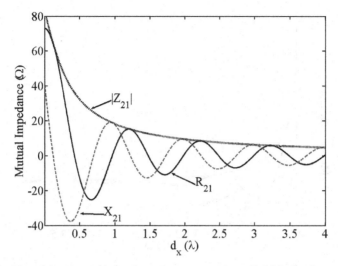

Figure 6.3. Plot of the mutual impedance between two parallel $\lambda/2$ dipoles as a function of separation distance in wavelengths.

$$w_5 = k\left(\sqrt{x_{AD}^2 + (y_{AD} + L_y)^2} + y_{AD} + L_y\right)$$

$$w_6 = k\left(\sqrt{x_{AD}^2 + (y_{AD} + L_y)^2} - y_{AD} - L_y\right)$$

Figure 6.3 is a graph of the real and imaginary parts of the mutual impedance ($Z_{21} = R_{21} + jX_{21}$) as well as the magnitude of the mutual impedance between two parallel y-oriented dipoles along the x axis. The phase is linear and equal to $-kx_{AB}$, so it was not included in the plot. The impedance quickly decreases as x_{AB} increases. When $x_{AB} = 0$, the mutual impedance becomes the self-impedance. Figure 6.4 is a graph of the real and imaginary parts of the mutual impedance as well as the magnitude of the mutual impedance between two collinear y-oriented dipoles along the y axis. The mutual impedance is much smaller than for the parallel dipole case as x_{AB} increases, since the dipole radiation in the y-direction is small. The impedance plots start at $y_{AC} = \lambda/2$ instead of $y_{AC} = 0$, because the dipoles overlap at smaller spacings. Extensive plots of the mutual impedance between two parallel dipoles of different lengths and separation distances in the x and y directions using (6.24) appear in reference 4.

Example. Find the self-, mutual, and driving point impedance of two parallel half-wavelength dipoles ($a = 0.001\lambda$) spaced $\lambda/2$ apart assuming $I_1 = I_2$.

Use (6.5), (6.21), and (6.22) to get $Z_{11} = 73.1309 + j42.5452$, $Z_{12} = -12.5324 - j29.9293$, and $Z_{d1} = 60.5985 + j12.6159$. If $I_1 \neq I_2$, then Z_{d1} will be different than Z_{d2}, as in an amplitude taper.

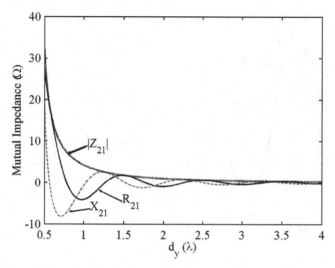

Figure 6.4. Plot of the mutual impedance between two collinear $\lambda/2$ dipoles as a function of separation distance in wavelengths.

A planar dipole of length L and width w is the complement of a slot of length L and width w. Booker's relationship introduced in Chapter 4 can also be applied here. Assuming there are two slots in an infinite ground plane, then the mutual admittance between the slots is related to the mutual impedance between two dipoles that are complementary to the slots by [5]

$$Y_{12}^{\text{slot}} = 4Z_{12}^{\text{dipole}}Y_0^2 \tag{6.25}$$

If the slot only radiates from one side of the ground plane, then the relationship is [5]

$$Y_{12}^{\text{slot}} = 2Z_{12}^{\text{dipole}}Y_0^2 \tag{6.26}$$

A slot of width, w, is also considered complementary to a dipole of radius a when the width and radius are related by [5]

$$w = 4a \tag{6.27}$$

Example. Find the self- and mutual admittances of two parallel half-wavelength slots that are complementary to a dipole with $a = 0.001\lambda$ and are spaced $\lambda/2$ apart assuming $I_1 = I_2$.

If $w \approx 0.004\lambda$ then the slot is complementary to the dipole. Substituting the dipole impedances in the previous examples into (6.25) results in

$$Y_{11}^{\text{slot}} = 4(73.13 + j42.55)/377^2 = (2.06 + j1.20) \times 10^{-3} \Omega^{-1}$$

$$Y_{12}^{\text{slot}} = 4(-12.53 - j29.93)/377^2 = -(0.35 + j0.84) \times 10^{-3}\,\Omega^{-1}$$

Note that the slot is assumed to radiate out both sides.

6.3. METHOD OF MOMENTS

Finding the mutual coupling between nondipole antennas requires numerical methods like the method of moments (MoM). The MoM most closely follows the analytical derivations presented thus far. It is a mathematical technique for solving electromagnetics equations of the form [6]

$$\mathcal{L}\{\mathcal{J}\} = \mathcal{V} \tag{6.28}$$

where $\mathcal{L}$ is a linear integrodifferential operator, $\mathcal{J}$ is the induced current, and $\mathcal{V}$ is the forcing function. In this equation, $\mathcal{V}$ is known, and the idea is to invert $\mathcal{L}$ to obtain the unknown function

$$\mathcal{J} = \mathcal{L}^{-1}\{\mathcal{V}\} \tag{6.29}$$

The procedure involves a technique that transforms (6.29) into a system of linear algebraic equations of the form

$$Z^{MM} I = V \tag{6.30}$$

where Z^{MM} is the MoM impedance matrix, I is the current column vector, and V is the induced voltage column vector. The steps in the MoM procedure are as follows:

- Expand I in into a series of basis functions with unknown coefficients.
- Take the inner product of both sides with a set of known testing functions.
- Apply the boundary conditions at N points to get a set of N equations and unknowns.
- Analytically or numerically solve the definite integral(s) to find Z^{MM}.
- Numerically solve the resultant matrix equation for I.

Once I is known, then the far-field, near-field, efficiency, polarization, or antenna impedance can be easily calculated. The MoM is a powerful technique but works best for

- Electrically small objects
- A perfect electrical conductor (PEC)
- Narrowband applications.

The MoM is widely used to calculate the impedance and radiated fields of many types of antennas.

A thin dipole antenna of radius a has current flowing along its length but not in any other direction. If the wire antenna is situated along the z axis, then [7]

$$E_z = -j\omega\mu \iiint_{v'} J_z(z') \frac{e^{-jkR}}{4\pi R} dv' - j\frac{1}{\omega\varepsilon} \frac{\partial^2}{\partial z^2} \iiint_{v'} J_z(z') \frac{e^{-jkR}}{4\pi R} dv' \qquad (6.31)$$

This equation may be rewritten as

$$E_z = \frac{1}{j\omega\varepsilon} \iiint_{v'} \left[k^2 \frac{e^{-jkR}}{4\pi R} + \frac{\partial^2}{\partial z^2}\left(\frac{e^{-jkR}}{4\pi R} \right) \right] J_z(z') dv' \qquad (6.32)$$

Since the wire is a PEC, the current lies only on the surface. If the wire has a radius a, then the volume integral in (6.32) reduces to a surface integral given by

$$E_z = \frac{1}{j\omega\varepsilon} \oint_s \left[k^2 \frac{e^{-jkR}}{4\pi R} + \frac{\partial^2}{\partial z^2}\left(\frac{e^{-jkR}}{4\pi R} \right) \right] J_z(z') dz' d\phi' \qquad (6.33)$$

If $R = \sqrt{(z-z')^2 + a^2}$ (implying that the observation point is on the surface of the wire) and the wire is very thin ($a \ll \lambda$), the resulting current is

$$I_z(z') = 2\pi J_z(z') \qquad (6.34)$$

and the surface integral of $J_z(z')$ becomes a line integral of $I_z(z')$.

$$E_z = \frac{1}{j\omega\varepsilon} \int_{-L/2}^{L/2} \left[k^2 \frac{e^{-jkR}}{4\pi R} + \frac{\partial^2}{\partial z^2}\left(\frac{e^{-jkR}}{4\pi R} \right) \right] I_z(z') dz' \qquad (6.35)$$

If a z-directed field is incident upon the wire, then the total tangential field at the wire surface must be zero or

$$E_z^{\text{total}} = E_z^i + E_z^{\text{scattered}} = 0 \qquad (6.36)$$

Substituting (6.35) for the scattered field into (6.36) yields an integral equation of the first kind.

$$\frac{j}{\omega\varepsilon} \int_{-L/2}^{L/2} \underbrace{\left[\frac{e^{-jkR}}{4\pi R} + k^2 \frac{\partial^2}{\partial z^2}\left(\frac{e^{-jkR}}{4\pi R} \right) \right]}_{K(z,\,z')} I_z(z') dz' = E_z^i \qquad (6.37)$$

Taking the derivative of inside the integral kernel, $K(z, z')$, results in

$$K(z,z')=\frac{e^{-jkR}}{4\pi R^5}\left[(1+jkR)(2R^2-3a^2)+(kaR)^2\right] \quad (6.38)$$

This equation is based on the following approximations:

- The wire is infinitely thin, so that only z-directed currents, not radial or circumfiential currents, exist.
- The wire is a PEC.
- The boundary condition is enforced at the wire surface, but the current, $I_z(z')$, is a filament at the center of the wire.
- The distance, R, is never zero.

  The current on the wire, $I(z')$ can be written as a sum of weighted basis functions

$$I(z')=\sum_{n=1}^{N} I_n B_n(z') \quad (6.39)$$

Next, inserting (6.39) into (6.37) yields [1]

$$\int_{-L/2}^{L/2} K(z,z')\sum_{n=1}^{N} I_n B_n(z')dz' = \sum_{n=1}^{N} I_n \int_{-L/2}^{L/2} K(z,z')B_n(z')dz' = -j\omega\varepsilon E_z^i(z) \quad (6.40)$$

The basis functions, B_n, are mathematical building blocks weighted by the coefficients, I_n. Equation (6.39) provides an interpolation of the current over the length of the wire. It is possible to solve this integral using Gaussian quadrature if the basis functions are chosen properly. Many different basis functions are used in practice. Subdomain basis functions exist over segmented portions of the wire, while entire domain basis functions exist over the length of wire. Three common examples of subdomain basis functions are [8]:

1. *Pulse*

$$B_n = \begin{cases} 1, & z'_{n-1} \le z' \le z'_n \\ 0, & \text{otherwise} \end{cases}$$

2. *Piecewise Linear*

$$B_n = \begin{cases} \dfrac{z'-z'_{n-1}}{z'_n-z'_{n-1}}, & z'_{n-1} \le z' \le z'_n \\ \dfrac{z'_{n+1}-z'}{z'_{n+1}-z'_n}, & z'_n \le z' \le z'_{n+1} \\ 0, & \text{otherwise} \end{cases}$$

3. *Cosine*

$$B_n = \cos\left[\frac{(2n-a)\pi z'}{L}\right]$$

Enforcing the boundary conditions at M points enables (6.40) can be written in matrix form as

$$\begin{bmatrix} \int\limits_{-L/2}^{L/2} B_1(z')K(z_1,z')dz' & \cdots & \int\limits_{-L/2}^{L/2} B_N(z')K(z_1,z')dz' \\ \vdots & \ddots & \vdots \\ \int\limits_{-L/2}^{L/2} B_1(z')K(z_M,z')dz' & \cdots & \int\limits_{-L/2}^{L/2} B_N(z')K(z_M,z')dz' \end{bmatrix} \begin{bmatrix} I_1 \\ \vdots \\ I_N \end{bmatrix} = \begin{bmatrix} E_z^i(z_1) \\ \vdots \\ E_z^i(z_M) \end{bmatrix} \quad (6.41)$$

The integrals in (6.41) must be solved before finding the coefficients of the basis functions. One way of simplifying the process is to take the inner product of each integral with a weighting function, so that a matrix row is represented by

$$\sum_{n=1}^{N} I_n \int \int\limits_{-L/2}^{L/2} W_m(z)K(z,z')B_n(z')dz'dz = -j\omega\varepsilon \int E_z^i W_m(z)dz \quad (6.42)$$

One common approach is to have the basis and testing functions the same. This technique is known as Galerkin's method [6]:

$$\sum_{n=1}^{N} I_n \int \int\limits_{-L/2}^{L/2} K(z,z')B_m(z)B_n(z')dz'dz = -j\omega\varepsilon \int E_z^i B_m(z)dz \quad (6.43)$$

If the weighting functions are delta functions, then (6.42) becomes

$$\sum_{n=1}^{N} I_n \int \int\limits_{-L/2}^{L/2} \delta(z-z_m)K(z,z')B_n(z')dz'dz = -j\omega\varepsilon \int E_z^i \delta(z-z_m)dz \quad (6.44)$$

which reduces to

$$\sum_{n=1}^{N} I_n \int\limits_{-L/2}^{L/2} K(z_m,z')B_n(z')dz' = -j\omega\varepsilon E_z^i(z_m) \quad (6.45)$$

This formulation is known as point matching [6] or collocation. If point matching is used with pulse basis functions, then applying the boundary conditions at N collocation points yields the matrix equation

$$\begin{bmatrix} \int_{-L/2}^{L/2+\Delta z} K(z_1,z')dz' & \cdots & \int_{L/2-\Delta z}^{L/2} K(z_1,z')dz' \\ \vdots & \ddots & \vdots \\ \int_{-L/2}^{L/2+\Delta z} K(z_M,z')dz' & \cdots & \int_{L/2-\Delta z}^{L/2} K(z_M,z')dz' \end{bmatrix} \begin{bmatrix} I_1 \\ \vdots \\ I_N \end{bmatrix} = \begin{bmatrix} E_z^i(z_1) \\ \vdots \\ E_z^i(z_M) \end{bmatrix} \quad (6.46)$$

As long as Δz is small, then the integral is accurately approximated by midpoint integration.

$$\begin{bmatrix} \Delta z K(z_1,z_1) & \cdots & \Delta z K(z_1,z_N) \\ \vdots & \ddots & \vdots \\ \Delta z K(z_N,z_1) & \cdots & \Delta z K(z_N,z_N) \end{bmatrix} \begin{bmatrix} I_1 \\ \vdots \\ I_N \end{bmatrix} = \begin{bmatrix} E_z^i(z_1) \\ \vdots \\ E_z^i(z_M) \end{bmatrix} \quad (6.47)$$

Figure 6.5 shows a diagram of part of the dipole. The wire is broken into segments of length $\Delta z = L/N_{\text{seg}}$. The points z_m are located at the midpoints on the surface of the wire, while the integration variable z' is along the filament at the center of the wire.

The next step after calculating the impedance matrix is to represent the right-hand side by either a plane wave (receive antenna) or a voltage source (transmit antenna). A normalized plane wave incident at each segment is [8]

$$E_z^i(z_n) = e^{jkz_n\cos\theta} \quad (6.48)$$

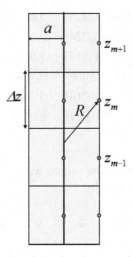

Figure 6.5. Dipole broken into equal segments.

The simplest voltage source is the delta-gap model that has the center segment set to V_0.

$$\begin{bmatrix} E_z^i(z_1) \\ \vdots \\ E_z^i(z_{(N-1)/2}) \\ E_z^i(z_{(N+1)/2}) \\ E_z^i(z_{(N+3)/2}) \\ \vdots \\ E_z^i(z_N) \end{bmatrix} = \begin{bmatrix} 0 \\ \vdots \\ 0 \\ V_0 \\ 0 \\ \vdots \\ 0 \end{bmatrix} \tag{6.49}$$

Example. Find the input impedance to a dipole with $L = 0.1\lambda$ and $a = 0.005\lambda$.

Solution: To easily display the results, only 7 segments are used for the wire. Point matching with pulse basis functions will be used to generate the impedance matrix. The MoM impedance matrix has values given by

$$\mathrm{Re}\{Z^{MM}\} = \begin{bmatrix} 11.26 & 11.25 & 11.22 & 11.18 & 11.12 & 11.04 & 10.94 \\ 11.25 & 11.26 & 11.25 & 11.22 & 11.18 & 11.12 & 11.04 \\ 11.22 & 11.25 & 11.26 & 11.25 & 11.22 & 11.18 & 11.12 \\ 11.18 & 11.22 & 11.25 & 11.26 & 11.25 & 11.22 & 11.18 \\ 11.12 & 11.18 & 11.22 & 11.25 & 11.26 & 11.25 & 11.22 \\ 11.04 & 11.12 & 11.18 & 11.22 & 11.25 & 11.26 & 11.25 \\ 10.94 & 11.04 & 11.12 & 11.18 & 11.22 & 11.25 & 11.26 \end{bmatrix}$$

$$\mathrm{Im}\{Z^{MM}\} = \begin{bmatrix} -1.0248 & 0.4197 & 0.0605 & 0.0181 & 0.0078 & 0.0041 & 0.0025 \\ 0.4197 & -1.0248 & 0.4197 & 0.0605 & 0.0181 & 0.0078 & 0.0041 \\ 0.0605 & 0.4197 & -1.0248 & 0.4197 & 0.0605 & 0.0181 & 0.0078 \\ 0.0181 & 0.0605 & 0.4197 & -1.0248 & 0.4197 & 0.0605 & 0.0181 \\ 0.0078 & 0.0181 & 0.0605 & 0.4197 & -1.0248 & 0.4197 & 0.0605 \\ 0.0041 & 0.0078 & 0.0181 & 0.0605 & 0.4197 & -1.0248 & 0.4197 \\ 0.0025 & 0.0041 & 0.0078 & 0.0181 & 0.0605 & 0.4197 & -1.0248 \end{bmatrix} \\ \times 10^5$$

The delta gap model is used for the source with $V_0 = 1/\Delta z = 70\,\mathrm{V}$, so the right-hand side is a 7×1 matrix given by

$$V = [0 \quad 0 \quad 0 \quad 70 \quad 0 \quad 0 \quad 0]^T$$

Solving for the current on the wire segments results in

$$\mathrm{Re}\{I\} = \begin{bmatrix} 0.3237 \\ 0.4950 \\ 0.5911 \\ 0.6224 \\ 0.5911 \\ 0.4950 \\ 0.3237 \end{bmatrix} \times 10^{-5} \quad \text{and} \quad \mathrm{Im}\{I\} = \begin{bmatrix} 0.4146 \\ 0.7495 \\ 1.1060 \\ 1.6920 \\ 1.1060 \\ 0.7495 \\ 0.4146 \end{bmatrix} \times 10^{-3}$$

$$Z_{in} = 2.17 - j591$$

The small real part and large imaginary part is indicative of a short dipole.

Example. Find the directivity and input impedance for a dipole that is 0.49λ long with $a = 0.000001\lambda$.

Solution: Figure 6.6 is a plot of dipole impedance as a function of the number of segments. The real part of the impedance shows little change as a function of the number of segments, while the imaginary part shows dramatic change. Segmentation is not as critical for calculating the directivity (Figure 6.7). The difference between the antenna pattern for the dipole broken into 5 segments and the one broken into 101 segments is barely noticeable in the plot in Figure 6.8. The moral is that a relatively higher segmentation rate is required for accurately calculating the dipole impedance than is required for calculating the antenna pattern.

The MoM also applies to surfaces and dielectrics [7]. For instance, a rectangular patch antenna is divided into small triangles, and the current is found

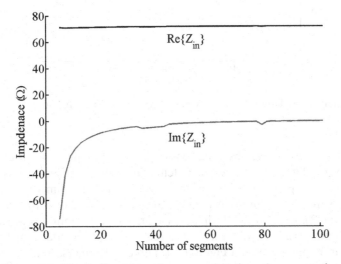

Figure 6.6. Dipole input impedance as a function of segmentation.

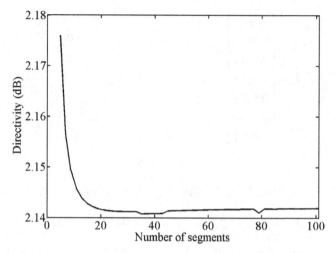

Figure 6.7. Dipole directivity as a function of segmentation.

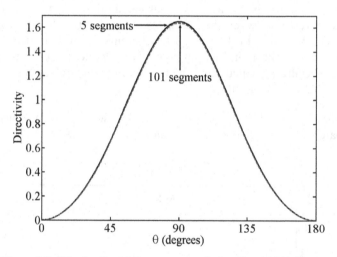

Figure 6.8. Dipole directivity versus angle for 5 and 101 segments.

on each of the triangles. Three-dimensional basis functions are used for the triangular grid. Figure 6.9 shows the patch with six different grid levels as listed in Table 6.1. The patch has dimensions of 37.8 mm × 50.2 mm with a pin feed that is 9.3 mm off center. The substrate has ε_r = 3.6 with dimensions 53.8 mm × 66.2 mm × 1.6 mm. Using coarse gridding, maximum triangle size of $\lambda/3$, produces terrible results for both directivity and impedance calculations as shown in Figures 6.10 and Figure 6.11. Directivity calculations for a grid as coarse as $\lambda/7$ are accurate, while $\lambda/5$ are very good. As the grid gets finer for

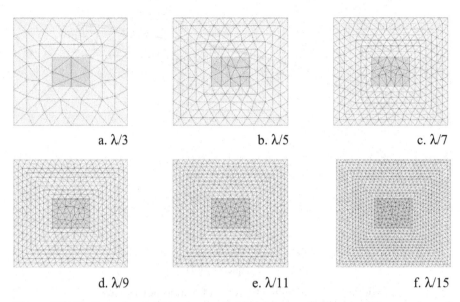

Figure 6.9. Rectangular patch antenna divided into small triangles. The number of unknowns quickly increases as the size of the triangles gets smaller.

TABLE 6.1. Number of Metal and Dielectric Triangles in the Patch Model for a Specified Maximum Triangle Edge Length

Max Triangle Edge Length	Number of Metal Triangles	Number of Dielectric Triangles	Relative Time
$\lambda/3$	94	144	1
$\lambda/5$	256	336	4
$\lambda/7$	498	612	15
$\lambda/9$	832	946	48
$\lambda/11$	1238	1378	118
$\lambda/15$	2276	2300	506

impedance calculations, the resonant frequency increases until it hits 2 GHz at $\lambda/11$. Increasing to $\lambda/15$ improves the depth of S_{11} at 2 GHz. As the number of triangles increases, the amount of time needed to find the solution increases, and the amount of memory needed also increases. The last column in Table 6.1 is the relative time needed to calculate the antenna pattern of the patch. When the maximum triangle size is $\lambda/15$, then the calculations take 506 times longer than when the maximum triangle size is $\lambda/3$.

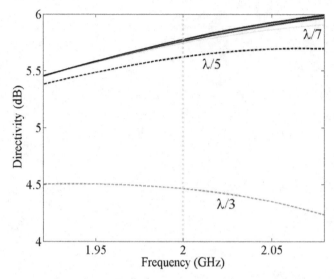

Figure 6.10. Patch directivity as a function of triangle size.

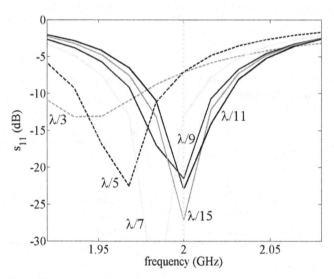

Figure 6.11. Patch S_{11} as a function of triangle size.

6.4. MUTUAL COUPLING IN FINITE ARRAYS

Equations (6.22), (6.23), and (6.24) apply to the mutual impedance between any two elements in an N element array with and without a ground plane. Consider the planar array in Figure 6.12. The diagonal elements of the imped- ance matrix are all equal to the self-impedance of a dipole and are found using

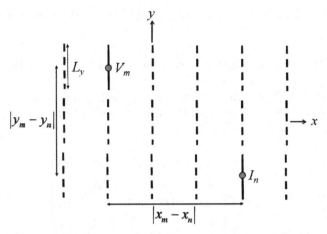

Figure 6.12. Planar array of dipoles on a rectangular grid.

(6.21). Depending upon their locations, the mutual impedance between element m and element n are found using (6.22), (6.23), or (6.24) with

$$
\begin{aligned}
x_{AB} = x_{AD} = |x_m - x_n| \\
y_{AB} = y_{AD} = |y_m - y_n|
\end{aligned}
\tag{6.50}
$$

Example. Find the impedance matrix and driving point impedances for 5 parallel $\lambda/2$ dipoles spaced $\lambda/2$ apart.

The diagonal elements of the impedance are all equal and found using (6.21). The mutual impedance between element m and element n are found using (6.22).

$$
\text{Re}\{Z\} =
\begin{bmatrix}
73.1 & -12.5 & 4.0 & -1.9 & 1.1 \\
-12.5 & 73.1 & -12.5 & 4.0 & -1.9 \\
4.0 & -12.5 & 73.1 & -12.5 & 4.0 \\
-1.9 & 4.0 & -12.5 & 73.1 & -12.5 \\
1.1 & -1.9 & 4.0 & -12.5 & 73.1
\end{bmatrix}
\tag{6.51}
$$

$$
\text{Im}\{Z\} =
\begin{bmatrix}
42.5 & -29.9 & 17.7 & -12.3 & 9.4 \\
-29.9 & 42.5 & -29.9 & 17.7 & -12.3 \\
17.7 & -29.9 & 42.5 & -29.9 & 17.7 \\
-12.3 & 17.7 & -29.9 & 42.5 & -29.9 \\
9.4 & -12.3 & 17.7 & -29.9 & 42.5
\end{bmatrix}
\tag{6.52}
$$

The driving point impedances for the elements when uniformly weighted are given by (6.9):

$$Z_d = [63.8 + j27.4 \quad 50.2 - j11.9 \quad 56.1 + j18.1 \quad 50.2 - j11.9 \quad 63.8 + j27.4]$$
$$(6.53)$$

Example. Find the impedance matrix and driving point impedances for 5 parallel $\lambda/2$ dipoles spaced λ apart.

The procedure is repeated for the increased element spacing. Note that the self-impedance does not change, but the mutual impedances are less than those for the same array with $\lambda/2$ spacing.

$$Re\{Z\} = \begin{bmatrix} 73.1 & 4.0 & 1.1 & 0.5 & 0.3 \\ 4.0 & 73.1 & 4.0 & 1.1 & 0.5 \\ 1.1 & 4.0 & 73.1 & 4.0 & 1.1 \\ 0.5 & 1.1 & 4.0 & 73.1 & 4.0 \\ 0.3 & 0.5 & 1.1 & 4.0 & 73.1 \end{bmatrix} \qquad (6.54)$$

$$Im\{Z\} = \begin{bmatrix} 42.5 & 17.7 & 9.4 & 6.3 & 4.8 \\ 17.7 & 42.5 & 17.7 & 9.4 & 6.3 \\ 9.4 & 17.7 & 42.5 & 17.7 & 9.4 \\ 6.3 & 9.4 & 17.7 & 42.5 & 17.7 \\ 4.8 & 6.3 & 9.4 & 17.7 & 42.5 \end{bmatrix} \qquad (6.55)$$

The driving point impedances for this array are given by

$$Z_d = [79.0 + j80.7 \quad 82.7 + j93.6 \quad 83.3 + j96.7 \quad 82.7 + j93.6 \quad 79.0 + j80.7]$$
$$(6.56)$$

More often than not, the dipoles are placed above a ground plane [9]. If the dipole array is above an infinite perfectly conducting ground plane, then the array/ground plane is replaced with the array and an image array at $z = -h$ (Figure 6.13). The currents in the image array are in the opposite direction of the currents in the actual array due to the boundary condition for the tangential electric field. Two impedance matrixes are calculated. The first contains the mutual impedances between all the elements as though there were no ground plane. The second, Z^i, contains the mutual impedances between each element and the N images. If the array is $z = h$ above the ground plane, then the image impedance matrix, Z^i, is found by solving (6.22), (6.23), or (6.24) using

$$x_{AB} = x_{AD} = \sqrt{|x_m - x_n|^2 + h^2} \qquad (6.57)$$

with y_{AB} and y_{AD} given by (6.50). Once the self- and mutual impedances are found, then the driving point impedance is given by

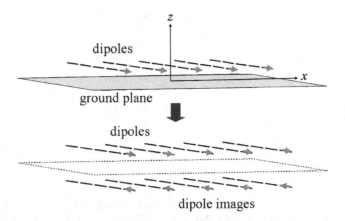

Figure 6.13. Planar array of dipoles on a rectangular grid placed above an infinite ground plane. The ground plane is replaced with an image of the dipoles h below the ground plane with currents opposite those of the array.

$$Z_{dn} = \sum_{m=1}^{N} \frac{I_m}{I_n}\left(Z_{nm} - Z_{nm}^i\right) \tag{6.58}$$

The image impedance elements have a negative sign, because the image current flows in the opposite direction of the element currents.

Example. Find the impedance matrix and driving point impedances for 5 parallel $\lambda/2$ dipoles spaced $\lambda/2$ apart and placed $h = \lambda/4$ above an infinite perfectly conducting ground plane.

The ground plane has a dramatic effect on the mutual and self-impedances in the impedance matrix.

$$\mathrm{Re}\{Z\} = \begin{bmatrix} 85.7 & 12.1 & -9.3 & 5.3 & -3.3 \\ 12.1 & 85.7 & 12.1 & -9.3 & 5.3 \\ -9.3 & 12.1 & 85.7 & 12.1 & -9.3 \\ 5.3 & -9.3 & 12.1 & 85.7 & 12.1 \\ -3.3 & 5.3 & -9.3 & 12.1 & 85.7 \end{bmatrix} \tag{6.59}$$

$$\mathrm{Im}\{Z\} = \begin{bmatrix} 72.5 & -30.7 & 8.1 & -2.9 & 1.3 \\ -30.7 & 72.5 & -30.7 & 8.1 & -2.9 \\ 8.1 & -30.7 & 72.5 & -30.7 & 8.1 \\ -2.9 & 8.1 & -30.7 & 72.5 & -30.7 \\ 1.3 & -2.9 & 8.1 & -30.7 & 72.5 \end{bmatrix} \tag{6.60}$$

$$Z_d = [90.5 + j48.3 \; 105.9 - j16.3 \; 91.3 + j27.3 \; 105.9 - j16.3 \; 90.5 + j48.3] \tag{6.61}$$

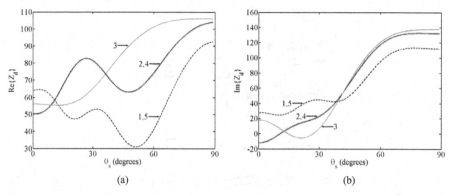

Figure 6.14. Plot of the driving point impedance as a function of scan angle for a 5-element array of parallel $\lambda/2$ dipoles spaced $\lambda/2$ apart. **(a)** Real part. **(b)** Imaginary part.

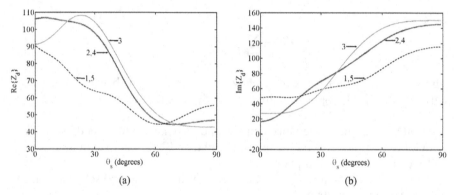

Figure 6.15. Plot of the driving point impedance as a function of scan angle for a 5-element array of parallel $\lambda/2$ dipoles spaced $\lambda/2$ apart and $\lambda/4$ above an infinite perfectly conducting ground plane. **(a)** Real part. **(b)** Imaginary part.

Steering the beam changes the mutual impedances between elements. For instance, steering the beam of an N element uniformly spaced linear array changes (6.9) to

$$Z_{dn} = \sum_{m=1}^{N} Z_{nm} \frac{|I_m| e^{-jk(m-1)d\sin\theta_s}}{|I_n| e^{-jk(n-1)d\sin\theta_s}} = \sum_{m=1}^{N} Z_{nm} \frac{|I_m|}{|I_n|} e^{jk(n-m)d\sin\theta_s} \qquad (6.62)$$

Figure 6.14 shows plots of the real and imaginary parts of the driving point impedance as a function of scan angle for a 5-element uniform array of $\lambda/2$ parallel dipoles with $x_{AB} = 0.5\lambda$. Small scan angles produce small changes in the impedance, while large scan angles produce large changes in the impedance. Figure 6.15 shows the driving point impedances when the array is placed

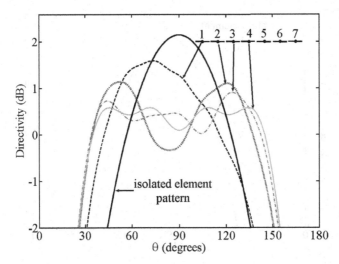

Figure 6.16. Element patterns of an isolated dipole compared to the first 4-element patterns in a 7-collinear-element array.

above a ground plane. The ground plane causes a significant change in the scan impedance compared to the case with no ground plane. In both the array with and without a ground plane, the scan impedances begin to significantly change from the impedance at $\theta_s = 0$ when $\theta_s > 30°$.

The mutual coupling in a small finite array affects the radiation of each element differently. For example, a 7-element linear array of uniformly weighted 0.47λ collinear dipoles spaced half a wavelength apart have element patterns that differ from the isolated element pattern as shown in Figure 6.16. All patterns experience a loss in directivity along with an increase in beamwidth. If all 7-element patterns are averaged, then the element pattern in Figure 6.17 results. This average element pattern is symmetric and has a directivity that is 1.5 dB less than the isolated pattern (Figure 6.17). The decrease in directivity corresponds to the increase in beamwidth from 78° to 126°. The fairly flat-top average element pattern helps keep the main beam constant over a wide scanning angle before the element pattern drops off. Some element patterns squint, while others have many oscillations. If all the element patterns are replaced by the average element pattern, then the array pattern is reasonably reproduced.

Changing the dipole orientation in the array to parallel produces dramatically different element patterns as shown in Figure 6.18. The isolated element pattern is omnidirectional in the plane of the array. The directivity of all the elements actually increases in this case while the beamwidth decreases. As a result, scanning the beam causes the main beam to decrease more than expected. The average element pattern shown in Figure 6.19 is an average of all 7 dipoles in the array. It has a beamwidth of 120°. The average element

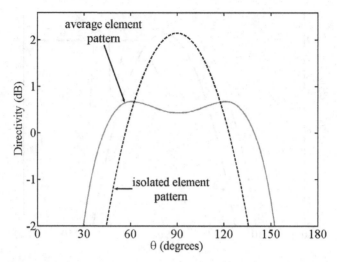

Figure 6.17. Average versus isolated element patterns in a 7-collinear-element array.

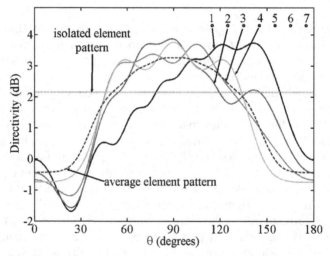

Figure 6.18. Element patterns of an isolated dipole compared to the first 4-element patterns in a 7-parallel-element array.

pattern has a higher gain than the isolated element pattern within ±39° of broadside and has a lower gain farther from broadside.

Mutual coupling not only changes the element patterns, but also changes the input impedance to each of the elements. Figure 6.20 shows the real and imaginary parts of the dipole input impedance for the first four elements in the parallel and collinear arrays when $\theta_s = 0°$. Element 1 is an edge element

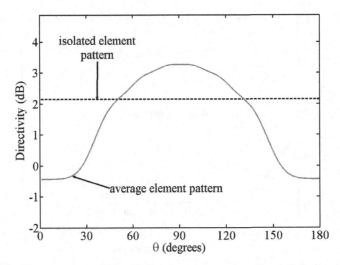

Figure 6.19. Average versus isolated element patterns in a 7-parallel-element array.

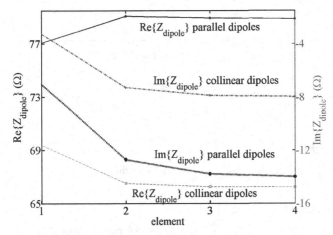

Figure 6.20. Real and imaginary input impedance of the first 4-element patterns in a 7-element array.

and has the most deviation from average. The element orientation does produce different element impedances.

Figures 6.21 and 6.22 show the real and imaginary impedances of the center element of an array with $N = 1, 3, 5, \ldots, 25$ elements. The single element array has the expected impedance of an isolated dipole. Adding one dipole to either side produces over a 10% change in the impedance of the center dipole. The input impedance of the parallel dipoles is more susceptible to mutual coupling than the collinear dipoles. After a while, adding more elements does not change the impedance. Consequently, a dipole surrounded by a large number of elements will have the same impedance as a dipole in an infinite array.

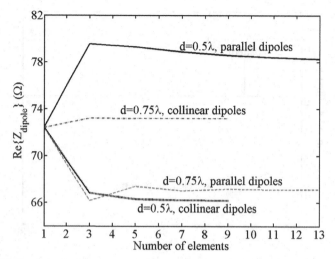

Figure 6.21. Real part of the center element impedance as a function of the number of element on either side of it.

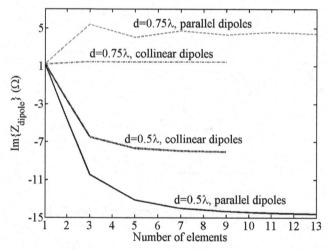

Figure 6.22. Imaginary part of the center element impedance as a function of the number of element on either side of it.

The center element pattern also changes as the number of elements in the array grows. Figure 6.23 shows the element pattern of the center element of a parallel array of dipoles spaced 0.5λ apart with $N = 1, 3, 5, \ldots, 25$. The number of ripples in the element pattern increases with N. Changes to the center element pattern decrease as N increases, because each additional element is farther away than the center element.

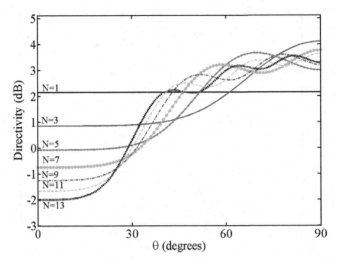

Figure 6.23. Center element pattern in an array of $N = 1, 3, 5, \ldots, 25$ parallel elements with $d = 0.5\lambda$.

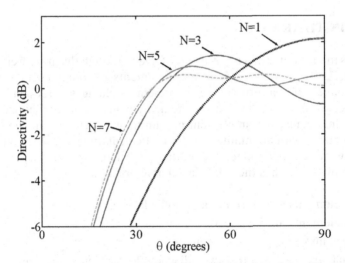

Figure 6.24. Center element pattern in an array of 1, 3, 5, and 7 collinear elements $d = 0.5\lambda$.

Element spacing plays a major role in the effect on the element patterns. When a collinear array of dipoles are spaced 0.5λ apart, then the center element pattern dramatically changes with the addition of elements as shown in Figure 6.24. Increasing the spacing to 0.75λ causes little change to the center element pattern (Figure 6.25).

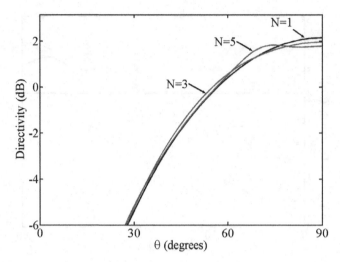

Figure 6.25. Center element pattern in an array of 1, 3, 5, and 7 collinear elements $d = 0.75\lambda$.

6.5. INFINITE ARRAYS

Elements in the center of a large array behave as though the array were infinite in extent. As shown previously, as more elements surround a center element, the impedance and pattern of the center element changes less. Calculating the impedance of an element in an infinite array is much easier than calculating the same impedance for an element in a finite array. The advantage of doing the calculations with an infinite array is that calculations are done on a single element with periodic boundary conditions applied.

An infinite array has the following characteristics:

1. All elements are placed in a periodic lattice.
2. All elements are identical.
3. It has uniform amplitude weights.
4. Relative phase between two adjacent elements is a constant.
5. It radiates a plane wave.
6. It has infinite directivity.
7. It has no sidelobes.

The infinite array derives its characteristics from a uniform array when $N \to \infty$. In the limit, the array factor becomes a Dirac delta function.

$$AF_{N\to\infty} = \frac{\sin(N\psi/2)}{\sin(\psi/2)} = \frac{\sin(N\psi/2)}{\psi/2} = \delta(\psi/2) = 2\delta(\psi) \qquad (6.63)$$

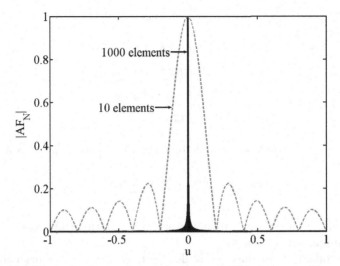

Figure 6.26. Normalized array factors for 10- and 1000-element uniform arrays.

Figure 6.26 is a visual demonstration of (6.63) as N increases from 10 to 1000 when $d = 0.5\lambda$ and the array factors are normalized. The 1000-element array factor looks very similar to a delta function. The main beam and grating lobes appear at locations given by the equations in Chapter 2. Since an infinite array transmits an infinite amount of power at one angle, the directivity is infinite.

All the elements in an infinite array see the same environment, so they all have the same reflection coefficient. Since the reflection coefficient, Γ, is a function of scan angle, then the realized gain for a lossless element is related to the gain of the element by [2]

$$G_{\text{realized}}(\theta, \phi) = G(\theta, \phi)[1 - \Gamma^2(\theta, \phi)] \tag{6.64}$$

It is important that Γ stays small within the scan range of the array. Thus, finding Γ or the element impedance is very important in array design.

6.5.1. Infinite Arrays of Point Sources

Consider the case of an infinite linear array of point sources spaced d along the x axis. If the elements have equal amplitude weights and a beam steering phase shift, then the current at element n is given by

$$I_n(x) = I_0 e^{-jndk_{xs}} \delta(x - nd)\delta(y)\delta(z) \tag{6.65}$$

when the beam is steered to θ_s and $k_{xs} = k \sin \theta_s$. The array factor is given by

$$AF = \sum_{n=-\infty}^{\infty} \int_{-\infty}^{\infty} \int_{-\infty}^{\infty} \int_{-\infty}^{\infty} e^{-jx'k_{xs}} \delta(x' - nd)\delta(y')\delta(z') e^{j(k_x x' + k_y y' + k_z z')} dx' dy' dz' \tag{6.66}$$

where

$$k^2 = k_x^2 + k_y^2 + k_z^2$$
$$k_x = k \sin\theta \cos\phi$$
$$k_y = k \sin\theta \sin\phi$$
$$k_z = k \cos\theta$$

Using the sifting property of the delta function, the array factor is written as

$$\text{AF} = \sum_{n=-\infty}^{\infty} e^{-jndk_{xs}} e^{jk_x nd} = \sum_{n=-\infty}^{\infty} e^{jnd(k_x - k_{xs})} \tag{6.67}$$

This equation looks like the array factor for a uniform array, except there are an infinite number of elements.

If a periodic function has slowly decaying terms, then it Fourier transform has terms that quickly decay. The Poisson sum relates an infinite periodic series to its infinite periodic Fourier transform and is given by

$$\frac{2\pi}{d} \sum_{n=-\infty}^{\infty} \tilde{f}\left(\frac{2\pi n}{d}\right) = \sum_{n=-\infty}^{\infty} f(nd) \tag{6.68}$$

where $\tilde{f}$ is the Fourier transform of f. Since $k = 2\pi/\lambda$ and the 2π is already contained in the variable, the following formulation for Fourier transforms is used:

$$f(y) = \frac{1}{2\pi} \int_{-\infty}^{\infty} e^{jk_y y} \tilde{f}(k_y) dk_y$$
$$\tilde{f}(k_y) = \int_{-\infty}^{\infty} e^{-jk_y y} f(y) dy \tag{6.69}$$

The infinite Fourier series in (6.67) can be replaced by an infinite sum of Dirac delta functions using the Poisson sum formula.

$$\text{AF} = \sum_{n=-\infty}^{\infty} e^{jnd(k_x - k_{xs})} = \frac{2\pi}{d} \sum_{n=-\infty}^{\infty} \delta\left(k_x - k_{xs} - \frac{2\pi n}{d}\right) \tag{6.70}$$

This expression for the array factor is called the grating lobe or Floquet series [2], because the location of the Dirac delta functions (Floquet modes) are the same as the location of the grating lobes (see Chapter 2).

$$k_x = k_{xs} + \frac{2\pi n}{d}$$
$$\sin\theta_g = \sin\theta_s + \frac{\lambda n}{d} \tag{6.71}$$

A similar derivation is possible for planar arrays. If the planar array in the x–y plane has rectangular spacing, then

$$
\text{AF} = \sum_{n=-\infty}^{\infty} \sum_{m=-\infty}^{\infty} \int_{-\infty}^{\infty} \int_{-\infty}^{\infty} \int_{-\infty}^{\infty} e^{-jx'k_{xs}} e^{-jy'k_{ys}} \delta(x'-nd_x)\delta(y'-md_y)\delta(z')
$$

$$
e^{j(k_x x' + k_y y' + k_z z')} dx' dy' dz' \tag{6.72}
$$

Using the Poisson sum, (6.72) becomes [10]

$$
\text{AF} = \sum_{n=-\infty}^{\infty} e^{jnd(k_x - k_{xs})} \sum_{m=-\infty}^{\infty} e^{jmd(k_y - k_{ys})}
$$

$$
= \frac{4\pi^2}{d_x d_y} \sum_{n=-\infty}^{\infty} \delta\left(k_x - k_{xs} - \frac{2\pi n}{d_x}\right) \sum_{m=-\infty}^{\infty} \delta\left(k_y - k_{ys} - \frac{2\pi m}{d_y}\right) \tag{6.73}
$$

The grating lobe series for point sources in an infinite array with triangular spacing is given by

$$
\text{AF} = \frac{4\pi^2}{d_x d_y} \sum_{n=-\infty}^{\infty} \delta\left(k_x - k_{xs} - \frac{2\pi n}{d_x}\right) \sum_{m=-\infty}^{\infty} \delta\left(k_y - k_{ys} + \frac{2\pi m}{d_y} - \frac{4\pi n}{d_y}\right) \tag{6.74}
$$

An interesting observation from these grating lobe series is that the antenna radiates a plane wave at very specific angles and nowhere else. In other words, there are no sidelobes. The grating lobe series is frequently referred to as a Floquet series. Each main lobe that radiates is known as a Floquet mode. Thus, the total power radiated is the sum of all the Floquet modes, because there are no sidelobes.

Example. Graph and label the Floquet modes for a 1000-element uniform array with $d = 1.0\lambda$ steered 30° from broadside. Figure 6.27 shows the plots. Since the 1000-element array factor looks like two delta functions, they can be treated as two Floquet modes.

6.5.2. Infinite Arrays of Dipoles and Slots

The driving point impedance of a $L_x \times w_y$ flat dipole or slot in an infinite array in a $d_x \times d_y$ rectangular lattice (Figure 6.28) can be calculated analytically using the following formulas [11]:

1. Dipoles in free space:

$$
Z_d = \frac{240 L_x^2}{\pi d_x d_y} \left[F_0^2 G_0^2 H_{00} - j \sum_{\substack{m=-\infty \\ m\neq 0}}^{\infty} \sum_{\substack{n=-\infty \\ n\neq 0}}^{\infty} F_n^2 G_m^2 \Psi_{mn} \right] \tag{6.75}
$$

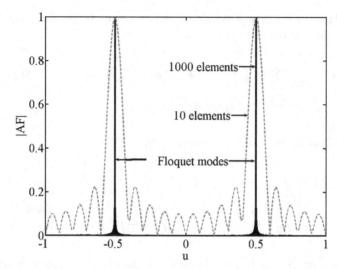

Figure 6.27. Normalized array factors for 10- and 1000-element uniform arrays with $d = 1.0\lambda$ and steered 30° from broadside.

2. Dipoles h above a ground plane:

$$Z_d = \frac{480L_x^2}{\pi d_x d_y}\left\{F_0^2 G_0^2 H_{00}\sin^2(kh\cos\theta) + \frac{j}{2}\left[F_0^2 G_0^2 H_{00}\sin(2kh\cos\theta) - \sum_{\substack{m=-\infty \\ m\neq 0}}^{\infty}\sum_{\substack{n=-\infty \\ n\neq 0}}^{\infty} F_n^2 G_m^2 \Psi_{mn}\right]\right\}$$

$$(6.76)$$

3. Slots in a $a \times b$ rectangular waveguide:

$$\frac{Y}{Y_r} = \frac{\lambda_g ab}{2\lambda d_x d_y}\left[\frac{1-(L_x/a)^2}{\cos(\pi L_x/2a)}\right]F_0^2 G_0^2 H_{00} + j\left\{\frac{4b}{\lambda_g}\left[\ln\csc\frac{\pi w_y}{2b} + \frac{1}{2}\left(\frac{b}{\lambda_g}\right)^2\cos^4\frac{\pi w_y}{2b}\right]\right.$$

$$-\frac{4b}{\lambda_g}\left(\frac{\lambda_g}{\lambda_{g3}}\right)^2\left[\frac{\cos(3\pi L_x/2a)(1-L_x^2/a^2)}{\cos(\pi L_x/2a)(1-9L_x^2/a^2)}\right]\left[1+\left(\frac{\pi w_y}{2\lambda_{g3}}\right)^2\right]\ln\left(\frac{4\lambda_{g3}}{\gamma\pi b'}\right)$$

$$\left.-\frac{\lambda_g ab}{2\lambda d_x d_y}\left[\frac{1-L_x^2/a^2}{\cos(\pi L_x/2a)}\right]^2\sum_{\substack{m=-\infty \\ m\neq 0}}^{\infty}\sum_{\substack{n=-\infty \\ n\neq 0}}^{\infty} F_n^2 G_m^2 \Psi_{mn}\right\} \qquad (6.77)$$

where

$$F_n(\theta,\phi) = \frac{\sin[k_y w_y/2 + n\pi w_y/d_y]}{k_y w_y/2 + n\pi w_y/d_y}$$

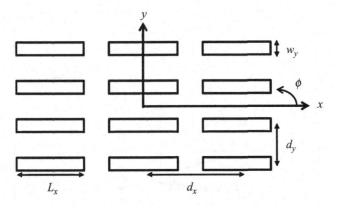

Figure 6.28. Infinite rectangular lattice of flat dipoles or slots.

$$G_m(\theta, \phi) = \frac{\cos(k_x L_x/2 + m\pi L_x/d_x)}{1 - (k_x L_x/\pi + 2m\pi L_x/d_x)^2}$$

$$H_{00}(\theta, \phi) = \frac{1 - (k_x/k)^2}{\cos\theta}$$

$$\Psi_{mn}(\theta, \phi) = \frac{(k_x/k + m\lambda/d_x)^2 - 1}{\sqrt{(k_x/k + m\lambda/d_x)^2 + (k_y/k + n\lambda/d_y)^2 - 1}}$$

$$k_x = k\sin\theta\cos\phi$$

$$k_y = k\sin\theta\sin\phi$$

$$\gamma = 1.781$$

$$\lambda_{g3} = \left| \frac{\lambda}{1 - (3\lambda/2a)^2} \right|$$

$$Y_r = \frac{\lambda}{\lambda_g Z_0}$$

$$\lambda_g = \frac{2\pi}{\sqrt{k^2 - (\pi/a)^2}}$$

These formulas are not sensitive to w_y.

Example. Plot the normalized driving point resistance for a half wavelength dipole in an infinite array with $d_x = 0.5\lambda$ and $d_y = 0.5\lambda$ in free space and 0.25λ above a ground plane.

Figure 6.29 is a plot of the real part of (6.75), and Figure 6.30 is a plot of the real part of (6.76). The array is scanned as a function of θ in the E plane ($\phi = 0°$), the diagonal plane ($\phi = 45°$), and the H plane ($\phi = 90°$). The plots

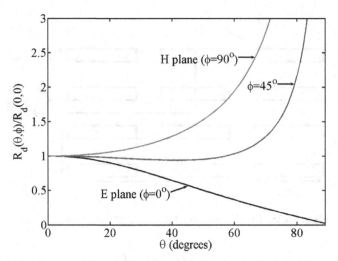

Figure 6.29. Real part of the driving point impedance of an infinite array of half-wavelength dipoles.

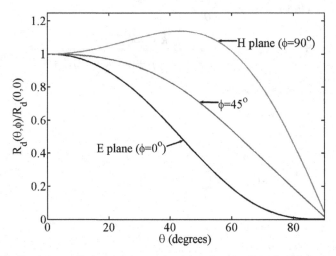

Figure 6.30. Real part of the driving point impedance of an infinite array of half-wavelength dipoles 0.25λ above a ground plane.

are nearly identical until about 20° when they begin to diverge. There is a 50% spread in the normalized resistance at 35° for the no ground plan case and at 37° for the ground plane case.

Another approach to the analysis of infinite arrays is to place an element inside a unit cell. As an example, Figure 6.31 shows a unit cell that contains a slot in an infinite array [3]. The unit cell extends from the array to infinity in

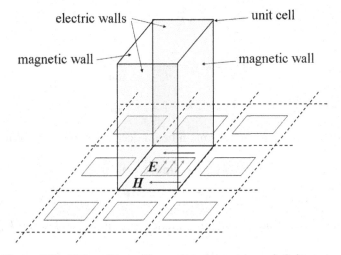

Figure 6.31. Unit cell bounding a slot element in an infinite array.

the z direction. The walls of the unit cell have either electric or magnetic boundary conditions, depending upon the direction of the electric or magnetic fields in the slots. The walls do not perturb the fields; but once the walls are in place, all fields outside the walls can be ignored. In a way, the unit cell behaves like a waveguide. As shown in Figure 6.31, the magnetic walls are parallel to the electric field in the slot while the electric walls are parallel to the magnetic field in the slot.

The dominant mode in the unit cell waveguide is TEM. Only one mode exists if the size of the waveguide (element spacing) is chosen properly for the operating frequency. If there are no higher-order modes, then there are no grating lobes either. When the array scans in the x–z plane, the amplitude of the fields remains the same, but a constant phase difference exists between consecutive slots. Opposite walls of the unit cell are identical except for a phase difference that is a function of the steering angle. In the unit cell, the walls parallel to the long slot dimension remain electric walls while the other walls differ by a phase shift. The dominant mode in the unit cell becomes TE since a longitudinal component of the field is present in the z direction. Scanning in the y–z plane produces a dominant TM mode. Scanning at arbitrary angles produces both TE and TM modes.

Assume the elements in the infinite array lie in a plane with

$$(x_{mn}, y_{mn}) = \left(md_x + \frac{nd_y}{\tan\gamma}, nd_y \right) \tag{6.78}$$

where $\gamma = 90°$ is rectangular spacing. All the Floquet modes associated with this element spacing have propagation constants given by

$$k_{xmn} = k_{x00} + \frac{2m\pi}{d_x} \tag{6.79}$$

$$k_{ymn} = k_{y00} + \frac{2n\pi}{d_y} - \frac{2m\pi}{d_x \tan\gamma} \tag{6.80}$$

$$k_{zmn} = \sqrt{k_0^2 - k_{xmn}^2 - k_{ymn}^2} \tag{6.81}$$

The dominant Floquet mode or main beam occurs when m and n are zero.

$$k_{x00} = k_0 \sin\theta_s \cos\phi_s \tag{6.82}$$

$$k_{y00} = k_0 \sin\theta_s \sin\phi_s \tag{6.83}$$

Floquet modes become propagating modes or grating lobes when $k_{zmn} \geq 0$ or

$$k_0^2 \geq k_{xmn}^2 + k_{ymn}^2 \tag{6.84}$$

Example. If a planar array is scanning in the x–z plane, find the maximum element spacing that still excludes the $m = 0$ and $n = 0$ Floquet mode.

The dominant mode occurs when $m = 0$ and $n = 0$. The first higher-order mode occurs when $m = 1$ and $n = 0$ leading to $k_z = |k\sin\theta - 2\pi/d_x|$. To keep this mode below cutoff, then $k < k_z$ and

$$d_x < \frac{\lambda}{1 + \sin\theta} \tag{6.85}$$

which is the same as the element spacing where no grating lobes appear.

6.6. LARGE ARRAYS

Some tricks can be played to reduce the computation time for calculating the characteristics of a large array. As the array gets larger, more elements have neighboring elements completely surrounding them, so the mutual coupling of these elements is similar. As such, they can be treated to have identical characteristics with small variations that tend to average out as the number of elements increases. The percent of edge elements in a rectangular planar array with a rectangular element grid is given by

$$\% \text{ edge elements} = \frac{2N_x + 2N_y - 4}{N_x N_y} \times 100 \tag{6.86}$$

A contour plot of (6.86) appears in Figure 6.32. The number of edge elements depends upon the total number of elements in the array and the ratio of N_x

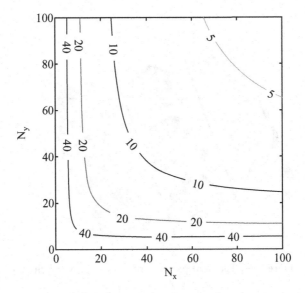

Figure 6.32. Percentage of edge elements in a rectangular planar array with a rectangular element grid.

1	2	3
4	5	6
7	8	9

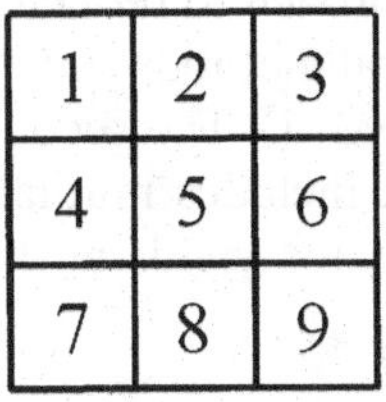

Figure 6.33. A 3×3 array with 8 edge elements and 1 center element.

to N_y. As an example, a 400-element rectangular planar array with a rectangular element grid has 24% edge elements when it is 40×10, and it has only 19% edge elements when it is 20×20. A square array always has fewer edge elements than a rectangular array. If the array increases in size until it is 40×40, then the percent of edge elements decreases to 9.75%. Other array contours, such as a hexagonal shape, have even more edge elements than predicted by (6.86).

A 3×3 square array on a square lattice (Figure 6.33) has 9 elements with 9 different element patterns. If the elements are dipoles, then the patterns can be calculated using the MoM as shown in Figure 6.34. Elements 1, 3, 7, and 9 have the same pattern but are rotated by $0°$, $90°$, $180°$, and $270°$, respectively. Element pattern 4 is flipped $180°$ relative to element 6. Element pattern 2 is flipped $180°$ relative to element 8. Element 5 is in the center and has a unique pattern that is symmetric about x and y.

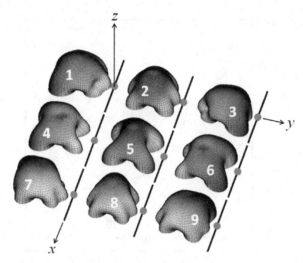

Figure 6.34. Element patterns of a 3 × 3 array of dipoles.

Calculating the array far-field pattern and mutual coupling when a large array has more complicated elements than dipoles requires a lot of computer time and memory. The first approach to reducing computer time and memory requirements is to use numerical methods suited to the problem at hand, such as the fast multiple method (FMM). The second approach is to find a representative element pattern that includes mutual coupling effects. Multiplying this pattern times the array factor yields an accurate approximation of the array far-field pattern.

6.6.1. Fast Multipole Method

The FMM was named one of the "Top Ten Algorithms of the Century" [12] and has found extensive use in electromagnetics. FMM accelerates the iterative solver in the MoM by replacing the Green's function using a multipole expansion that groups sources lying close together and treats them like a single source [13]. Figure 6.35 shows that the sources lying in region Q that independently interact with P can be grouped together and replaced by a single source that interacts with P. Both MoM and FMM use basis functions to model interactions between triangles and segments, but unlike MoM, FMM groups basis functions and computes the interaction between groups of basis functions, rather than between individual basis functions. Figure 6.36 is a diagram of the hierarchical spatial decomposition that separates the simulation space into regions that are far enough apart to interact according to the approximate grouping shown in Figure 6.35. As the distance increases, larger regions of space can be grouped into single approximations. By treating the interactions between far-away basis functions using the FMM, the corresponding matrix

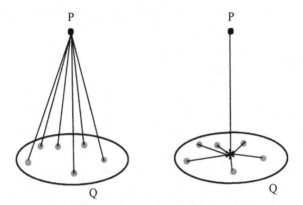

Figure 6.35. Fast multipole method.

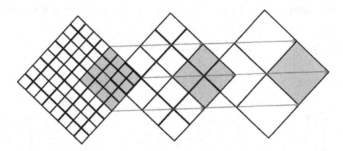

Figure 6.36. Diagram of the spatial decomposition used in FMM.

elements do not need to be explicitly stored, resulting in a significant memory requirement [13]. The FMM can reduce the matrix–vector product in an iterative solver from $O(N^2)$ to $O(N\log N)$ in memory [14].

The FMM often creates incredible memory and time savings when applied to antennas, such as an array of dipole antennas. Consider a square array of N dipoles in a square grid with element spacing $d = \lambda/2$. As shown in Figure 6.37, the amount of memory required for FMM increases linearly with the number of elements, and Figure 6.38 shows a similar graph for time. Memory and time requirements for the MoM increase extremely fast. In fact, for a 21×21 array, the MoM takes 1700 MB and 2977 s, while the FMM only takes 75.5 MB (22.5 times less) and 79 s (37.7 times less). In this case, the FMM models a much larger array than does the MM alone.

The FMM is not always superior to the MoM, though. If the elements in an array are spaced far apart, then the calculations using FMM are very slow compared to MoM. Calculating the currents and far field of a 5×20 array of dipoles spaced $\lambda/2$ apart is done much more efficiently using FMM. In that case, the calculation only requires 24 s versus 314 s using MoM, which is 13

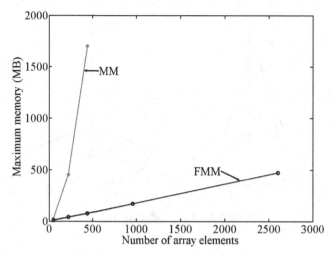

Figure 6.37. Maximum memory used to calculate the currents on an array of N dipoles.

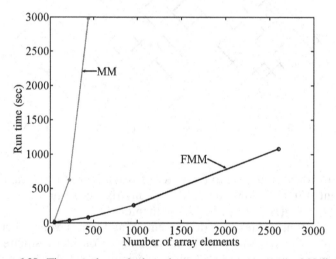

Figure 6.38. Time used to calculate the currents on an array of N dipoles.

times faster. On the other hand, increasing the element spacing to 10λ increases the FMM calculation to 5719 s while not changing the MoM calculation time. This means the FMM was 18 times slower. Thus, FMM becomes slow when the sources are spread far apart in space.

As another example, the patch antenna meshed in Table 6.1 is modeled using FMM and MM. Table 6.2 shows the time and memory requirements for this problem using MM and FMM when the maximum triangle length is $\lambda/5$ and $\lambda/15$. In this case, the FMM takes more memory and more time than the MoM. Although FMM can offer exceptional speed and memory use improve-

TABLE 6.2. Time and Memory Needed to Calculate the S_{11} of a Patch Antenna Using Both MoM and Fast Multipole Method

Max Triangle Edge Length	MoM		FMM	
	Time (seconds)	Memory (MB)	Time (seconds)	Memory (MB)
$\lambda/5$	85	23	375	63
$\lambda/15$	10,124	1380	48,396	2480

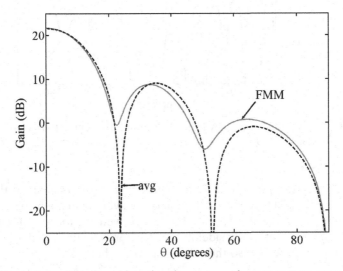

Figure 6.39. Array pattern found by using the average element pattern compared with the FMM solution at $\phi = 0$.

ments in many problems, there are situations where FMM does not work well, and the user should beware.

6.6.2. Average Element Patterns

An average element pattern is the average of all the element patterns in the array. Replacing all the element patterns with the average element pattern and then multiplying by the array factor results in an accurate reproduction of the array pattern. The element pattern for each element in the array is calculated when all other elements are terminated in a matched load. Their average then replaces each element in the array. Since the pattern contains mutual coupling effects, the resulting array pattern is more accurate than when an isolated element pattern is used. As an example, consider a planar array of x-oriented half-wavelength dipoles spaced $\lambda/2$ apart with 20 elements in the y direction and 5 elements in the x direction. A 25-dB Taylor taper is placed across the y direction while a uniform taper is in the x direction. Figure 6.39

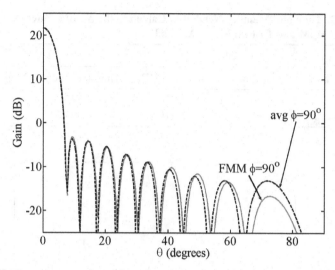

Figure 6.40. Array pattern found by using the average element pattern compared with the FMM solution at $\phi = 90°$.

is a cut of the array pattern at $\phi = 0$, while Figure 6.40 is a cut of the array pattern at $\phi = 90°$. The FMM solution and the average element pattern solution agree well for the $\phi = 90°$ cut, but not so well for the $\phi = 0°$ cut. Even though the $\phi = 90°$ cut is low sidelobe, it has more elements and a lower percentage of edge elements, so the averaging works better.

6.6.3. Representative Element Patterns

Another approach is to calculate the element patterns for each element in a 3×3 array as shown in Figure 6.33. There are 9 different element patterns, only 4 if rotations are not included. These element patterns replace similar element patterns in larger arrays that have the same element grid as the 3×3 array. Thus, for a $N_x \times N_y$ array, the element patterns of elements 1, 3, 7, and 9 of the 3×3 array are placed at the corners of the $N_x \times N_y$ array. Elements 2, 4, 6, and 8 replace the elements on the edges of the larger array. Finally, element 5 of the 3×3 array replaces all interior elements of the $N_x \times N_y$ array. Table 6.3 lists the number of element patterns from the 3×3 array that are used in an $N_x \times N_y$ array. For instance, the element patterns of a 5×20 element array of dipoles spaced $d_x = d_y = 0.5\lambda$ in a square grid can be represented by the elements in the 3×3 as shown in Figure 6.41. Using this approach results in the array patterns shown in Figures 6.42 and 6.43. At least for this array, the average element pattern worked better than the representative element pattern approach. The results deviate at the peak of the main beam as well as in the sidelobes. Even though the average element pattern results are better, calculating the average element pattern of a large array takes considerable

TABLE 6.3. Number of Elements from the 3 × 3 Used in a $N_x \times N_y$ Array

Element:	1	2	3	4	5	6	7	8	9
Number:	1	$2(N_x-2)$	1	$2(N_y-2)$	(N_x-2) (N_y-2)	$2(N_y-2)$	1	$2(N_x-2)$	1

1	2	2	2	2	2	2	2	2	2	2	2	2	2	2	2	2	2	2	3
4	5	5	5	5	5	5	5	5	5	5	5	5	5	5	5	5	5	5	6
4	5	5	5	5	5	5	5	5	5	5	5	5	5	5	5	5	5	5	6
4	5	5	5	5	5	5	5	5	5	5	5	5	5	5	5	5	5	5	6
7	8	8	8	8	8	8	8	8	8	8	8	8	8	8	8	8	8	8	7

Figure 6.41. A 5 × 20 array is built from the 9 element patterns.

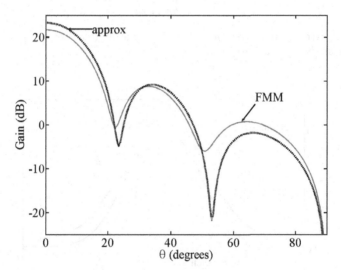

Figure 6.42. Array pattern ($\phi = 0°$) for a 5 × 20 array found by using the element patterns from the 3 × 3 array.

more time than calculating the 9-element patterns of a 3 × 3 array (only 4-element patterns are needed if symmetry is invoked).

6.6.4. Center Element Patterns

Figure 6.44 shows the element patterns for each element in a 3-element array of parallel dipoles lying in the x direction, spaced $\lambda/2$ apart along the y axis. Averaging these element patterns produces the lower-gain, symmetric pattern labeled "3 avg". Calculating the array pattern by multiplying this element pattern times the array factor is much faster than calculating the array pattern

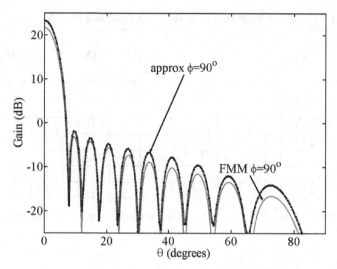

Figure 6.43. Array pattern ($\phi = 90°$) for a 5×20 array found by using the element patterns from the 3×3 array.

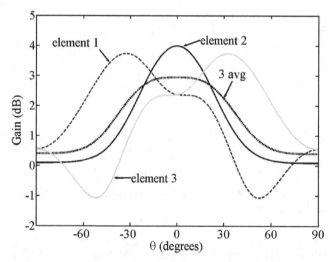

Figure 6.44. Element patterns from a 3-element array of parallel dipoles lying in the x direction, spaced $\lambda/2$ apart along the y axis with their average element pattern.

using a full wave electromagnetic solver for the large array. The question is whether such an approach yields accurate results.

Figure 6.44. The element patterns from a 3-element array of parallel dipoles lying in the x direction, spaced $\lambda/2$ apart along the y axis with their average element pattern.

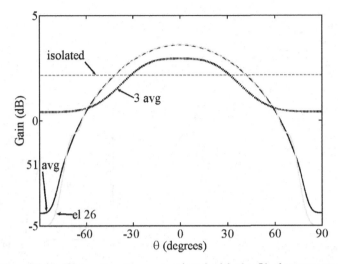

Figure 6.45. Element patterns associated with the 51-element array.

Consider a 51-element array of parallel dipoles lying in the x direction, spaced $\lambda/2$ apart along the y axis. The average element pattern for this array along with the center element pattern (element 26) are graphed in Figure 6.45 along with the isolated element pattern and the "3 avg" pattern from Figure 6.44. The average element pattern and the element pattern of element 26 are nearly identical and would be very close to the element pattern in an infinite array. Multiplying the element patterns in Figure 6.45 by the array factor for a 51-element array results in the antenna patterns shown in Figure 6.46. The FMM solution for the full array is also shown for comparison. The biggest discrepancy between these patterns occurs at the main beam and the far-out sidelobes. The expanded areas in Figure 6.46 show that the center element and the average element pattern for the 51-element array provide excellent matches to the FMM solution. Since calculating 51-element patterns, averaging them, and then multiplying by the array factor takes longer than the FMM solution, the best approach is to use the element pattern of the center element of the large array to represent all element patterns in the array. Calculating the element pattern for an infinite array or the center element pattern is easier and faster than calculating the element pattern of the center element of a large array, so it is the preferred approach.

Example. Compare the center and average element patterns of a 3×3 array of rectangular patches to the element pattern of the patch in an infinite array (Figures 6.47–6.50).

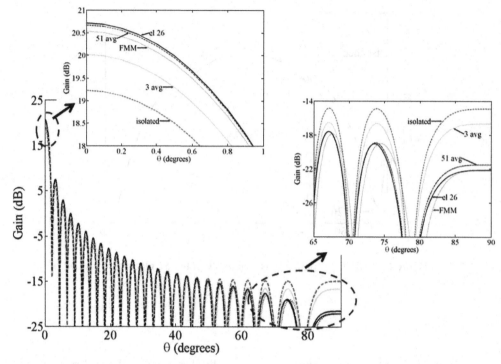

Figure 6.46. Array pattern for the 51-element array calculated using FMM and four approaches to multiplying an element pattern times the array factor.

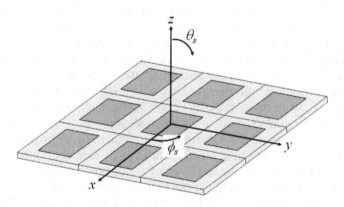

Figure 6.47. A 3×3 array of rectangular patches.

6.7. ARRAY BLINDNESS AND SCANNING

Arrays must be designed for their entire scanning range, not just for broadside. Wheeler showed that for any infinite array, the impedance at a scan angle, θ_s, is approximately equal to [15]

Figure 6.48. Element patterns of the 3 × 3 array of rectangular patches.

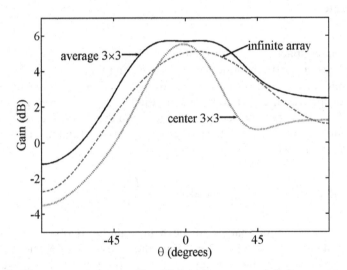

Figure 6.49. Element pattern cuts ($\phi = 0°$) for the average, infinite array, and center element patterns.

$$\frac{Z_{\theta=\theta_s}}{R_{\theta=0}} = \begin{cases} 1/\cos\theta_s, & H \text{ plane} \\ \cos\theta_s, & E \text{ plane} \end{cases} \qquad (6.87)$$

In an E-plane scan, the impedance approaches infinity toward endfire, while for the H-plane scan, the impedance approaches zero at endfire. This result is independent of element spacing. The emergence of grating lobes, which is a function of element spacing, also causes extreme impedance mismatch. As a

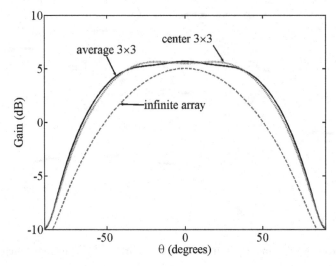

Figure 6.50. Element pattern cuts ($\phi = 90°$) for the average, infinite array, and center element patterns.

result, element spacing must be small at the highest frequency and maximum scan angle to preclude the emergence of a grating lobe.

Lechtreck discovered another scan limiting phenomenon that causes a large impedance mismatch closer to broadside than the angle where grating lobes emerged [16]. The mismatch caused a large dip in the E-plane element pattern due to a surface wave on the face of the array. He postulated that this blindness angle occurs when

$$|\sin \theta_{\text{blind}}| = \left(\frac{\lambda_0}{d}\right) - \left(\frac{c_0}{v_s}\right) \tag{6.88}$$

where d is the element spacing in the scan plane, v_s is phase or slow wave velocity of a surface wave traveling along the array, c_0 is the velocity of light in free space, and λ_0 is the wavelength in free space. Note that this equation reduces to the equation for predicting grating lobes when $v_s = c_0$, or all the radiation travels away from the array. Lechtreck verified (6.88) by measuring a significant dip in the center element pattern at $\pm67°$ of a rectangular 65-element array with hexagonal spacing when $d = 0.506\lambda$ and $v_s = 0.93c_0$ (measured). Decreasing the surface wave velocity causes the blindness to occur closer to broadside. Decreasing the element spacing moves the blindness angle away from broadside [17].

Scan blindness occurs when an array neither transmits nor receives power (100%—or at least a very large amount—is reflected) at certain scan angles. It occurs when the steering phase of the elements matches the phase propagation of a surface wave on the face of the antenna. Thus, an electromagnetic wave excited by the elements travels as a surface wave rather than propagating away from the face of the array. In order to induce scan blindness, a structure that can support a slow wave must be near or on the array face [18]. Waveguide arrays with dielectric covers, microstrip arrays over a thick dielectric substrate, and corrugated surfaces are examples where waves become trapped inside the dielectric slab and propagate parallel to the elements. Scan blindness becomes worse as the array size increases and is not a severe problem for small arrays. Complete scan blindness when 100% of the power is reflected from the elements only occurs in infinite arrays [19].

Scanning to the blind angle means that k_z becomes imaginary while a surface wave mode propagates along the surface of the array. Blindness occurs when the imaginary part of the impedance becomes extremely large even when the real part may be nonzero [20]. In order for blindness to occur, k_z must be zero, and the wave propagation constant must equal the surface wave propagation constant.

$$k_{sw} = \sqrt{k_{xmn}^2 + k_{ymn}^2} \tag{6.89}$$

Surface waves exist in a dielectric substrate of thickness h when for the TE mode when [21]

$$k_{zd} \cos k_{zd} h + j k_{z0} \sin k_{zd} h = 0 \tag{6.90}$$

and for the TM mode when

$$\varepsilon_r k_{z0} \cos k_{zd} h + j k_{zd} \sin k_{zd} h = 0 \tag{6.91}$$

where 0 in the subscript indicates free space and d indicates dielectric. For substrates with $h < \lambda_0/4/\sqrt{\varepsilon_r - 1}$ only the lowest order TM surface wave mode exists. If there are only currents in the x direction (e.g., x-oriented printed dipoles), then polarization mismatch prevents a TM surface wave for H-plane scanning. The surface wave propagation constant should include the presence of the microstrip patches, but for most practical purposes, the patches can be ignored and the unloaded surface wave propagation constant used with an error in blindness angle of much less than one degree [22].

Schaubert found that E-plane scan blindness in single polarized arrays of tapered slot antennas with a ground plane can be predicted by approximating a corrugated surface parallel to the x–z plane with corrugations t thick and h tall (length of TSA) and separated by the element spacing, d_x. He found that blindness occurs when [23]

$$\tan k_{zd} h - \frac{\alpha_z}{k_{zd}} = 0 \tag{6.92}$$

where

$$\alpha_z = k_0 \sqrt{\sin^2 \theta_{\text{blind}} + \left(\frac{\lambda_0}{d_x - t} \right)^2 - 1} \qquad (6.93)$$

$$k_z = k_0 \sqrt{1 - \sin^2 \theta_{\text{blind}} - \left(\frac{\lambda_0}{d_x - t} \right)^2} \qquad (6.94)$$

If $a - t > \lambda/2$, then a TM_z mode with the electric field in the y–z plane exists between the parallel plates. The dielectric thickness and permittivity of a TSA do not greatly affect the blindness for the thin dielectric substrates used to make them. The scan blindness is very dependent on the H-plane separation and the length of the depth of the corrugation. Blindness moves toward endfire as the H-plane spacing, frequency, or antenna depth increases.

Example. Plot the blind angle over a frequency range of 3.8 to 4.2 GHz for an infinite planar array of TSAs that scan in the E plane. Assume $h = 3.5$ cm, $d_x = d_y = 4.43$ cm.

Figure 6.51 shows the plot of the blind angle as a function of frequency. Increasing the frequency moves the blind angle away from broadside.

Usually, the dielectric slab is very thin, so only the lowest-order TM mode propagates. The first TM surface wave propagation constant is bounded by $k_0 < k_{sw} < \sqrt{\varepsilon_r} k_0$. A surface wave resonance occurs when k_{sw} equals a Floquet mode propagation constant or [10]

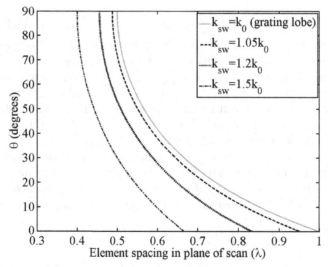

Figure 6.51. Scan blindness as a function of frequency for the E-plane TSA array.

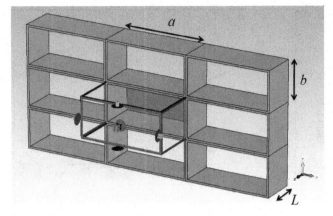

Figure 6.52. Infinite array of open-ended rectangular waveguides.

$$k_{zdmn} = \sqrt{k_d^2 - k_{xdmn}^2 - k_{ydmn}^2} \qquad (6.95)$$

$$k_{z0mn} + \frac{jk_{zdmn}\tan(k_{zdmn}h)}{\varepsilon_r} = 0 \qquad (6.96)$$

$$k_{z0mn} = \sqrt{k_0^2 - k_{pmn}^2} \qquad (6.97)$$

$$k_{zdmn} = \sqrt{\varepsilon_r k_0^2 - k_{pmn}^2} \qquad (6.98)$$

$$k_{pmn}^2 = k_{xmn}^2 + k_{ymn}^2 \qquad (6.99)$$

When $k_0 h \sqrt{\varepsilon_r}$ is small, then

$$k_{ps} = k_0 \left\{ 1 + 0.5[k_0 h(1 - 1/\varepsilon_r)]^2 \right\} \qquad (6.100)$$

Example. An infinite planar array of X-band rectangular waveguides are in a rectangular lattice in the x–y plane (Figure 6.52). Assume $a = 2.29$ cm, $b = 1.02$, and the wall thickness is 1 mm. Use the unit cell approximation to plot the E-plane and H-plane element patterns at 10 GHz.

Figure 6.53 shows the plot of the E-plane and H-plane element patterns of the array.

6.8. MUTUAL COUPLING REDUCTION/COMPENSATION

Changing the array environment changes the mutual coupling. Adding a ground plane to a dipole array improves the H-plane scan but deteriorates the E-plane scan. In order to improve the E-plane scan without affecting the H-plane scan, a periodic array of conducting baffles is placed between the dipoles.

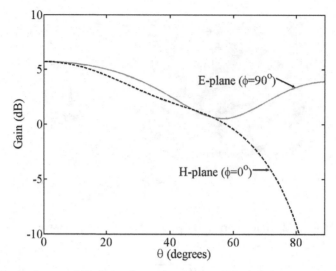

Figure 6.53. *E*-plane and *H*-plane element patterns of the infinite array of rectangular waveguides.

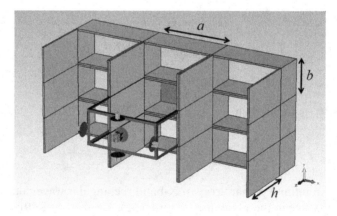

Figure 6.54. Infinite array of open-ended rectangular waveguides with fences in the *y–z* plane.

If the baffles are thin, perpendicular to the ground plane, and running between the dipoles or slots, the baffles will not change the *H*-plane scanning behavior. The height of the baffles determines their effect on the *E*-plane scan [24].

Example. The infinite planar array of X-band rectangular waveguides in Figure 6.52 are modified to have metallic fences in the (a) *E* plane (Figure 6.54) and (b) *H* plane (Figure 6.55). Show how the height of the fences changes the element patterns at 10 GHz.

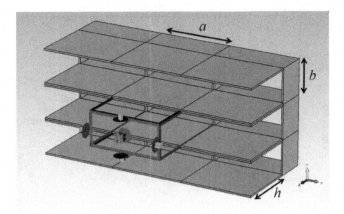

Figure 6.55. Infinite array of open-ended rectangular waveguides with fences in the *x–z* plane.

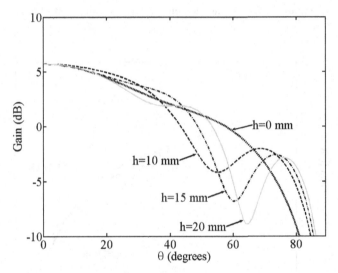

Figure 6.56. *H*-plane element pattern as a function of *h*.

As can be seen in Figure 6.56, the *E*-plane fences induce a null in the *H*-plane element pattern. The null gets deeper and moves away from broadside as the fence gets taller. Figure 6.57 shows that the *E*-plane element pattern has a higher gain out to about 60° due to the fences. Fences in the *x–z* plane have no effect on the *H*-plane element pattern (Figure 6.58) but move the dip in the *E*-plane pattern away from broadside as well as increase its depth (Figure 6.59). The *h* = 10 fence induces a very flat element pattern from 0° to 45°. Another dip in the pattern begins to form at 37° for *h* = 15 mm. It also gets deeper and moves to a higher angle when *h* = 20 mm.

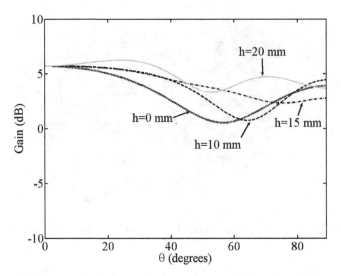

Figure 6.57. *E*-plane element pattern as a function of *h*.

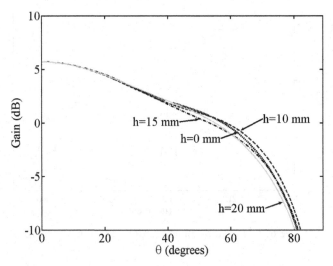

Figure 6.58. *H*-plane element pattern as a function of *h*.

The easiest way to reduce the deleterious effects of mutual coupling is to pack the elements closer together, so that grating lobes do not enter real visible space [25]. Unfortunately, moving the elements closer together also implies the use of narrowband elements. Changing the impedance of the ground plane between elements by using corrugations [26] allows larger element spacings before blindness occurs.

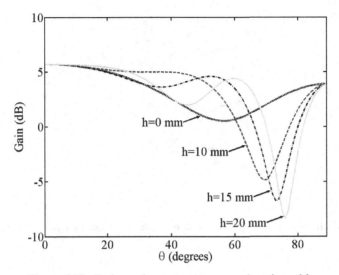

Figure 6.59. *E*-plane element pattern as a function of *h*.

Mutual coupling in a transmit only array can be dramatically reduced by using a circulator with one port terminated with a matched load [27]. This approach cannot be used as a receive array, because the received signal would also be sent to the matched load as well.

A tunable element has an impedance that is a function of the scan angle of the array. This approach is complex and expensive and limits bandwidth.

The position of the element null is a function of the dielectric substrate height and permittivity. Decreasing the height of the dielectric substrate moves the scan blindness closer to endfire. Decreasing the dielectric constant of the substrate will also move the scan blindness closer to endfire.

Example. Demonstrate the effects of dielectric cover thickness and permittivity on the infinite planar array of X-band rectangular waveguides in Figure 6.52.

Figure 6.60 shows the infinite array with a dielectric cover ($\varepsilon_r = 2$) h thick. Figures 6.61 and 6.62 are the *H*- and *E*-plane element patterns for several different values of *h*. The dip in the *E*-plane pattern becomes more prominent as the thickness increases. Figures 6.63 and 6.64 are the *H*- and *E*-plane element patterns for several different values of ε_r when $h = 10$ mm. The dips in both patterns significantly increase with an increase in ε_r.

Surface waves in a microstrip patch substrate can be suppressed by the proper choice of thicknesses for given material properties [28]. The dominant TM surface wave associated with a circular patch can be suppressed by properly selecting the patch size. Unfortunately, the optimum size is not resonant.

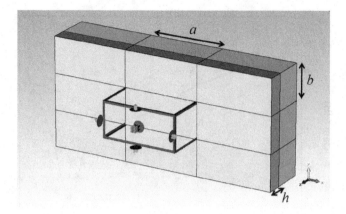

Figure 6.60. Infinite array of open-ended rectangular waveguides with a dielectric cover.

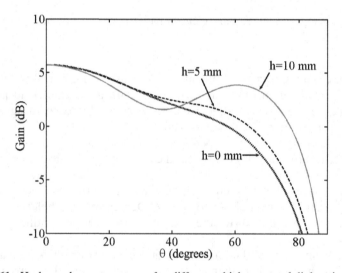

Figure 6.61. *H*-plane element patterns for different thicknesses of dielectric cover.

Two ways to reduce the circular patch size while eliminating the surface wave are [29]:

1. Place a cylindrical core (radius less than the radius of the patch) of dielectric material with a dielectric constant less than the substrate beneath the circular patch.
2. Make the patch an angular ring with several equally spaced shorting pins.

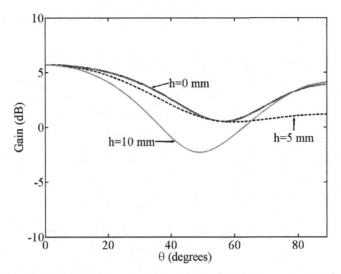

Figure 6.62. *E*-plane element patterns for different thicknesses of dielectric cover.

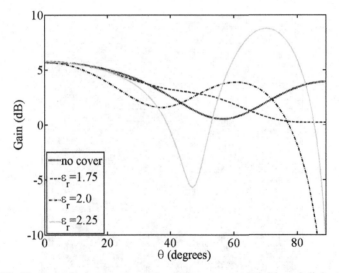

Figure 6.63. *H*-plane element patterns for different permittivities of dielectric cover.

More recent approaches to mitigating the effects of mutual coupling use electromagnetic bandgap (EBG) materials to eliminate surface waves in microstrip arrays. As the substrate dielectric constant increases, the bandwidth decreases. The increased thickness encourages the propagation of unwanted surface waves. Placing a mushroom style EBG structure between the patches has been experimentally shown to reduce coupling between adjacent patches

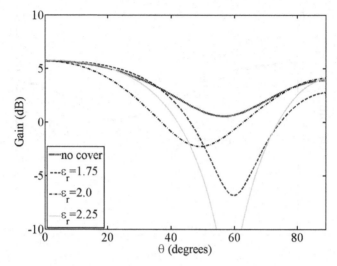

Figure 6.64. *E*-plane element patterns for different permittivities of dielectric cover.

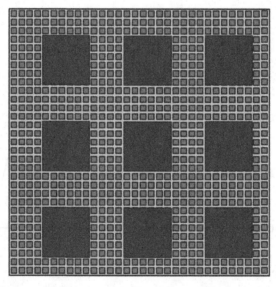

Figure 6.65. Rectangular microstrip patches surrounded by EBG structure.

by up to 8 dB [30]. This approach was extended to eliminate scan blindness in an infinite array of rectangular patches [31]. Several other mutual coupling/ scan blindness mitigation techniques using EBG structures have also been proposed [32–35]. Figure 6.65 is a diagram of an array of rectangular microstrip patches in the midst of an EBG structure.

REFERENCES

1. W. L. Stutzman and G. A. Thiele, *Antenna Theory and Design*, 2nd ed., New York: John Wiley & Sons, 1998.

2. J. L. Allen and B. L. Diamond, *Mutual Coupling in Array Antennas*, Technical Report EDS-66-443, Lincoln Laboratory, 1966.

3. S. Edelberg and A. Oliner, Mutual coupling effects in large antenna arrays: Part 1–Slot arrays, *IRE Trans. Antennas Propagat.*, Vol. 8, No. 3, 1960, pp. 286–297.

4. H. King, Mutual impedance of unequal length antennas in echelon, *IRE Trans. Antennas Propagat.*, Vol. 5, No. 3, 1957, pp. 306–313.

5. R. S. Elliott, *Antenna Theory and Design*, revised edition, Hoboken, NJ: John Wiley & Sons, 2003.

6. R. F. Harrington and IEEE Antennas and Propagation Society, *Field Computation by Moment Methods*, Piscataway, NJ: IEEE Press, 1993.

7. A. F. Peterson, S. L. Ray, and R. Mittra, *Computational Methods for Electromagnetics*, New York: IEEE Press, 1998.

8. C. A. Balanis, *Advanced Engineering Electromagnetics*, New York: John Wiley & Sons, 1989.

9. P. Carter, Jr., Mutual impedance effects in large beam scanning arrays, *Antennas Propagation, IRE Trans.* Vol. 8, No. 3, 1960, pp. 276–285.

10. A. Bhattacharyya, *Phased Array Antennas and Subsystems: Floquet Analysis, Synthesis, BFNs, and Active Array Systems*, Hoboken, NJ: Wiley-Interscience, 2006.

11. A. A. Oliner and R. G. Malech, Mutual coupling in infinite scanning arrays, in *Microwave Scanning Antennas*, R. C. Hansen, ed., New York: Academic Press, 1966, pp. 195–335.

12. J. Dongarra and F. Sullivan, Guest Editors Introduction to the top 10 algorithms, *Comput. Sci. Eng.*, Vol. 2, No. 1, 2000, pp. 22–23.

13. J. Board and L. Schulten, The fast multipole algorithm, *Comput. Sci. Eng.*, Vol. 2, No. 1, 2000, pp. 76–79.

14. A. T. Ihler, *An Overview of Fast Multipole Methods*, Cambridge: MIT, 2004.

15. H. Wheeler, The grating-lobe series for the impedance variation in a planar phased-array antenna, *IEEE Trans. Antennas Propagat.*, Vol. 14, No. 6, 1966, pp. 707–714.

16. L. Lechtreck, Effects of coupling accumulation in antenna arrays, *IEEE Trans. Antennas Propagat.*, Vol. 16, No. 1, 1968, pp. 31–37.

17. H. A. Wheeler, A systematic approach to the design of a radiator element for a phased-array antenna, *Proc. IEEE*, Vol. 56, No. 11, 1968, pp. 1940–1951.

18. D. Pozar and D. Schaubert, Scan blindness in infinite phased arrays of printed dipoles, *IEEE Trans. Antennas Propagat.*, Vol. 32, No. 6, 1984, pp. 602–610.

19. G. H. Knittel, A. Hessel, and A. A. Oliner, Element pattern nulls in phased arrays and their relation to guided waves, *Proc. IEEE*, Vol. 56, No. 11, 1968, pp. 1822–1836.

20. A. K. Bhattacharyya, Floquet-modal-based analysis for mutual coupling between elements in an array environment, *IEE Proc. Microwaves Antennas Propagat.*, Vol. 144, No. 6, 1997, pp. 491–497.

21. D. M. Pozar, Analysis and design considerations for printed phased-array antennas, in *Handbook of Microstrip Antennas*, J. R. James and P. S. Hall, eds., Stevenage, UK: Peregrinus, 1989, 693–754.

22. D. Pozar and D. Schaubert, Analysis of an infinite array of rectangular microstrip patches with idealized probe feeds, *IEEE Trans. Antennas Propagat.*, Vol. 32, No. 10, 1984, pp. 1101–1107.

23. D. H. Schaubert, A class of *E*-plane scan blindnesses in single-polarized arrays of tapered-slot antennas with a ground plane, *IEEE Trans. Antennas Propagat.*, Vol. 44, No. 7, 1996, pp. 954–959.

24. S. Edelberg and A. Oliner, Mutual coupling effects in large antenna arrays II: Compensation effects, *IRE Trans. Antennas Propagat.*, Vol. 8, No. 4, 1960, pp. 360–367.

25. A. W. Rudge, *The Handbook of Antenna Design*, 2nd ed., London: P. Peregrinus on behalf of the Institution of Electrical Engineers, 1986.

26. E. DuFort, Design of corrugated plates for phased array matching, *IEEE Trans. Antennas Propagat.*, Vol. 16, No. 1, 1968, pp. 37–46.

27. R. C. Hansen, *Microwave Scanning Antennas*, New York: Academic Press, 1964.

28. N. Alexopoulos and D. Jackson, Fundamental superstrate (cover) effects on printed circuit antennas, *IEEE Trans. Antennas Propagat.*, Vol. 32, No. 8, 1984, pp. 807–816.

29. D. R. Jackson, J. T. Williams, A. K. Bhattacharyya et al., Microstrip patch designs that do not excite surface waves, *IEEE Trans. Antennas Propagat.*, Vol. 41, No. 8, 1993, pp. 1026–1037.

30. Y. Fan and Y. Rahmat-Samii, Microstrip antennas integrated with electromagnetic band-gap (EBG) structures: A low mutual coupling design for array applications, *IEEE Trans. Antennas Propagat.*, Vol. 51, No. 10, 2003, pp. 2936–2946.

31. F. Yunqi and Y. Naichang, Elimination of scan blindness in phased array of microstrip patches using electromagnetic bandgap materials, *Antennas Wireless Propagat. Lett. IEEE*, Vol. 3, 2004, pp. 63–65.

32. Z. Iluz, R. Shavit, and R. Bauer, Microstrip antenna phased array with electromagnetic bandgap substrate, *IEEE Trans. Antennas Propagat.*, Vol. 52, No. 6, 2004, pp. 1446–1453.

33. G. Donzelli, F. Capolino, S. Boscolo et al., Elimination of scan blindness in phased array antennas using a grounded-dielectric EBG material, *Antennas Wireless Propagat. Lett. IEEE*, Vol. 6, 2007, pp. 106–109.

34. E. Rajo-Iglesias, O. Quevedo-Teruel, and L. Inclan-Sanchez, Mutual coupling reduction in patch antenna arrays by using a planar EBG structure and a multilayer dielectric substrate, *IEEE Trans. Antennas Propagat.*, Vol. 56, No. 6, 2008, pp. 1648–1655.

35. Z. Lijun, J. A. Castaneda, and N. G. Alexopoulos, Scan blindness free phased array design using PBG materials, *IEEE Trans. Antennas Propagat.*, Vol. 52, No. 8, 2004, pp. 2000–2007.

7 Array Beamforming Networks

An array feed or beamforming network takes the signals at all the elements and combines them to form a receive beam or, conversely, takes a transmitted signal and distributes it to the elements in the array to form a transmit beam. This chapter presents an assortment of analog and digital beamforming networks for creating one or more beams. A corporate feed consists of a series of power splitters/combiners that distribute the signals. One input/output is connected to all the elements. Couplers sample the signals from the elements that are combined to form multiple beams. Examples include the Blass matrix and the Butler matrix. Another approach distributes the signals from one or more antennas to the elements of an array. Bootlace and Rotman lenses are good examples. Finally, the most versatile and advanced beamforming nework is called the digital beamformer. This approach places an RF analog-to-digital converter at each element, so the signal at each element goes directly to the computer.

7.1. TRANSMISSION LINES

Most feed networks are made from (a) transmission lines, such as coaxial cables, stripline, and microstrip, (b) passive devices, such as couplers and power splitters, or (c) active components, such as amplifiers and receivers. Impedance matching is extremely important to ensure efficient power transfer through the feed network. Most devices and transmission lines have a characteristic impedance of $50\,\Omega$, so most antennas are designed to have 50-Ω input impedance as well. The value of $50\,\Omega$ was arrived at through a compromise between power handling and low loss for air-dielectric coaxial cable [1]. The insertion loss has a minimum around $77\,\Omega$, and the maximum power handling occurs at $30\,\Omega$ for any coaxial cable with air dielectric. A nice round number between these two optima is $50\,\Omega$, so almost all coaxial cables have a 50-Ω impedance with a few having $75\,\Omega$ for dipole antennas.

The cross section of a coaxial cable is shown in Figure 7.1. The characteristic impedance of a coaxial cable is found from [2]

Antenna Arrays: A Computational Approach, by Randy L. Haupt
Copyright © 2010 John Wiley & Sons, Inc.

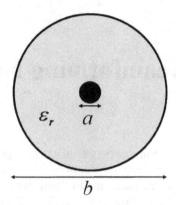

Figure 7.1. Cross section of a coaxial cable.

$$Z_c = \frac{1}{2\pi}\sqrt{\frac{\mu}{\varepsilon}}\ln\frac{b}{a} \qquad (7.1)$$

where b is the outside diameter and a is inside diameter. Oftentimes, a dielectric material like polytetrafluoroethylene (PTFE), also known as Teflon, fills the space between the inner and out conductors. Loss in the dielectric material filling the cable and the resistive losses in the conductors attenuate the signal propagating inside the transmission line. The losses get bigger as the frequency increases. Skin effect losses decrease as the diameter of the cable increases.

Example. A commercial coaxial cable has $Z_c = 50\,\Omega$ with $a = 0.036$ in. and $b = 0.125$ in. What is ε_r of the dielectric inside the cable?
    Inserting these values into (7.1) yields

$$\varepsilon_r = \left(\frac{1}{2\pi Z_c}\sqrt{\frac{\mu_0}{\varepsilon_0}}\ln\frac{b}{a}\right)^2 = \left(\frac{1}{2\pi(50)}(377)\ln\frac{0.125}{0.036}\right)^2 = 2.23$$

which is the dielectric constant of Teflon.

    Stripline consists of a thin, narrow conductor inside a dielectric substrate between two large ground planes. Figure 7.2 is a cross section of a stripline of width W suspended halfway between two ground planes separated by h. The characteristic impedance is a function of W, h, and ε_r [3].

$$Z_c = \begin{cases} \dfrac{30\pi h}{\sqrt{\varepsilon_r}(W+0.441h)}, & W \geq 0.35h \\[2ex] \dfrac{30\pi h}{\sqrt{\varepsilon_r}(2W-0.091h)}, & W < 0.35h \end{cases} \qquad (7.2)$$

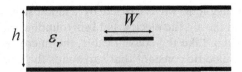

Figure 7.2. Cross section of stripline.

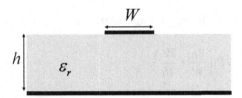

Figure 7.3. Cross section of a microstrip line.

Since the characteristic impedance is 50 Ω and the dielectric constant is known, then the width of the strip is found from [3]

$$
W = \begin{cases} \dfrac{0.6\pi h}{\sqrt{\varepsilon_r}} - 0.441h, & \sqrt{\varepsilon_r} \leq 2.4,\ Z_c = 50\ \Omega \\[3mm] 0.85h - h\sqrt{0.6 - \dfrac{0.6\pi}{\sqrt{\varepsilon_r}} - 0.441h}, & \sqrt{\varepsilon_r} > 2.4,\ Z_c = 50\ \Omega \end{cases} \tag{7.3}
$$

Example. Find the width of a 50-Ω stripline in a RT duroid substrate ($\varepsilon_r = 2.2$) 6 mm thick. Substituting into (7.3) results in

$$
W = \frac{30\pi(6)}{50\sqrt{2.2}} - 0.441(6) = 5.0 \text{ mm}
$$

Microstrip is a transmission line made on a printed circuit board (PCB). A conducting strip lies on top of a dielectric slab known as a substrate with a conducting ground plane beneath the substrate (Figure 7.3). Microstrip is cheap, lightweight, and compact. Another major advantage of microstrip over stripline is that all active components are mounted on top of the board. Microstrip also has its disadvantages. It radiates, causing unintended coupling with other components, and it is dispersive, so different frequencies travel at different speeds. In addition, it cannot handle high power. FR4 is a popular low-cost substrate used below a few gigahertz (no exact cutoff because loss increases with frequency). The substrate has a thin copper clad on both sides. One side serves as the ground plane while the circuit design is etched or routed from the other side. At microwave frequencies, FR4 losses are high and the dielectric constant tolerance is large, causing degraded performance. Since part of the electromagnetic wave carried by a microstrip line exists in the substrate and part in the air, the electric field surrounding the microstrip exists in two different media. Like the coaxial cable, the microstrip line appears to be in an effective dielectric constant that surrounds the microstrip line [3].

$$\varepsilon_{\text{eff}} = \frac{\varepsilon_r + 1}{2} + \frac{\varepsilon_r - 1}{2} \frac{1}{\sqrt{1 + 12h/W}} \tag{7.4}$$

The characteristic impedance of a microstrip line is given by [3]

$$Z_c = \begin{cases} \dfrac{60}{\sqrt{\varepsilon_{\text{eff}}}} \ln\left(\dfrac{8h}{W} + \dfrac{W}{4h}\right), & W \leq h \\[4mm] \dfrac{120\pi}{\sqrt{\varepsilon_{\text{eff}}}\left[W/h + 1.393 + 0.667\ln\left(W/h + 1.444\right)\right]}, & W > h \end{cases} \tag{7.5}$$

The width of the microstrip can be calculated given the characteristic impedance [3].

$$W = \begin{cases} \dfrac{8he^A}{e^{2A} - 2}, & W \leq 2h \\[4mm] \dfrac{2h}{\pi}\left\{B - 1 - \ln(2B - 1) + \dfrac{\varepsilon_r - 1}{2\varepsilon_r}\left[\ln(2B - 1) + 0.39 - \dfrac{0.61}{\varepsilon_r}\right]\right\}, & W > 2h \end{cases} \tag{7.6}$$

where h is substrate thickness, W is the width of a microstrip line, and

$$A = \frac{Z_c}{60}\sqrt{\frac{\varepsilon_r + 1}{2}} + \frac{\varepsilon_r - 1}{\varepsilon_r + 1}\left(0.23 + \frac{0.11}{\varepsilon_r}\right)$$

$$B = \frac{377\pi}{2Z_c\sqrt{\varepsilon_r}}$$

Example. Find the width of a 50-Ω microstrip line on a quartz substrate ($\varepsilon_r = 3.8$) 2 mm thick.

Using (7.6), the width is found to be 4.6 mm.

The transmission lines in the feed networks have many bends in order to guide the signals to/from the elements. A 90° bend in a microstrip line produces a large reflection from the end of the line. Some signal bounces around the corner, but a large portion reflects back the way the signal traveled down the line. If the bend is an arc of radius at least three times the strip width, then reflections are minimal [4]. This large bend takes up a lot of real estate compared to the 90° bend. A sharp 90° bend behaves as a shunt capacitance

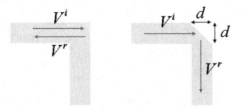

Figure 7.4. Mitered bend in microstrip.

between the ground plane and the bend. In order to create a better match, the bend is mitered to reduce the area of metallization and remove the excess capacitance. The signal is no longer normally incident to microstrip edge, so it reflects from the end down the other arm. Figure 7.4 shows the difference between a straight bend in the microstrip (left) and a mitered bend (right). Douville and James experimentally determined the optimum miter for a wide range of microstrip geometries [5]. Figure 7.4 shows an isosceles triangular piece of metal removed from the corner with side length given by

$$d = 2W\left(0.52 + 0.65e^{-1.35W/h}\right), \qquad W \geq 0.25h, \, \varepsilon_r \leq 25 \qquad (7.7)$$

Since most lines are 50 Ω, this formula simplifies to

$$d = 2W\left(0.46 + 0.025\varepsilon_r\right), \qquad \varepsilon_r \leq 16, \, Z_c \approx 50 \, \Omega \qquad (7.8)$$

Example. Find d for a 50-Ω microstrip line with a 1.5-mm alumina substrate ($\varepsilon_r = 9.4$).

First, find W from (7.6): $W = 1.87$ mm. Next, find d from (7.7): $d = 2.39$ mm. Finally, find d from (7.8): $d = 2.59$ mm.

7.2. S PARAMETERS

Scattering parameters, or S parameters, are the reflection and transmission coefficients between the incident and reflected voltage waves of a multiport RF device. Consider the N-port device in Figure 7.5. Parts of an RF signal incident on one port exit all the ports. The first number in the subscript refers to the output port, while the second number refers to the input port. Thus S_{23} is the ratio of the output voltage at port 2 (V_2^-) due to the input voltage at port 3 (V_3^+). The voltage signal exiting port n is written as

$$V_n^- = S_{n1}V_1^+ + \cdots + S_{nn}V_n^+ + \cdots + S_{nN}V_N^+ \qquad (7.9)$$

where

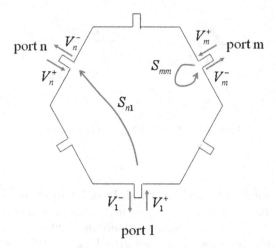

Figure 7.5. Scattering parameters for an N-port network.

$$S_{nm} = \frac{V_n^-}{V_m^+}\bigg|_{V_p^+=0 \text{ for } p \neq m}$$ (7.10)

A similar equation can be written for each port. All N equations can be combined into a matrix equation given by

$$\begin{bmatrix} S_{11} & \cdots & S_{1N} \\ \vdots & \ddots & \vdots \\ S_{N1} & \cdots & S_{NN} \end{bmatrix} \begin{bmatrix} V_1^+ \\ \vdots \\ V_N^+ \end{bmatrix} = \begin{bmatrix} V_1^- \\ \vdots \\ V_N^- \end{bmatrix}$$ (7.11)

where the $N \times N$ matrix is known as the S-parameter matrix. A loss-free network means that no power is dissipated or $\sum_{n=1}^{N} V_n^+ = \sum_{n=1}^{N} V_n^-$. In this case, the S-parameter matrix is unitary when S times its complex conjugate transpose equals the identity matrix. A lossy passive network has $\sum_{n=1}^{N} V_n^+ > \sum_{n=1}^{N} V_n^-$. The magnitudes of the reflection coefficients are always less than one. If there is active amplification, then the transmission coefficients may be greater than one. A network is reciprocal when $S_{mn} = S_{nm}$, and the S-parameter matrix equals its transpose. Networks with amplifiers or anisotropic materials, such as ferrites, are usually nonreciprocal.

Example. A circulator satisfies the matched and lossless conditions at the expense of reciprocity. Through the use of ferrites, a circulator either passes a signal from port 1 to port 2, from port 2 to port 3, and from port 3 to port 1 or in the reverse. Circulators are commonly used in monostatic radar systems to isolate the transmit and receive networks. Find the S-parameter matrix in both directions. The S parameters for a circulator that passes signals in the 1–2–3–1 direction are given by

$$[S] = \begin{bmatrix} 0 & 0 & 1 \\ 1 & 0 & 0 \\ 0 & 1 & 0 \end{bmatrix} \tag{7.12}$$

and in the opposite direction are given by

$$[S] = \begin{bmatrix} 0 & 1 & 0 \\ 0 & 0 & 1 \\ 1 & 0 & 0 \end{bmatrix} \tag{7.13}$$

Insertion loss is the loss of power that results from inserting a device in a transmission line or waveguide. Low insertion loss is very important when putting devices, such as phase shifters, in an array. The insertion loss (in decibels) is calculated as the ratio of the output power to the input power.

$$IL = 10 \log_{10} \frac{P_2}{P_1} \tag{7.14}$$

7.3. MATCHING CIRCUITS

A matching circuit reduces the reflection coefficient between two impedances, Z_1 and Z_3. One of the most commonly used matching circuits is the quarter-wave transformer shown in Figure 7.6. At the center frequency, a section of transmission line $\lambda_0/4$ long placed between the two transmission lines will eliminate the reflection coefficient if its impedance is

$$Z_2 = \sqrt{Z_1 Z_3} \tag{7.15}$$

The bandwidth of this transformer is calculated from [2]

$$BW = 2 - \frac{4}{\pi} \cos^{-1} \left(\frac{2 \Gamma_m Z_2}{|Z_3 - Z_1| \sqrt{1 - \Gamma_m^2}} \right) \tag{7.16}$$

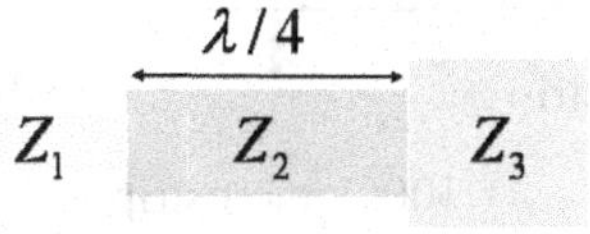

Figure 7.6. Quarter-wave transformer.

where Γ_m is the maximum reflection coefficient. The bandwidth decreases as the difference between Z_1 and Z_3 increases.

If the bandwidth of the match is not adequate, then additional quarter-wave sections can be added between the two original impedances. The impedances are found through a number of approaches like binomial, Chebyshev, and so on [2]. Extending the bandwidth in this manner takes additional space, which is usually at a premium.

Example. Find the quarter-wave transformer in microstrip that matches a 50-Ω microstrip line to a 100-Ω microstrip line when the substrate is 1.6 mm thick and has $\varepsilon_r = 2.2$.

$$Z_2 = \sqrt{5000} = 70.7\,\Omega$$

The strip width is found from (7.6)

$$W = 2.04 \text{ mm}$$

7.4. CORPORATE AND SERIES FEEDS

Corporate and series feeds are the most common methods of signal distribution to elements in an array. A corporate or parallel feed uses couplers and power dividers to distribute transmit and/or receive signals in the array. Couplers/power dividers funnel off signals from a transmission line or route the signal to other transmission lines. Elements connected to a series feed all tap the same transmission line or waveguide.

A T junction splits or combines signals in a corporate feed. It is a three-port network that splits the signal from port 1 or combines the signals from ports 2 and 3 (Figure 7.7). From a transmit point of view, the signal enters port 1 and then splits with half (3 dB) of the power sent to one arm and the other

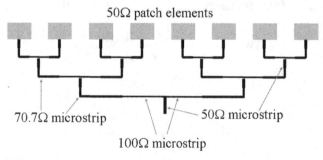

Figure 7.7. An 8-element corporate-fed microstrip array.

front   **back**

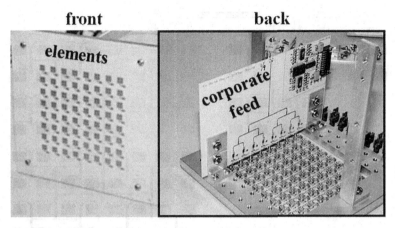

Figure 7.8. Picture of 8 × 8 planar array with brick architecture. (Courtesy of Ball Aerospace & Technologies Corp.)

half to the other arm. A T junction is a lossless, reciprical power divider that is matched at the input port but not matched at the two output ports. If port 1 is matched, then ports 2 and 3 are not. Thus, when the receive signals enter ports 2 and 3, portions are reflected from the mismatch. Figure 7.7 is a picture of an 8-element corporate-fed microstrip uniform array. The tree-like structure of the feed appropriately combines/distributes the signals from/to the elements. A quarter-wave transformer appears at the splits in order to match the lines of different impedances. The 50-Ω input line splits into two 100-Ω lines. If the microstrip line continued to split like this, then the lines feeding the elements would be 400 Ω. Not only would the microstrip line be very thin, but the element impedance would have to be very high for matching. Thus, the 100-Ω line is converted back to 50 Ω using a quarter-wave transformer of 70.7 Ω. This process is done for each arm of the corporate feed as shown in Figure 7.7.

When the corporate feed for a planar array is orthogonal to the array face, then it is called a brick architecture. Each column of 8 elements in Figure 7.8 has a corporate feed. The 8 × 8 array of circularly polarized patch elements lies on a square grid with element spacing of 1 cm over a 14- to 15-GHz operating bandwidth.

Examples of planar array feed networks (tile architecture) that lie in the x–y plane are shown in Figures 7.9–7.11. A square-planar array having the number of elements equal to a multiple of four has symmetry such that the feed network has a tree-type architecture such as the one shown in Figure 7.9 [6]. The patches have an impedance of 200 Ω, so the thin microstrip lines also have an impedance of 200 Ω while the thicker lines have an impedance of 100 Ω. If the number of elements has $2^m 3^n$ elements, then the following steps are used to build the feed [6]:

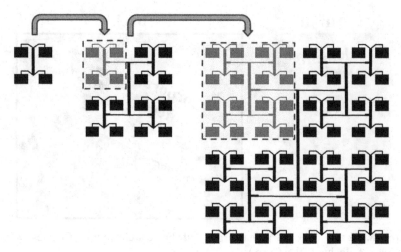

Figure 7.9. Tree structure feed for a square-planar array.

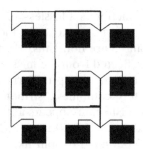

Figure 7.10. Planar feed network for a 3 × 3 array.

1. Start with a 2^m design like Figure 7.9.
2. Deleted external rows and columns until the desired $2^m 3^n$ is reached.
3. Modify feed-line impedances—replace 1:1 splitters with 2:1 splitters.
4. Modify impedance transformers.

Figure 7.10 is an example of a 3 × 3 array derived from a 4 × 4 array [6]. Other architectures, such as concentric ring arrays, require very complex feed structures as demonstrated by the feed network in Figure 7.11 [6]. The amplitude and phase to each element fed from port 1 must be the same, so the lengths and impedances of all the lines and matches must be carefully calculated. All the power splitters in this case are equal (1:1).

A resistive divider is reciprocal and matched at the three ports but is lossy. In this case, the signal loss is the result of heat dissipation rather than reflections as in the T junction. An N-way resistive divider consists of N arms each with a resistance of

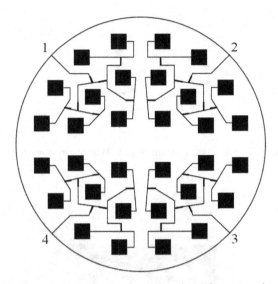

Figure 7.11. Four feed networks that comprise a 3-concentric-ring array.

$$R = Z_c \frac{(N-1)}{(N+1)} \tag{7.17}$$

A two-way divider requires $R = 2Z_c/3$, while a three-way divider needs resistors of $R = Z_c/2$.

Resistive divider efficiency decreases as N increases. The power transferred to each arm is $1/N^2$, which is much less than the $1/N$ of a lossless divider. Isolation in the resistive divider is equal to its insertion loss.

The Wilkinson power divider splits an input signal into two equal phase output signals or combines two equal-phase signals into one in the opposite direction [3]. It is lossless when all the output ports are matched. A diagram of a three port Wilkinson divider is shown in Figure 7.12. Quarter-wave transformers match ports 2 and 3 to port 1. The resistor (R_w) matches all ports and isolates port 2 from port 3 at the center frequency. Since the resistor adds no loss to the power split, an ideal Wilkinson divider is 100% efficient. A signal enters port 1 then splits into equal amplitude and phase signals at output ports 2 and 3. When the signals are recombined at port 3, they have equal amplitude and are 180° out of phase, since CCW signal travels half a wavelength further than the CW signal. The signals cancel at port 3. Since each end of the isolation resistor between ports 2 and 3 is at the same potential, no current flows through it, so it is decoupled from the input. The two output port terminations add in parallel at the input. In order to combine to Z_c, a quarter-wave transformer is placed in each leg. If the quarter-wave lines have an impedance of $\sqrt{2}Z_c$, then the input is matched when ports 2 and 3 are terminated in Z_c. Figure 7.13 is a picture of a three-port Wilkinson power divider. Note that this

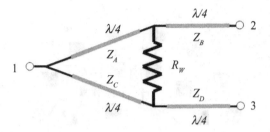

Figure 7.12. Diagram of a three-port Wilkinson power divider.

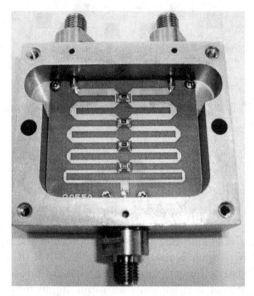

Figure 7.13. Picture of a three-port Wilkinson power divider.

Wilkinson power divider has four stages. The four different width curves have different impedances designed to make it very wide band 0.5 to 4 GHz. This type of power divider can be extended to make an 8-to-1 power combiner (Figure 7.14). Its tree-like structure takes up a lot of room because it is very broadband.

So far, only corporate feeds that are based on a power of two have been presented. If the number of elements in the array is not a power of two, then power dividers with ratios other than 2:1 are needed. Also, when there is an amplitude taper, the power division needs to be unequal [6]. In order to make the unequal split, the quarter-wave sections must be of different impedances, to encourage more of the signal to travel into or out of the lower-impedance arm. In addition, a second set of quarter-wave sections are needed to transform the arm impedances back to 50 Ω. This configuration looks similar to a

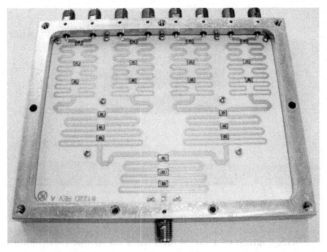

Figure 7.14. Picture of an 8-to-1 Wilkinson power combiner.

two-stage Wilkinson without the second isolation resistor. The following set of equations ensure that all the ports are matched and ports 2 and 3 are isolated [7]:

$$Z_A = Z_c\left(C^{-0.5} + C^{-1.5}\right)$$

$$Z_B = Z_c(C)^{0.25}\sqrt{1+C}$$

$$Z_C = Z_c C^{-0.25}$$

$$Z_D = Z_c C^{0.25} \qquad (7.18)$$

$$R_w = Z_c\left(C^{0.5} + C^{-0.5}\right)$$

$$C = \frac{P_2}{P_3}$$

Example. Design the feed network for a 4-element Chebyshev array with 20-dB sidelobes. The amplitude weights for the Chebyshev taper are

$$w = [0.5761 \quad 1.0 \quad 1.0 \quad 0.5761]$$

which equates to power at the elements of

$$P = [0.3319 \quad 1.0 \quad 1.0 \quad 0.3319]$$

$$C = \frac{1}{0.3319} = 3.01$$

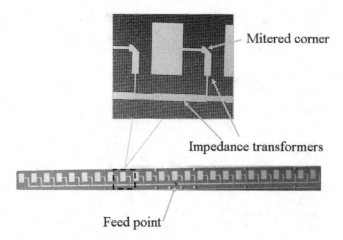

Figure 7.15. Picture of a series feed for a linear microstrip antenna.

$$Z_A = 38.39\Omega$$

$$Z_B = 131.88\Omega$$

$$Z_C = 37.96\Omega$$

$$Z_D = 65.86\Omega$$

$$R_w = 50\left(\sqrt{3.01} + \sqrt{\frac{1}{3.01}} \right) = 116.43\Omega$$

The elements of a series fed array all share the same transmission line/waveguide. An example is shown in Figure 7.15 of a single microstrip line that feeds a long line of rectangular patches. The microstrip line to an element taps the signal from the main microstrip line. A quarter-wave transformer with a mitered corner provides the match to the patch. A very common example of a series-fed array is the Hawk radar waveguide array shown in Figure 7.16. The signal enters the port, splits with half traveling up and the other half traveling down. Since the signal reaches the top and bottom waveguides first and the center waveguides last, the waveguides that tap the signal from the feed waveguide and wrap around to feed the slotted waveguide in the front of the array become increasingly shorter from top to bottom.

The array in Figure 7.17 is an example of a combination feed where the corporate feed distributes the signal to several series-fed patches. Some interesting corporate and series feeds can be found in reference 8.

A 32-element array of pyramidal horns with a corporate waveguide feed was designed to operate at 76.6 GHz [9]. The array measures 1.3 cm tall by 17.8 cm wide by 12.7 cm deep with horns that are 1.3λ (0.39 cm) in the H plane (array plane) and 2.0λ (0.79 cm) in the E plane. A 5-stage H-plane waveguide

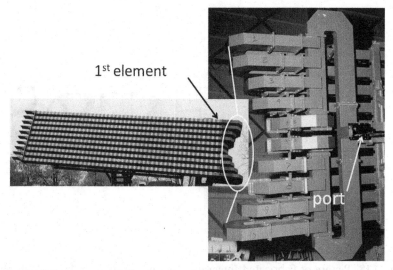

1st element

port

Figure 7.16. Picture of a series-fed Hawk radar waveguide array. (Courtesy of the National Electronics Museum.)

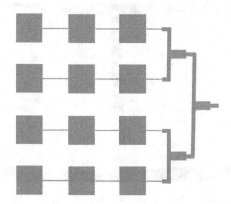

Figure 7.17. Picture of a corporate-series-fed microstrip array.

power divider uses equal power splits in stages 1 and 5 and unequal power splits in stages 2, 3, and 4 to produce a 40-dB Taylor distribution. Element spacing did not leave enough room for unequal power splitting in stage 5. Figure 7.18 shows the fields modeled inside the waveguide feed (right side only) with a blow-up of the 93:7 split at stage 3. The array was milled in two pieces—top and bottom—using numerical control techniques with very tight tolerances (Figure 7.19). Antenna patterns were measured at 76.3, 76.6, and 78 GHz as shown in Figure 7.20. The pattern at 76.6 GHz has the highest gain. A blow-up of this pattern appears in the upper right of the figure.

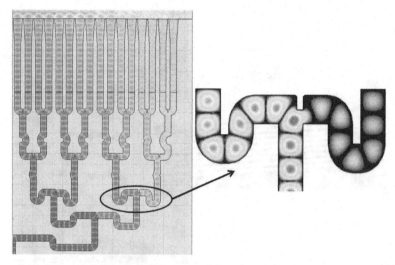

Figure 7.18. Picture of Fabricated Integrated Receive Array with the top half on the left and the bottom half on the right. (Courtesy of Amir Zaghloul.)

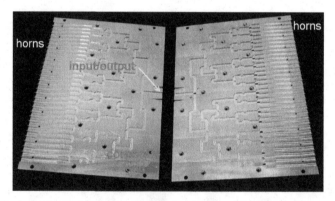

Figure 7.19. Array half with Taylor distribution power split of 93:7. (Courtesy of Amir Zaghloul.)

7.5. SLOTTED WAVEGUIDE ARRAYS

Cutting holes in a rectangular waveguide is a relatively simple way to create a series-fed linear array. The elements are typically narrow rectangular slots cut in the waveguide as shown in Figure 7.21. Slotted waveguide arrays are often used in place of reflector antennas, because they are very thin and are easily designed for low sidelobes. The Joint Surveillance Target Attack Radar System (Joint STARS) is a long-range, air-to-ground surveillance system designed to locate, classify, and track ground targets in all weather conditions.

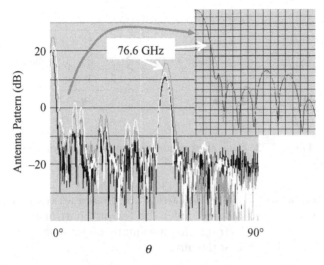

Figure 7.20. Measured antenna patterns of the 32-element array of horns. (Courtesy of Amir Zaghloul.)

Figure 7.21. Slots cut into side of waveguide for the PAVE Mover antennas of Joint STARS. (Courtesy of Northrop Grumman and available at the National Electronics Museum.)

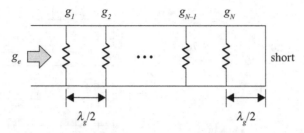

Figure 7.22. Equivalent circuit of a longitudinal slot array.

The E-8C, a modified Boeing 707, carries this phased-array radar antenna in a 26-foot canoe-shaped radome under the forward part of the fuselage [10, 11]. The antenna scans electronically in azimuth, and it scans mechanically in elevation from either side of the aircraft.

7.5.1. Resonant Waveguide Arrays

A waveguide terminated with a short has a standing wave with peaks that start $\lambda_g/4$ from the short and occur every $\lambda_g/2$ afterwards (see Chapter 4). The most common form of a resonant waveguide array has longitudinal slots in the broad face of the waveguide. The current on the waveguide wall reverses direction every $\lambda_g/2$ and flows in the opposite direction on either side of a centerline. In order to have consecutive elements in phase, they are placed $\lambda_g/2$ apart and offset on opposite sides of the centerline by Δ_n. Thus, these alternating offsets add an additional π radians of phase shift that compensates for the π radians of phase shift due to the $\lambda_g/2$ slot spacing in order to have all the slots radiate in phase. The offsets in a TE rectangular waveguide of cross section $a \times b$ are found from [12]

$$\Delta_n = \frac{a}{\pi} \sin^{-1} \sqrt{\frac{\lambda_0 b g_n}{2.09 \lambda_g a \cos^2\left(\dfrac{\pi \lambda_0}{2 \lambda_g}\right)}} \tag{7.19}$$

Figure 7.22 shows the equivalent circuit of a longitudinal resonant slot array with N conductances spaced $\lambda_g/2$ apart. Since the slots are spaced $\lambda_g/2$ apart, the input conductance to the array is the sum of all the normalized slot conductances.

$$g_e = \sum_{n=1}^{N} g_n \tag{7.20}$$

Ideally, all the power input to the waveguide should be radiated through the slots. The power radiating from each slot should sum to one when the input power is normalized [13].

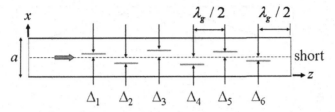

Figure 7.23. Diagram of the 6-element longitudinal slot array with a 20-dB Chebyshev taper.

$$\sum_{n=1}^{N} P_n = 1 \qquad (7.21)$$

A voltage V applied across slot n radiates the power

$$P_n = 0.5V^2 g_n = 0.5KV^2 a_n^2 \qquad (7.22)$$

where a_n is the slot amplitude and K is a constant. As a result, $a_n = g_n$, so the offsets are found from (7.19) for a desired amplitude taper.

Example. Find the slot offsets for a longitudinal offset slot resonant array of 6 elements with a Chebyshev taper having 20-dB sidelobes. Assume the array operates at 10 GHz. First, find the desired amplitude taper:

$$w = [0.5406 \quad 0.7768 \quad 1 \quad 1 \quad 0.7768 \quad 0.5406]$$

Then find K:

$$K = 1 \Big/ \sum_{n=1}^{6} w_n = 0.2158$$

The conductances for each slot are found using (7.22):

$$g = [0.0771 \quad 0.1592 \quad 0.2638 \quad 0.2638 \quad 0.1592 \quad 0.0771]$$

Finally, the offsets are found from (7.20):

$$\Delta_1 = \Delta_6 = 0.2293$$
$$\Delta_2 = \Delta_5 = 0.3295$$
$$\Delta_3 = \Delta_4 = 0.4242$$

Since $\lambda_g = 3.98$ cm for a waveguide with dimensions 2.286 by 1.016 cm, the element spacing is $\lambda_g/2 = 1.99$ cm. Figure 7.23 is a diagram of the 6-element slot array design. The array has a gain of 13.9 dB, and the first sidelobe is 24.9 dB below the main beam (Figure 7.24).

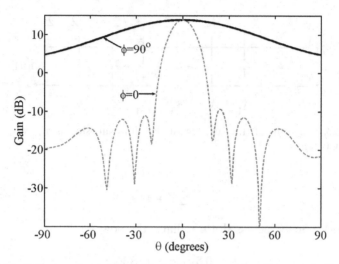

Figure 7.24. Orthogonal cuts in the far-field pattern of the slot array in Figure 7.23.

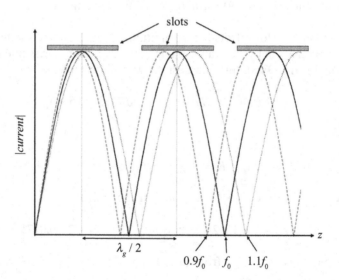

Figure 7.25. As the frequency changes, the slots are not longer at the same position on the standing wave inside the waveguide.

A resonant slot array has a narrow bandwidth. As the frequency changes, the peaks of the standing wave move (dashed line in Figure 7.25). Thus, the field distribution in a slot changes more dramatically, the farther it is from the short. Long resonant slot arrays are narrowband because the impedance bandwidth is approximately equal to 50%/N [14].

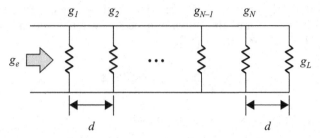

Figure 7.26. Equivalent circuit of a longitudinal nonresonant slot array.

In general, the mutual coupling for longitudinal broadwall slots in a single waveguide (linear array) can be ignored. Sidewall slots have high mutual coupling, so more care must be taken when waveguides are stacked to make a planar array [15].

7.5.2. Traveling-Wave Waveguide Arrays

A traveling-wave slot array has an element spacing either greater than or less than $\lambda_g/2$. The waveguide ends in a matched load that reduces the reflections that causes the standing wave. The element spacing is chosen to create a beam at the angle $\theta = \theta_s$ when $\phi = 0$. Figure 7.26 shows the equivalent circuit of a longitudinal nonresonant slot array with N conductances spaced d apart. The slots are resonant (have real admittance) and the mutual coupling is negligible.

The array factor when $\phi = 0$ is

$$\mathrm{AF} = \sum_{n=1}^{N} w_n e^{j(k\sin\theta - k\sin\theta_s)nd + jn\pi} \tag{7.23}$$

The main beam points at θ_s when

$$(k_g - k\sin\theta_s)d = 2m\pi - \pi, \qquad m = 0, \pm1, \pm2, \ldots; n = 1 \tag{7.24}$$

If θ_s is known, then the element spacing is found from (7.24):

$$d = \frac{(2m-1)\lambda_0\lambda_g}{2(\lambda_0 - \lambda_g\sin\theta_s)} \tag{7.25}$$

If d is known, then θ_s is found from (7.24):

$$\theta_s = \sin^{-1}\left[\frac{\lambda}{\lambda_g} - \frac{(2m+1)\lambda}{2d}\right] \tag{7.26}$$

Assuming that only one beam is desired, the minimum spacing occurs when $\theta_s = -90°$ for $m = 0$ and the maximum occurs when a second main beam emerges at $\theta_s = -90°$ for $m = 1$ [12].

$$\frac{0.5\lambda_0\lambda_g}{\lambda_g + \lambda_0} < d < \frac{0.5\lambda_0\lambda_g}{\lambda_g - \lambda_0} \tag{7.27}$$

If d is known, then the scan angle is given by

$$\theta_s = \sin^{-1}\left[\frac{2d\lambda - (2m-1)\lambda\lambda_g}{\lambda_g}\right] \tag{7.28}$$

The conductance of the resonant slots when q is the fraction of incident power dissipated in the matched load is given by

$$g_n = \frac{P_n}{q - \sum\limits_{m=n}^{N} P_m} = \frac{P_n}{1 - \sum\limits_{m=1}^{n-1} P_m} \tag{7.29}$$

This works when $N \geq 12$ and $\theta_s < 0$.

As with the resonant waveguide array, all the power input to the waveguide should radiate through the slots. If the input power is normalized, then all the powers radiating from the slots should sum to one as in (7.21). If a voltage V is applied across slot n, then the power radiated by that slot is

$$P_n = 0.5V^2 g_n = 0.5KV^2 w_n^2 \tag{7.30}$$

where w_n is the slot amplitude and K is a constant.

Example. Find the slot offsets for a longitudinal offset slot nonresonant array of 12 elements with an $\bar{n} = 3$ Taylor taper having 25-dB sidelobes. The main beam points at $\theta_s = -10°$ at 10 GHz.

First, find the desired amplitude taper:

$$w = [0.3726 \quad 0.4693 \quad 0.6276 \quad 0.7957 \quad 0.9286 \quad 1.0000$$
$$1.0000 \quad 0.9286 \quad 0.7957 \quad 0.6276 \quad 0.4693 \quad 0.3726]$$

Then find K:

$$K = (1-q)\bigg/ \sum_{n=1}^{12} w_n = 0.1154$$

The power at each slot is

$$P = [0.0160 \quad 0.0254 \quad 0.0455 \quad 0.0731 \quad 0.0995 \quad 0.1154$$
$$0.1154 \quad 0.0995 \quad 0.0731 \quad 0.0455 \quad 0.0254 \quad 0.0160]$$

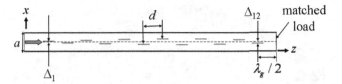

Figure 7.27. Diagram of a longitudinal slot array.

The conductances for each slot are found using (7.29):

$$g = [0.0160 \quad 0.0258 \quad 0.0474 \quad 0.0800 \quad 0.1185 \quad 0.1559$$
$$0.1847 \quad 0.1954 \quad 0.1783 \quad 0.1350 \quad 0.0872 \quad 0.0602]$$

Finally, the offsets are found from (7.20):

$$\Delta = [0.1046 \quad 0.1328 \quad 0.1799 \quad 0.2337 \quad 0.2844 \quad 0.3262$$
$$0.3550 \quad 0.3651 \quad 0.3488 \quad 0.3035 \quad 0.2440 \quad 0.2028]$$

Assuming $\lambda_g = 3.98\,\text{cm}$, the element spacing for $\theta_s = 30°$ is

$$d = \frac{(2m-1)\lambda_0\lambda_g}{2(\lambda_0 - \lambda_g \sin\theta_s)} = \frac{3(3.98)}{2(3+3.98\sin 10^o)} = 1.62\,\text{cm}$$

Figure 7.27 is a diagram of the 12-element slot array design.

Air-filled slotted waveguides have small scan angles in order to avoid grating lobes. Placing a dielectric inside the waveguide increases the scan range by decreasing λ_g, which in turn increases θ_s as calculated by (7.28).

So far, only the design of linear slot arrays has been described. Planar slot arrays are much more difficult, because the mutual coupling between parallel slots becomes significant. A good starting point is to design a linear slot array and stack them. This serves as an initial guess for a numerical optimization algorithm that will vary the offset and length of the slots to achieve the design goals [10].

The Airborne Warning and Control System (AWACS) uses an S-band planar slot array with over 4000 edge slots (Figure 7.28) to detect and track targets from the air [16]. It sits atop a Boeing 707 inside a radome (Figure 7.29). The radome and antenna rotate in azimuth at 10 revolutions per minute while the antenna uses 28 ferrite phase shifters to scan in elevation. Figure 7.30 shows the measured antenna pattern. The design goal for average side-lobe level was −37 dB while the actual measured average sidelobe level is −45 dB.

Figure 7.28. AWACS antenna array. (Courtesy of the National Electronics Museum.)

Figure 7.29. Picture of the AWACS in flight. (Courtesy of the US Air Force.)

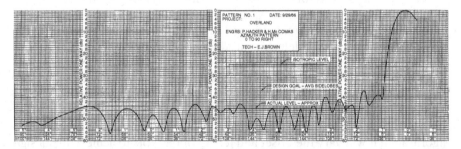

Figure 7.30. Far-field pattern of AWACS antenna. (Courtesy of the National Electronics Museum.)

Example. In reference 10, the design of a 2 × 4 resonant slotted waveguide array is presented at 8.933 GHz. The waveguide is 23.5 mm × 3.12 mm. Slot lengths and displacements are given in Table 7.1 when the slot width is 1.63 mm. A computer model of the array is shown in Figure 7.31 with the E- and H-plane pattern cuts in Figure 7.32.

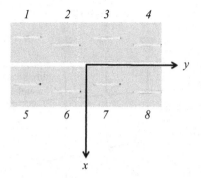

Figure 7.31. A 2 × 4 planar slot array.

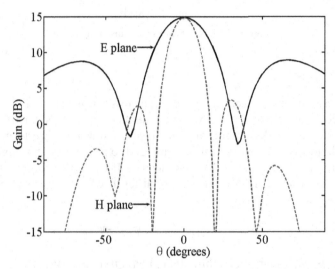

Figure 7.32. Orthogonal cuts in the antenna pattern of the 2 × 4 slot array.

TABLE 7.1. Slot Characteristics for a 2 × 4 Planar Array

Element:	1	2	3	4	5	6	7	8
Δ_n (mm):	−2.51	1.54	−3.10	1.54	−1.54	3.10	−1.54	2.51
Slot Length (mm):	17.60	16.94	17.98	16.99	16.99	17.98	16.94	17.60

7.6. BLASS MATRIX

A Blass matrix is a multibeam feed network that has M beams created from N elements [17]. In a receive antenna, each beam port couples part of the signal from each element using a series feed line as shown in Figure 7.33. The

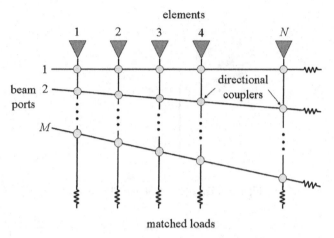

Figure 7.33. Diagram of a Blass matrix.

couplers are equally spaced along the transmission line in such a way as to create a constant phase shift between elements in order to steer the beam to a desired angle. The beam and element transmission lines are terminated in matched loads. These terminations are needed to prevent reflections but have the unwanted side effect of reducing efficiency. If $\psi_{m,n}$ is the phase length from beam port m to element n, then the phase difference between any two adjacent element is

$$\Delta\psi_m = \psi_{m,n+1} - \psi_{m,n} = \psi_{m,n} - \psi_{m,n-1} \tag{7.31}$$

This progressive linear phase shift steers beam m to an angle of

$$u_s = -\frac{\Delta\psi_m}{kd} \tag{7.32}$$

The Blass feed in Figure 7.33 is just one of many possible configurations [18].

7.7. BUTLER MATRIX

The Butler matrix [19] is a hardware version of the fast Fourier transform (FFT) [20] that was invented several years prior to the FFT. It transforms spatial signal samples taken by elements in a uniform linear array to samples in angular space which correspond to peaks of the main beams of array factors. A Butler matrix has $N = 2^M$ beam ports and $N = 2^M$ element ports that are interconnected via passive four-port 3-dB quadrature hybrid power couplers and fixed phase shifts.

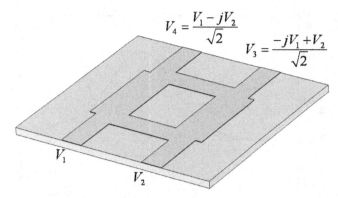

Figure 7.34. Computer model of a 3-dB quadrature hybrid coupler or two-port Butler matrix.

The simplest version of a Butler matrix is also better known as a 3-dB quadrature hybrid coupler. The 3-dB quadrature hybrid coupler is a four-port device with two inputs and two outputs [3]. An input signal splits equally between the two outputs, but one of the outputs has a 90° phase shift due to the additional distance it has to travel. Figure 7.34 is a computer model of a 3-dB quadrature hybrid in microstrip. If antennas were attached to port 3 and 4, then an input signal, V_1, at port 1 results in an output at port 4 of $V_1/\sqrt{2}$ and at port 3 of $-jV_1/\sqrt{2}$. An input signal at port 2, V_1, results in an output at port 4 of $-jV_2/\sqrt{2}$ and at port 3 of $V_2/\sqrt{2}$. The total signals at ports 3 and 4 are

$$V_3 = \frac{V_1 - jV_2}{\sqrt{2}}$$
$$V_4 = \frac{-jV_1 + V_2}{\sqrt{2}} \tag{7.33}$$

Figure 7.35 and Figure 7.36 show the amplitude and phase of the coupling coefficients as a function of frequency. At the design frequency of 6 GHz, the two ports receive equal signal amplitude but are 90° out of phase.

Larger Butler matrixes have $N = 2^m$, $m = 1, 2, \ldots$, input and output ports. The N beams are evenly spaced in angle such that each array factor beam n has a progressive linear phase shift given by

$$\delta_{sn} = (2n-1)\pi/N \qquad \text{for } n = 1, 2, \ldots, N \tag{7.34}$$

Therefore, the beam peaks occur when

$$\psi_x = 0 \text{ or } AF_n = \frac{\sin[N(kdu + \delta_{sn})/2]}{\sin[(kdu + \delta_{sn})/2]} \tag{7.35}$$

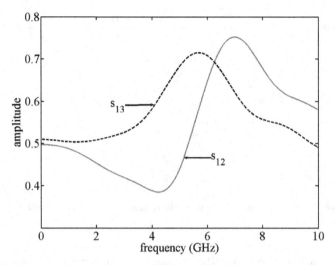

Figure 7.35. Amplitude of the coupling coefficients vs. frequency for the hybrid coupler in Figure 7.34.

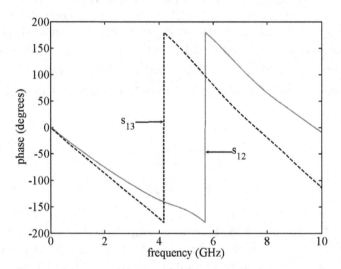

Figure 7.36. Phase of the coupling coefficients versus frequency for the hybrid coupler in Figure 7.34.

Since this is a uniform array factor, the peak sidelobes of each of the patterns are at about 13.2 dB. The angular coverage (bandwidth) is the angular separation between the two outer beams [21].

$$\text{angular coverage} = 2\sin^{-1}\left[\frac{\lambda}{2d}\left(1-\frac{1}{N}\right)\right] \tag{7.36}$$

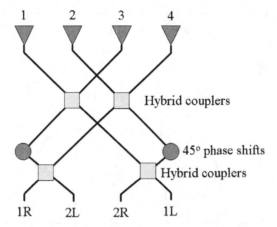

Figure 7.37. Diagram of a four-port Butler matrix.

The beam crossover level is found by substituting (7.36) into (7.35) to get

$$\frac{1}{N \sin\left(\dfrac{\pi}{2N}\right)} \tag{7.37}$$

which is approximately $2/\pi$ when N is large. The orthogonal beams overlap at the –3.9-dB level and have the full gain of the array.

A basic component of the FFT is the twiddle factor, W. The twiddle factor equates to a constant phase shift defined by the steering phase difference between two adjacent elements.

$$W = e^{j2\pi/N} \tag{7.38}$$

The array factor at port m of an N element uniform linear array along the x axis is given by

$$\mathrm{AF}_m = \sum_{n=0}^{N-1} e^{j\pi mn/N} s_n = \sum_{n=0}^{N-1} W^{mn} e^{jkndu}, \, m = \pm 1, \pm 2, \dots, \pm N/2 \tag{7.39}$$

where s_n is a sample of the incident plane wave at element n. Large Butler matrixes are lossy, but diagrams of up to 32 elements are shown in the literature [18]. Outputs can be combined to form beams, and amplitude tapers can be placed on the elements [22].

Example. Design a 4-element Butler matrix and plot the beams for uniform and low-sidelobe tapers.

Figure 7.37 is a diagram of the Butler matrix with $\lambda/2$ element spacing. Figure 7.38 is a plot of the 4 beams using a uniform amplitude taper and (7.39).

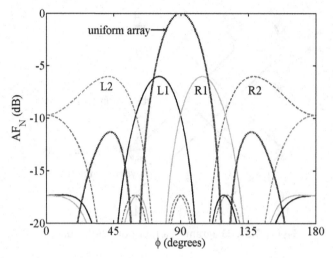

Figure 7.38. The four beams associated with the four-port Butler matrix.

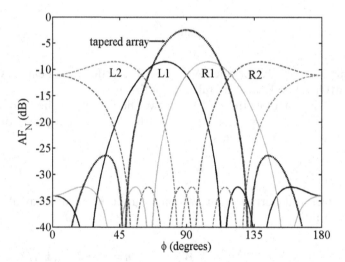

Figure 7.39. Butler matrix with amplitude taper.

An element amplitude taper of produces the low-sidelobe beams shown in Figure 7.39.

7.8. LENSES

A waveguide lens antenna consists of a feed placed at the focal point of a lens made from stacking waveguides. The shape of the lens and the length of the

Figure 7.40. Stepped waveguide lens antenna for the Nike AJAX MPA-4 radar. (Courtesy of the National Electronics Museum.)

waveguides are such that the phase path from the focal point to the front end of the waveguide is a constant. Since the phase velocity is higher inside the waveguide, the lens has the opposite curvature of a typical dielectric lens. Figure 7.40 is the zoned waveguide lens antenna for the Nike AJAX MPA-4 radar. By 1956, 80 Nike Ajax anti-aircraft missiles were deployed in the United States. An unconstrained lens [23] has curvature in the E plane (assuming linear polarization), while a constrained lens has curvature in the H plane. The unconstrained lens obeys Snell's law on transmit and focusing occurs in the plane of the electric field. The constrained lens does not obey Snell's law and focusing occurs normal to the electric field [24]. Thus, the signal radiated from each open-ended waveguide element has the same phase. Two more advanced versions of the waveguide lens are the Bootlace and Rotman lenses. These types of lenses are not limited to open-ended waveguides as elements [25].

7.8.1. Bootlace Lens

The bootlace lens consists of two back-to-back arrays with elements facing in opposite directions (Figure 7.41) [26]. An element on one side is connected to an element on the other side through a transmission line, phase shifter, amplifier, and so on. The elements on one side of the lens receives from/ transmits to a feed placed at the focal point. The shape of the input and output arrays, the length of the transmission lines between the input and output elements, and the amplitude and/or phase weights between the receive and transmit elements determine the performance of the lens. Transmission lines in the lens are not as limited in length as are the waveguides in a waveguide lens. Thus, a bootlace lens can be designed to have an on-axis focal point like

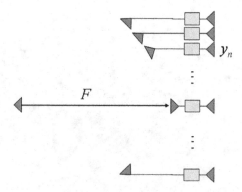

Figure 7.41. Diagram of the bootlace lens.

the waveguide lens, but it will also have two conjugate focal points. The feed has either one antenna placed at a focus or multiple antenna (multiple beams) placed along a focal arc. Gent found that the if the lens has a radius of $2R$, then the focal points lie on a circle having radius R [27]. A basic single-beam bootlace lens has one antenna at the feed, and the lens has a spherical shape on the side facing the feed and a flat face on the opposite side.

$$\delta_{cn} = k\left(\sqrt{F^2 + y_n^2} - F\right) \tag{7.40}$$

If the side of the lens facing the feed has a spherical shape while the other side has a flat face, then equal length transmission lines between the elements on the two faces produce a plane wave output. Since the transmission lines are the same length, it is independent of frequency. This lens can have up to 4 focal points that are symmetric about the lens axis. A feed placed on a curve that runs through these focal points will result in reasonably focused beams. Perfect focus only occurs when the feed is placed at one of the focal points. It is possible to extend the design of a bootlace lens to three dimensions [25]. The Patriot radar, AN/MPQ-53, is a space-fed array approximately 2.5 m in diameter with over 5000 elements (Figure 7.42) [28]. Elements in front of the array are circular waveguides that are dielectrically loaded. They are connected to the rear rectangular waveguide dielectrically loaded via a ferrite phase shifter. Its feed consists of horns that are 2.5 m in back of the array.

7.8.2. Rotman Lens

A Rotman lens [29] is a type of bootlace lens with three focal points. It forms simultaneous multiple beams for an antenna array as shown in Figure 7.43. M elements at the beam ports (x_{bm}, y_{bm}) are placed on a curved arc. They radiate to or receive from the curved back of the lens. The elements along the curved portion of the lens are at the array ports (x_{pn}, y_{pn}). Array elements are placed along the outside portion of the lens along a straight line (x_n, y_n). Each beam

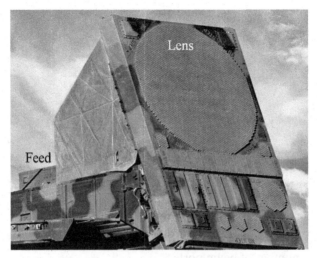

Figure 7.42. Patriot radar lens antenna array. (Courtesy of Raytheon Company.)

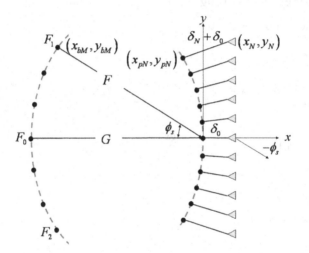

Figure 7.43. Diagram of a Rotman lens.

port creates one array factor from the array elements. The beams have their peaks steered to predetermined angles based on the geometry. Because it has no moving parts or phase shifters it offers an inexpensive alternative to corporate feed networks. The Rotman lens is very broadband, unlike the Butler matrix. A switching mechanism at the output can be used to select the appropriate beam in the desired look direction. In addition, interpolation can be used in direction finding to locate signals between adjacent beams.

The original design equations for the Rotman lens appeared in reference 29. These equations have been modified to include the permittivity of the lens substrate and the permittivity of the transmission lines and are given by [30]

$$x_n = \delta_0$$
$$y_n = -(N-1)d/2 : d : (N-1)d/2 \tag{7.41}$$

$$x_{pn} = w_n \frac{g-1}{c_\alpha(c_\alpha-g)}\sqrt{\frac{\varepsilon_{rt}}{\varepsilon_{rs}}} + \frac{y_n^2 s_\alpha^2}{2\varepsilon_{rs}(c_\alpha-g)F}$$

$$y_{pn} = \frac{y_n}{\sqrt{\varepsilon_{rs}}}\left(1 - w_n\sqrt{\frac{\varepsilon_{rt}}{\varepsilon_{rs}}}\right) \tag{7.42}$$

$$x_{bn} = -R\cos\gamma_m - G + R$$
$$y_{bn} = R\sin\gamma_m \tag{7.43}$$

where

N = number of array ports and elements

M = number of beam ports

d = element spacing

$Aw_n^2 + Bw_n + C = 0$

$$A = \left[\frac{(g-1)^2}{(c_\alpha-g)^2} + \frac{y_n^2}{\varepsilon_{rs}} - 1\right]\frac{\varepsilon_{rt}}{\varepsilon_{rs}}$$

$$B = \frac{s_\alpha^2(g-1)y_n^2}{(c_\alpha-g)^2}\sqrt{\frac{\varepsilon_{rt}}{\varepsilon_{rs}^3}} - 2y_n^2\sqrt{\frac{\varepsilon_{rt}}{\varepsilon_{rs}^3}} + 2\frac{g(g-1)}{c_\alpha-g}\sqrt{\frac{\varepsilon_{rt}}{\varepsilon_{rs}}} + 2g\sqrt{\frac{\varepsilon_{rt}}{\varepsilon_{rs}}}$$

$$C = \left[1 + \frac{s_\alpha^4 y_n^2}{4\varepsilon_{rs}(c_\alpha-g)^2} + \frac{gs_\alpha^2}{c_\alpha-g}\right]\frac{y_n^2}{\varepsilon_{rs}}$$

$$w_n = \frac{-B+\sqrt{B^2-4AC}}{2A}$$

$g = G/F$

$c_\alpha = \cos\alpha$

$s_\alpha = \sin\alpha$

ε_{rs} = relative dielectric constant of substrate

ε_{rt} = relative dielectric constant of transmission lines

$$w_n = \frac{\delta_n - \delta_0}{F}$$

$$R = \frac{(Fc_\alpha-G)^2 + F^2 s_\alpha^2}{2(G-Fc_\alpha)}$$

$$\gamma_m = -\alpha + \frac{2\alpha(m-1)}{M-1}$$

The original paper has an error in the B term which has unfortunately propagated through some papers published in the literature. These equations assume the lens feeds an N-element uniform linear array. There are M feed ports that are placed along a circular arc of radius R. The M beams point at the angles γ_m which are measured from broadside. Transmission line lengths inside the lens are found from w_n.

$$\delta_n = Fw_n + \delta_0 \tag{7.44}$$

where δ_0 is the smallest length at the center of the lens.

Example. Find the locations for the point sources for a Rotman lens with $G = 4.24$, $F = 4$, $d = 0.6$, $\alpha = 30°$, $N_{\text{beam}} = 7$, $N = 11$, $\varepsilon_{rs} = 1$, $\varepsilon_{rt} = 1$, and $\delta_0 = 0$.

Substituting these values into (7.41), (7.42), and (7.43), results in the design shown in Figure 7.44. The locations of the ports and elements are in Table 7.2.

Figure 7.45 is the design of an 8-beam Rotman lens with 8-element ports at a center frequency of 10 GHz. The matched loads on the sides absorb power incident on them in order to reduce unwanted reflections. This design was

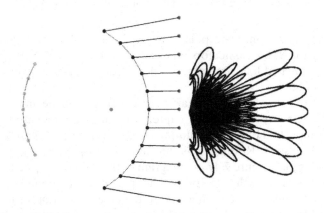

Figure 7.44. Rotman lens design with corresponding array factors.

TABLE 7.2. Element and Port Locations in λ for the Rotman Lens Design

n	1	2	3	4	5	6	7	8	9	10	11
x_{bn}	−3.84	−4.06	−4.19	−4.24	4.19	4.06	3.84	—	—	—	—
y_{bn}	−1.48	−1.01	−0.52	0	0.52	1.01	1.48	—	—	—	—
x_{pn}	−1.50	−0.96	−0.54	−0.24	−0.06	0	0.06	0.24	0.54	0.96	1.5
y_{pn}	−2.56	−2.18	−1.71	−1.17	−0.60	0	0.60	1.17	1.71	2.18	2.56
x_n	1	1	1	1	1	1	1	1	1	1	1
y_n	−3.0	−2.4	−1.8	−1.2	−0.6	0	0.6	1.2	1.8	2.4	3.0

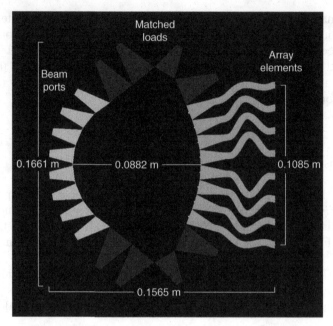

Figure 7.45. Computer design of a Rotman lens [31].

then implemented in microstrip as shown in Figure 7.46. Array elements are connected to the SMA connectors on top while receivers are connected to the beam ports. Matched loads are placed on the side SMA connectors to reduce internal reflections.

An array that has a full 360° azimuth coverage can be made around a polygon or circle as described in Chapter 5. Beam scanning with these structures is accomplished by a commutating feed and/or beam scanning. An alternate approach uses a Rotman lens that has overlapping beams covering all azimuth angles [32]. Figure 7.47 is a diagram of the concept applied to a square having 4 elements on each side. Every other port in the Rotman lens is a beam port while the other half are connected to the elements. In order to work, the focal arc and the curvature of the inside of the lens must be identical. A switching network connects the desired beam to the receiver. This concept can be extended to an array with more faces as shown in Figure 7.48b or to a circular array as in Figure 7.48c. Figure 7.49 is an experimental prototype of a 360° Rotman lens for a square array with 6 elements per face.

7.9. REFLECTARRAY

Reflect arrays are a cross between a reflector antenna and an antenna array. The feed antenna transmits to/receives from a flat to slightly curved surface covered with radiating elements as shown in Figure 7.50. Radiating elements

Array elements

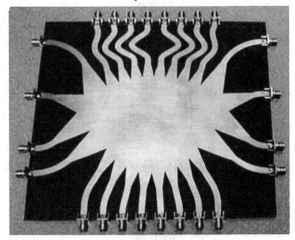

Beam ports

Figure 7.46. Experimental microstrip Rotman lens. (Courtesy of Pennsylvania State University.)

elements

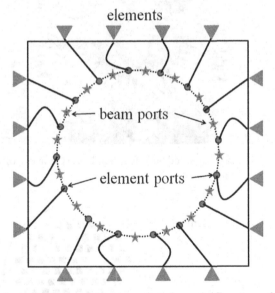

Figure 7.47. The beam and element ports are interleaved. (Courtesy of Amir Zaghloul.)

are open-ended waveguides, patches, dipoles, and so on. The elements focus the signal at the feed for receive or in the far field for transmit. Element type, orientation, and size determine the phase and/or polarization of the reflected signal. The first reflectarray used shorted rectangular waveguides [33] whose

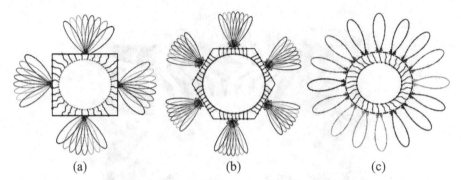

Figure 7.48. 360-degree Rotman lens designs. (**a**) Four sides. (**b**) Six sides. (**c**) Circular.

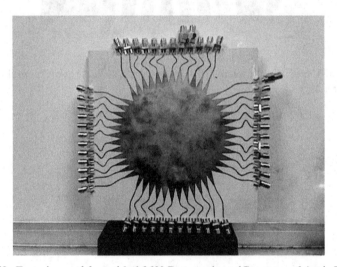

Figure 7.49. Experimental four-sided 360° Rotman lens. (Courtesy of Amir Zaghloul.)

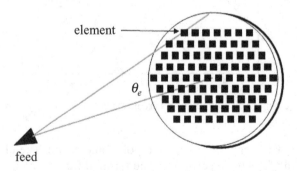

Figure 7.50. Diagram of the spatial phase delay of a reflectarray.

length determined the signal phase shift. These waveguide reflectarrays were too heavy to be practical. Reflectarrays became much more practical when they were made from printed elements.

Like a parabolic reflector, the reflect array has a very high efficiency, because it avoids lossy power dividers. Unlike the reflector, the main beam of the reflect array can be steered to large angles from broadside. The disadvantage of the reflect array is that it is narrowband. Bandwidth is limited by both the elements and the differential phase delay. The differential phase delay can be reduced as follows [34]:

1. Reduce the size of the reflecting surface.
2. Increase the distance of the feed from the reflecting surface.
3. Curve the reflecting surface.
4. Use time delays instead of phase shifts at the elements.

Aperture efficiency is found from [34]

$$\eta_T = \frac{\{[(1+\cos^{q+1}\theta_e)/(q+1)]+[(1-\cos^q\theta_e)/q]\}}{2\tan^2\theta_e[(1-\cos^{2q+1}\theta_e)/(2q+1)]} \qquad (7.45)$$

As with reflector antennas, there is also a spillover efficiency given by

$$\eta_s = 1 - \cos^{2q+1}\theta_e \qquad (7.46)$$

where θ_e is the angle between feed normal and line from feed to array edge, and $\cos^q\theta$ is the feed element pattern.

The RARF (Reflected Array Radio Frequency) used in the AN/APQ-140 Radar was manufactured by Raytheon in 1969 [35]. It has 3500 phase shifting modules that work for any polarization (Figure 7.51). Electronic phase controls allows $\pm 60°$ scanning in elevation and azimuth. This is a passive array with no gain in the modules.

7.10. ARRAY FEEDS FOR REFLECTORS

Phased array feeds have better illumination control of a reflector surface than does a single antenna element. This control aids in beam steering [36], adaptive nulling [37], beam shaping [38], and generating multiple beams [39]. Large arrays have more degrees of freedom to shape the pattern; but as the size of the array increases, the feed blockage and weight increases as well. The array pattern of a large uniform array has a narrow main beam that only illuminates a small portion of the reflector. In order to maximize antenna gain, the reflector is calibrated by adjusting the phase of the elements in the feed array. A circular planar array feed with triangular spacing has been used to compensate

Figure 7.51. Picture of RARF reflectarray. (Courtesy of the National Electronics Museum.)

for spherical reflector aberrations using a mean square minimization technique [40]. Approaches to synthesizing phased array weights to maximize gain usually assume perfect analog array weights. Closed-form solutions for the array feed excitations exist that optimize the directivity of an ideal offset reflector [41]. Compensating for reflector surface distortions improves the main beam gain as well as the sidelobe levels [42].

An array at the focal point of a front-fed reflector causes blockage that severely degrades the main beam of the reflector antenna. It is possible to adjust the phase shifters of the elements using a genetic algorithm until the maximum reflector gain is obtained at $\phi = 0°$ [43]. Optimizing the amplitude taper of the feed array contributes very little to the calibration, so it is ignored. Consider a 5-element uniform array with elements spaced half a wavelength apart placed at the focal point of a two-dimensional parabolic cylinder reflector with a diameter of 20λ and a focal distance of 10λ ($F/D = 0.5$). Its geometry with the compensated and uncompensated feed patterns appear in Figure 7.52. Compensated and uncompensated reflector antenna far-field patterns are shown in Figure 7.53. Calibration increases the main beam at $\phi = 0°$ by 4.7 dB and lowers the sidelobes adjacent to the main beam by 0.8 dB.

Using an offset reflector reduces feed blockage, but the array still needs calibration for small F/D to ensure maximum gain at $\phi = 0°$. An offset reflector with an F/D of 0.8 places the feed about 10λ from the reflector surface. At this F/D ratio, a uniform array works well, and the angle of the feed is not

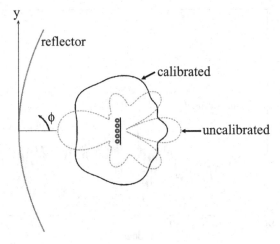

Figure 7.52. Calibrated and uncalibrated 5-element phased array feed for the reflector antenna.

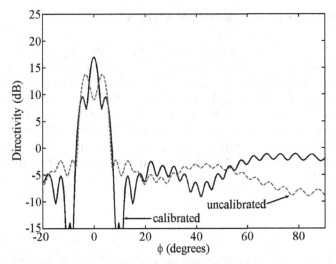

Figure 7.53. Front-fed reflector far-field patterns when the 5-element array feed is calibrated and uncalibrated.

very sensitive. Reducing the F/D to 0.3 brings the feed much closer to the reflector surface and necessitates calibration. Calibration for the offset feed includes adjusting the tilt or pointing direction of the feed. Figure 7.54 shows the antenna patterns of the calibrated and uncalibrated offset reflectors. Calibration provides a 4-dB improvement to the main beam at boresight.

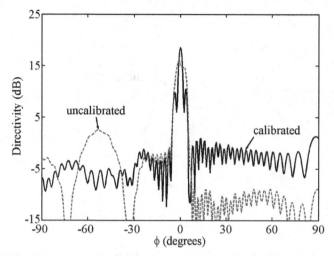

Figure 7.54. Offset-fed reflector far-field patterns when the 5-element array feed is calibrated and uncalibrated.

7.11. ARRAY FEEDS FOR HORN ANTENNAS

Many systems need high-gain antennas that electronically scan over a limited angular range, although the range may be several hundreds of beamwidths. Large, array-fed reflectors are high gain but have a scan that is limited to several tens of beamwidths. Large conventional arrays are high gain and capable of scanning over a large angular extent but are prohibitively expensive due to the large number of elements. A hybrid array-fed horn antenna is a viable and affordable alternative to full phased array antennas for applications such as space-based radar. The minimum number of elements in an array is given by [44]

$$N_{\min} = \frac{\Omega_s}{\theta_{3\,\mathrm{dB}}} \tag{7.47}$$

where Ω_s is the solid scan angle (field of view) and $\theta_{3\mathrm{dB}}$ is the array beamwidth. Dividing the number of elements in an array by $N_{\min}$ gives the element use factor [21]. For a scan sector of 60° or more, the element use factor is close to unity. If the scan is limited to ±20°, the planar array element use factor is very high, and conventional phased arrays with one control per element over a full aperture are not very efficient.

Figure 7.55 explains how the hybrid reflector-array antenna uses the Winstone Cone light concentrator to funnel all wavelengths passing through the entrance aperture out through the exit aperture to a planar array feed [45]. The two side reflectors were optimized for maximum ray intercept over the

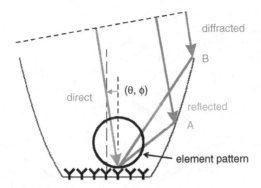

Figure 7.55. Diagram of the array-fed horn antenna. (Courtesy of Boris Tomasic, USAF AFRL.)

Figure 7.56. Picture of the experimental array-fed horn antenna. (Courtesy of Boris Tomasic, USAF AFRL.)

prescribed angular range [46]. Off-axis rays hitting the reflectors take multiple bounces before they reach the feed array. Reflectors are shaped so that incident waves within a ±20° sector only bounce once before reaching the feed array.

A 43-element feed with $d = 0.45\lambda_0$ at 10 GHz with 20% bandwidth was built for a parallel-plate flared waveguide shown in Figure 7.56 [44]. A diagram of the element probes and reflector outline appears in Figure 7.57. The parallel plates are separated by $0.37\lambda_0$, and the probe height is $0.23\lambda_0$. The antenna aperture is $40\lambda_0$ (1.2 m), while the array feed is $20\lambda_0$ (0.6 m). If the array scans ±20°, then the array element spacing must be less than 0.7λ. If the array is 1.2 m long, then it would have 57 elements. Thus, the array-fed horn antenna has an element use factor of 43/28 = 1.53, which is better than the full array use factor of 54/28 = 2.03 [44].

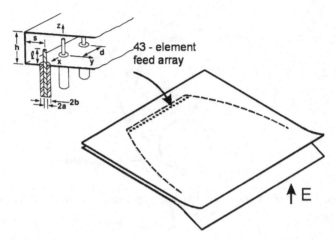

Figure 7.57. Diagram of the elements and array-fed horn antenna. (Courtesy of Boris Tomasic, USAF AFRL.)

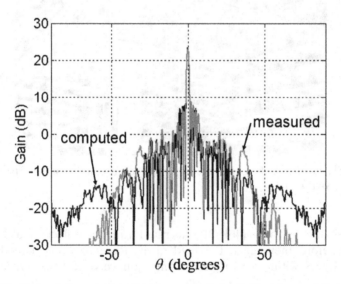

Figure 7.58. Plots of the calculated and measured far field pattern of the array-fed horn antenna at broadside. (Courtesy of Boris Tomasic, USAF AFRL.)

The array-fed horn antenna was modeled on a computer using physical optics (PO), and the experimental antenna was measured in an anechoic chamber. Figure 7.58 shows that the measured and computed far-field patterns agree well within $\theta = \pm 30°$. No grating lobes appear when the beam is steered to $\theta_s = -20°$ (Figure 7.59).

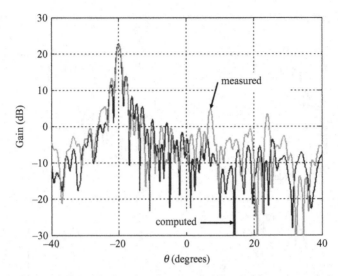

Figure 7.59. Plots of the calculated and measured far field pattern of the array-fed horn antenna at $\theta_s = -20°$. (Courtesy of Boris Tomasic, USAF AFRL.)

All the element patterns were calculated using the uniform theory of diffraction (UTD) and compared with measured element patterns (Figure 7.60 has four examples). Agreement is quite good over $\theta = \pm90°$. An accurate representation of the far-field pattern is obtained using the measured far-field element patterns in the 43-element array as was done in Chapter 6. The amplitude and phase weights (Figure 7.61) were found that limit the maximum sidelobe level to below 30 dB using the method of alternating projections [47]. Figure 7.62 is the resulting far-field pattern.

7.12. PHASE SHIFTERS

Phase shifters change the phase of a signal in a transmission line or waveguide by inducing a phase delay between 0° and 360° (in time delay terms, between 0 and λ/c). Ideally, the amplitude of the signal passing through the phase shifter is the same for all phase states. Some phase-shifting technologies have a high insertion loss and must be integrated with an amplifier to be useful. Most phase shifters are designed to be reciprocal (even when they use nonreciprocal components), in order to have identical characteristics in the transmit and receive modes. Phase shifters used in phased arrays must have digital interfaces (including those that use analog technology to cause the phase delay), so they can be computer-controlled. Phase shifters work by either forcing the signal to take a longer/shorter path or increasing/decreasing the phase velocity over a set distance.

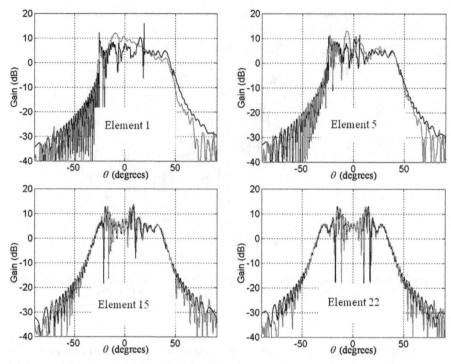

Figure 7.60. Plots of the calculated and measured far field element patterns of the array-fed horn antenna. (Courtesy of Boris Tomasic, USAF AFRL.)

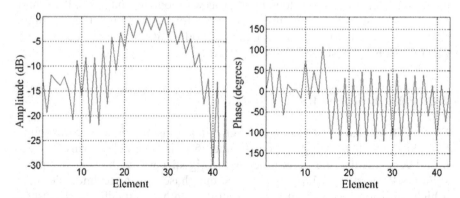

Figure 7.61. Low sidelobe element weights synthesized using the measured element patterns. (Courtesy of Boris Tomasic, USAF AFRL.)

Switched line phase shifters add line lengths to increase the phase of a signal. When bit n of an N-bit phase shifter is a 1, the signal is delayed by traveling an additional $180°/n$ in phase. Thus, a three-bit phase shifter with input [1 1 0] has an additional phase delay of $270°$ (see Figure 7.63). Switches

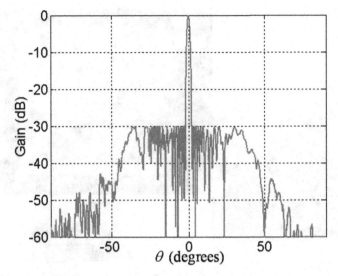

Figure 7.62. Low-sidelobe patttern using the measured element patterns and the synthesized weights. (Courtesy of Boris Tomasic, USAF AFRL.)

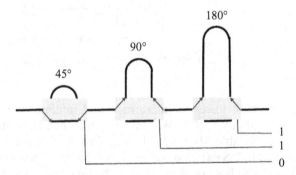

Figure 7.63. Diagram of a switched line phase shifter.

are fast and lightweight but have low power-handling capability and are possibly lossy. A number of different type of switches used in phase shifters technology include PIN diode and MEMS (micro-electromechanical systems).

The PIN diode has heavily doped p-type and n-type regions separated by an intrinsic region [3]. A forward bias results in a very low resistance to high frequencies, while a reverse bias acts like an open circuit with a small capacitance. PIN diodes have a low frequency limit of about 1 MHz due to carrier lifetime. A few milliamps of DC current can cause the PIN diode to switch one or more amps of RF current. These switches have the disadvantage of being current controlled rather than voltage controlled.

MEMS switches have low losses, high isolation, high linearity, small size, low power consumption, and low cost and are broadband. On the other hand,

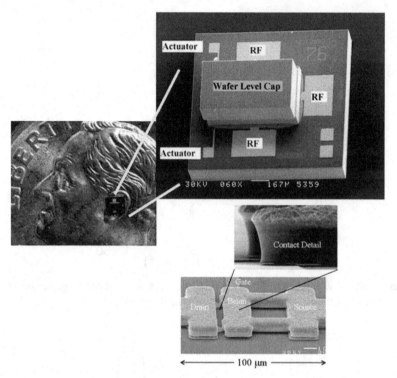

Figure 7.64. Picture of a Radant MEMS switch. (Courtesy of Radant Technologies, Inc.)

they tend to have high losses at microwave and millimeter waves, limited power-handling capability (~100 mW), and they may need expensive packaging to protect the movable MEMS bridges against the environment [16]. An example of an RF MEMS switch developed by Radant Technologies is shown in Figure 7.64 [48]. The base of the cantilever beam connects to the source to its right. The free-standing tip of the beam suspends above the drain. An electrostatic force pulls the beam down when a voltage is applied to the gate. Closing the switch results in an electrical path between the beam and the drain. MEMS switches have high isolation when open, but they have low insertion loss when closed while consuming a small amount of power. At first, MEMS switches had low reliability, but recent advances have made them competitive with other technologies.

Ferrite phase shifters use current to magnetize or demagnetize a piece of ferrite in a waveguide [3]. When it is demagnetized, the phase is zero. When it is magnetized, the phase is a function of the length of the ferrite. Ferrite phase shifters are reciprocal when the variable differential phase shift through the device is the same in either direction. A latching phase shifter makes use of the hysteresis curve for a ferrite. A current pulse controls whether the ferrite is in the positive or negative saturation state. The length of the ferrite

TABLE 7.3. Phase Shifter Characteristics [16]

Type	Ferroelectric	Semiconductor/MMIC	Ferrite	MEMS
Cost	Low	High	Very high	Low
Reliability	Good	Very good	Excellent	Good
Power handling	>1 W	>10 W	kW	<50 mW
Switch speed	ns	ns	10–100 µs	10–100 µs
DC power consumption	Low	Low	High	Negligible
Size	Small	Small	Large	Small

is chosen so that the desired differential phase shift between the two saturation states corresponds to phase bit n where the phase for bit n is $360°/2^n$. A phase shifter has independently biased ferrite sections of decreasing size in series. Latching ferrite phase shifters have the advantage of not requiring a continuous bias current and have been made from about 2 GHz to 94 GHz [49].

The rotary-field phase shifter, on the other hand, induces a phase shift by rotating a fixed amplitude magnetic bias field [49]. It is inherently low phase error since the phase shift is controlled by the angular orientation of the bias field rather than the magnitude of the bias field. The angular orientation is a function of the current fed to a pair of orthogonal windings. Rotary-field phase shifter operate from S-band through Ku-band (2–18 GHz).

Ferroelectric materials have a tunable dielectric constant. For phase shifters made from a ferroelectric material like barium strontium titanate (BST), part or all of the substrate of the phase shifting circuit is made from ferroelectric material [50]. As the fields pass through the ferroelectric layer, the phase velocity can be controlled by changing the permittivity of the ferroelectric [51].

Table 7.3 lists the characteristics of several types of phase shifters. Ferroelectrics and MEMS dominate the low cost. Ferrite phase shifters are needed for high-power applications, but also consume a lot of DC power consumption. Ferrites and MEMS tend to have slow switching speeds.

Figure 7.65 is a picture of a waveguide-fed linear array of slots. Parts on the left are assembled to make the array on the right. The phase shifters fit between the waveguide feed and the slots. Electronic controls lie on top of the phase shifters.

7.13. TRANSMIT/RECEIVE MODULES

The transmit/receive (T/R) module has five functions [52]:

1. Amplify the transmit signal.
2. Amplify and provide low-noise figure for the receive signal.
3. Switch between transmit and receive mode.

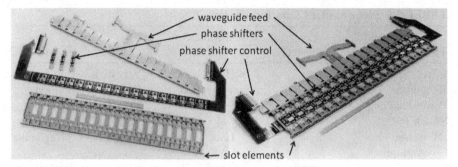

Figure 7.65. Diagram of a linear slot array with phase shifters. (Courtesy of Northrop Grumman and available at the National Electronics Museum.)

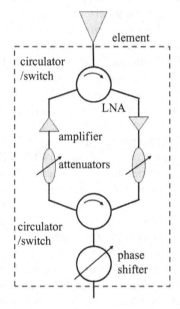

Figure 7.66. Diagram of a T/R module.

4. Phase control of transmit and receive signals.
5. Protect low-noise amplifier.

A diagram of a T/R module is shown in Figure 7.66 [53]. The phase shifter at the bottom of the T/R module performs beam steering and calibration. A circulator or switch at either end isolates the transmit and receive channels. Communications systems usually operate in full duplex mode (simultaneously transmit and receive), while radar systems usually operate in half-duplex mode (sequentially transmit and receive). Care must be taken to isolate the transmit and receive channels, as well as protect the transmit amplifier from reflections that result from impedance mismatches while beam scanning. An attenuator appears in both paths in order to calibrate the paths and implement low-sidelobe tapers. The attenuator only reduces gain, so the gain at each element starts at the minimum attenuation and is adjusted down rather than set at the mean level and adjusted up or down. Attenuators throw away signal power and create heat, so they need to be applied wisely to avoid overheating and loss of signal power. The receive channel typically has 10–20 dB of gain from the low-noise amplifier (LNA) in order to establish a low-noise figure before phase shifting and traveling through the feed network. Power amplifiers in the transmit channel typically have about 30 dB of gain in order to compensate for the losses in the feed network [53].

Module calibration ensures desired performance at each element [53]. Ideally, when calibrating the array, the phase shifter's gain remains constant as the phase settings are varied, but the attenuator's insertion phase can vary as a function of the phase setting. In this situation, the calibration process only has to be done at the broadside beam location. The first step of this calibration process adjusts the attenuators for uniform gain at the elements. The phase shifters are then adjusted to compensate for the insertion phase differences at each element. Steering the beam just adds a linear phase shift to the calibration phase. This calibration should be done across the bandwidth and range of operating temperatures. If the phase shifter's gain varies as a function of setting, then the attenuators need to be compensated as well. The process of adjusting the phase and amplitude iterates until the error is minimized. All the calibration settings are saved and applied at the appropriate times.

The transmitter feeding a T/R module can be either centralized at the array or subarray output or distributed at each element. Centralized power amplification has the following disadvantages [52]:

1. The insertion loss of the beamformer and phase shifters occurs after the transmit amplifier or before the receiving amplifier. These losses degrade the array's output power and noise figure. Attenuators also degrade a centralized system's output power and noise figure.
2. The power-handling requirement for centralized phase shifters and attenuators is significantly higher than a distributed system.
3. A centralized amplifier conveys the same phase noise to each antenna element, creating correlated errors in the aperture.
4. If the amplifier is a single device, then its failure will lead to failure of the phased array. A centralized amplifier can be made by power combining several lower-power amplifiers to minimize the impact of a single amplifier's failure.
5. Output power and efficiency degradation created by the insertion loss of the output power combiner is a significant drawback for solid-state centralized amplifiers.

Figure 7.67. Picture of the MERA T/R module. (Courtesy of the National Electronics Museum.)

The main disadvantage of distributed T/R modules is the increased cost and complexity associated with filters for reducing electromagnetic interference (EMI) [52]. In a centralized system, only one EMI filter is required and does not have the size restriction of fitting behind an element. Distributed amplification requires a filter for every T/R module. Distributed filtering typically has higher insertion loss than centralized filtering due to packaging size restrictions. The filter's loss may degrade system noise figure and output power depending on its location in the array.

T/R module development started in 1964 with the Molecular Electronics for Radar Applications (MERA) Program [54]. A photograph of the brick module including the radiating element (dipole) is shown in Figure 7.67. The monolithic microwave integrated circuit (MMIC) was fabricated on high-density alumina and high-resistivity silicon, using thin-film techniques. Circuits are located on both sides of the module. The side shown in the photograph contains a four-stage S-band transmit amplifier at 2.25 GHz. The amplifier is followed by a times four frequency multiplier plus filters in order to transmit the 9-GHz signal. Also shown in the upper left-hand corner is a balanced mixer and a 500-MHz lumped element IF amplifier. Located on the opposite side is the four-bit phase shifter and local oscillator network. Each module in the array transmitted a peak power of 0.6 W at 9.0 GHz. A noise figure of 12 dB was obtained with a gain of 14 dB. The received signal is output from the module at 2.125 GHz. The module is $7.1 \times 2.5 \times 0.8$ cm, and it weighs 0.95 ounces. A T/R switch was to be used at X band to separate the transmitter and receiver functions.

In 1987, the Advanced Tactical Fighter (ATF) program started the use of GaAs, MMIC technology in order to reduce the size and cost of T/R modules (Figure 7.68) [55]. The AN/APG-77 radar antenna is a phased array with 2000 T/R modules capable of scanning 120 degrees.

As MMIC and heat dissipation technology improved, arrays using many T/R modules migrated from a brick design to a tile design. Figure 7.69 is a diagram of one tile in a 17-GHz three-tile multibeam active array that was launched into geostationary orbit in 2008 [56]. Each tile has 33 RF modules, 16 low-noise amplifiers, a tile controller, and a power supply. The element

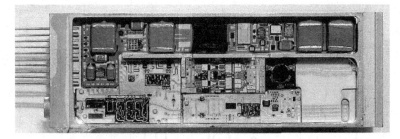

Figure 7.68. Picture of the ATF T/R module. (Courtesy of the National Electronics Museum.)

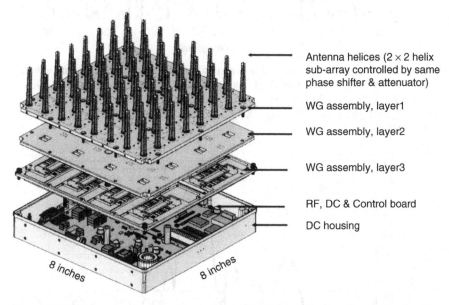

Antenna helices (2 × 2 helix sub-array controlled by same phase shifter & attenuator)

WG assembly, layer1

WG assembly, layer2

WG assembly, layer3

RF, DC & Control board

DC housing

8 inches

8 inches

Figure 7.69. Example of an 8 × 8 array tile. (Courtesy of the Lockheed Martin, Corp. E. Lier and R. Melcher, A modular and lightweight multibeam active phased receiving array for satellite applications: Design and ground testing, *IEEE Antennas Propagat. Mag.*, Vol. 51, No. 1, 2009, pp. 80–90.)

spacing is one inch, with each element having a directivity of 14.4 dBi. A 2 × 2 subarray of helical elements are connected to a 4-to-1 combiner. A plot of the antenna pattern for the tile is shown in Figure 7.70. Steering to the maximum scan angle of 5° results in the pattern shown in Figure 7.71.

T/R modules have not enjoyed the mass production of other commercial technologies until recently. Computer, wireless, and solar cell initiatives are pushing substrate manufacturing that benefit T/R module design. Figure 7.72 is a diagram of power-handling capability of different T/R module substrates versus frequency [53, 57]. Gallium nitride (GaN) and silicon carbide MMIC chips have the potential to increase the T/R module power by one or two orders of magnitude. SiC technology offers 1–10 kW in the low-GHz frequency band

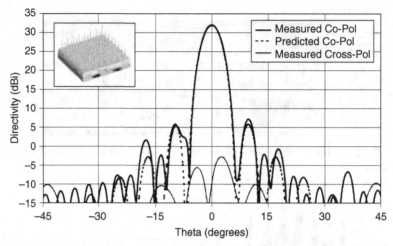

Figure 7.70. Predicted and measured antenna patterns of the tile when the main beam points at broadside. (Courtesy of the Lockheed Martin, Corp. E. Lier and R. Melcher, A modular and lightweight multibeam active phased receiving array for satellite applications: Design and ground testing, *IEEE Antennas Propagat. Mag.*, Vol. 51, No. 1, 2009, pp. 80–90.)

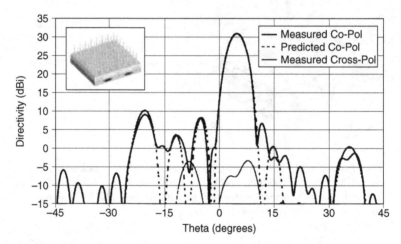

Figure 7.71. Predicted and measured antenna patterns of the tile when the main beam points at 5°. (Courtesy of the Lockheed Martin, Corp. E. Lier and R. Melcher, A modular and lightweight multibeam active phased receiving array for satellite applications: Design and ground testing, *IEEE Antennas Propagat. Mag.*, Vol. 51, No. 1, 2009, pp. 80–90.)

with better thermal concuctivity than GaAs. GaN modules work from a little less than 10 kW below 10 GHz to a little less than 100 W above 100 GHz. It is capable of reduced chip size, broad bandwidth, and higher operating voltage. SiGe and CMOS offer low-power, low-cost, high-performance T/R modules.

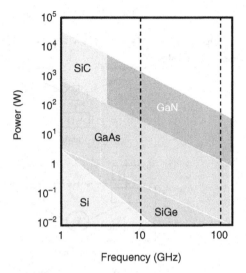

Figure 7.72. Solid state power handling versus frequency for several substrate materials.

7.14. DIGITAL BEAMFORMING

Digital beamforming (DBF) [58, 59] is similar to the use of T/R modules, except the T/R module contains an analog-to-digital (AD) converter that outputs a digital signal directly to the computer. It has the advantages of improved dynamic range, ease of forming multiple beams, adaptive nulling, and low sidelobes. Like T/R modules, the AD converter can be either central-ized or distributed [16]. The corporate feed in Figure 7.73a performs analog beamforming and converts the signal to digital once all the signals are combined. Figure 7.73b converts the analog signals to digital signals at each element. Figure 7.73c shows an array architecture in which the subarrays perform analog beamforming and the signals from the subarrays are converted to digital signals. This approach can increase the bandwidth of the array through digital time delay instead of hardware time delays. DBF is frequently implemented at the subarray level because the digital receivers are expensive, large, and heavy and require calibration. A full DBF array has the most flex-ibility and the least amount of feed hardware, because all weighting and combining is done in software. Calibration of the DBF can be done by cou-pling a calibration signal into the feed line from the element (Figure 7.74) [60]. Calibration can also be done using near-field or far-field sources [61]. Compensation for the amplitude and phase errors can be done in software. DBF is expensive and requires a lot of space behind each element, so it is usually limited to small arrays. DBF systems are moving up in frequency and becoming more common [62].

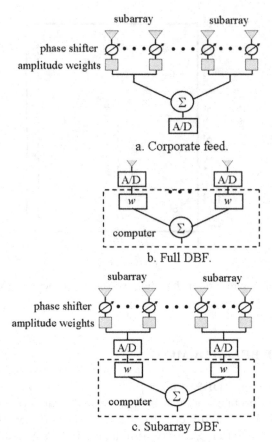

Figure 7.73. Digital beamforming versus corporate feed. (**a**) Corporate feed. (**b**) Subarray corporate feed and digital beamforming at the subarray ports. (**c**) Digital beamforming at each element.

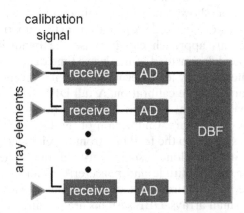

Figure 7.74. Diagram of a digital beamforming array.

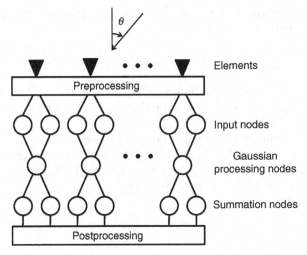

Figure 7.75. Diagram of a neural beamforming array.

7.15. NEURAL BEAMFORMING

An artificial neural network is ideally suited to form beams given the signals at each element of an array. Unlike traditional beamforming algorithms, which require either extremely well-matched channels or extensive calibration, the neural network beamformer is trained *in situ* and learns to perform the required function regardless of array element degradations, large manufacturing tolerances, or scattering from the platform. The neural beamformer architecture consists of antenna measurement input preprocessing, an artificial neural network, and output postprocessing (Figure 7.75). In some special cases, the neural beamformer is similar to a Butler matrix [63].

The input nodes receive preprocessed antenna data that are sent to hidden layer (or Gaussian processing) nodes. Element signals, $x_1 = (x_{l1}, x_{l2}, ..., x_{ln})$, $l = 1, 2, ..., L$, are measured over the antenna field of view ($\pm 60°$). Each element signal vector contains seven phase differences between adjacent elements in the array. The hidden layer of nodes have Gaussian activation functions. A processing node has an output that depends on the radial distance between the element signal vector and the node center, as well as on the spread parameter, σ. When the network receives the element input vector, the output of hidden layer node i is ϕ_{li} [64]:

$$\phi_{li} = e^{-\sum_{k=1}^{n} \frac{|x_{lk} - m_{ik}|^2}{2\sigma^2}} \tag{7.48}$$

where $i = 1, 2, ..., q$ for q hidden layer nodes. An interpolation matrix, Φ, contains the processing node outputs, ϕ_{li}. Row l of this matrix is the outputs of the q hidden layer nodes corresponding to input vector x_l.

The output matrix, Y, has components y_{lj} that are a weighted sum of the hidden layer outputs.

$$y_{lj} = \sum_{i=1}^{q} \phi_{li} w_{ij} \tag{7.49}$$

where $j = 1, 2, \ldots, r$ for r output nodes and w_{ij} are the components of the weight matrix, W. The weights come from training the network on a subset of the L measured input vectors.

An 8-element array is trained using $r = 13$ output nodes, centered at $10°$ intervals from $-60°$ to $60°$. If the AOA (angle of arrival) is between two training angles, the nodes have an output between 0 and 1. The AOA, θ, for a wave between nodes j and $j + 1$, is found from [65]

$$\theta = \theta_j + 10° \frac{y_{j+1}}{y_j + y_{j+1}} \tag{7.50}$$

where θ_j is the training angle for node j, and y_j and y_{j+1} are the node output values produced by the network at nodes j and $j + 1$, respectively.

The network is trained using supervised learning (the network receives input vectors and the corresponding desired outputs), and backpropagation is used to find the weights.

After training, the desired output matrix, Y_d, is a 13×13 identity matrix. The training input vectors are a subset of the L measured input vectors corresponding to waves from the training angles. The desired output matrix is the product of the training matrix times the weight vector.

$$Y_d = \Phi_t W \tag{7.51}$$

Since the number of hidden layer nodes equals the number of training angles, the weight matrix is found from

$$W = \Phi_t^{-1} \tag{7.52}$$

since Y_d is an identity matrix.

In order to determine the AOA, the node with the largest response and the largest of its one or two nearest neighbors are selected. The final confidence of the interpolated AOA is obtained by adding the values from the two nodes. If this value exceeds a predetermined threshold, then a source is considered detected.

The threshold for the false alarm rate is determined through training the neural network by generating many noise samples, preprocessing them, and stimulating the network [64]. The network's outputs are processed with no threshold. All interpolated outputs are considered detected sources. These purposely generated false alarms are used to determine a detection threshold. The number of noise samples and quantity of false alarms depend on the required the level of the detection threshold. A threshold for a probability of false alarm of 10^{-3} typically takes 100,000 noise samples. The 100 signal detections with the largest confidence values and the smallest value becomes our threshold; that is, any confidence value over that threshold would be a false alarm, and these 100 out of 100,000 would give a false alarm probability of 10^{-3}.

7.16. CALIBRATION

A phased array must be calibrated before it generates an optimum coherent beam. Calibration involves tuning the attenuators, phase shifters, receivers, and so on, such that peak gain is available for the desired low-amplitude taper for transmit and/or receive operations. A phased array is calibrated by adjusting the phase shifters until the signal path to each element is identical, thus ensuring the maximum gain of the array. The phase settings are stored for beam steering angles and select frequencies within the bandwidth of the array. Calibration tables may have to be updated on a periodic basis, depending upon the environment of the array. Age and temperature cause component characteristics to drift over time, so the array requires recalibration.

Methods for performing array calibration include [66]:

1. Near-Field Scan. A planar near field scanner positioned very close to the array moves a probe directly in front of each element to measure the amplitude and phase of that element when all other elements are turned off.
2. Far-Field Gain. An antenna is placed at boresight in the far field. The amplitude and phase of an element is measured when all other elements are turned off.

Usually, array calibration consists of pointing a beam at a transmit or receive calibration source in the far field. Radio telescope arrays often point the beam to a region of the sky with a single strong point source [67]. Calibration with near-field sources requires that distance and angular differences be taken into account. If the calibration source is in the far field, then the phase shifters are set to steer the beam in the direction of the source. Each element is then toggled through all of its phase settings until the phase setting that yields the maximum signal strength is found. The difference between the steering phase and the phase that yields the maximum signal is the calibration phase. The complex electric field at the element is found by measuring the maximum power, the minimum power, and the phase shift corresponding to the maximum power.

Making power measurements for every element in an array for every phase setting is extremely time-consuming. Calibration techniques that measure both amplitude and phase of the calibrated signal tend to be much faster.

Accurately measuring the signal phase is reasonable in an anechoic chamber but difficult in the operational environment. Measurements at four orthogonal phase settings yield sufficient information to obtain a maximum likelihood estimate of the calibration phase [66]. The element phase error is calculated from power measurements at the four phase states, and the procedure is repeated for each element in the array. Additional measurements improve the signal-to-noise ratio, and the procedure can be repeated to achieve desired accuracy within resolution of the phase shifters, since the algorithm is intrinsically convergent.

Another approach uses amplitude-only measurements from multiple elements to find the complex field at an element [68]. The first step measures the power output from the array when the phases of multiple elements are successively shifted with the different phase intervals. Next, the measured power variation is expanded into a Fourier series to derive the complex electric field of the corresponding elements. The measurement time reduction comes at the expense of increased measurement error.

REFERENCES

1. Why 50 Ohms? Jul 30, 2009; http://www.microwaves101.com/encyclopedia/why50ohms.cfm.

2. R. E. Collin, Foundations for Microwave Engineering, *IEEE Press Series on Electromagnetic Wave Theory*, New York: IEEE Press, 2001.

3. D. M. Pozar, *Microwave Engineering*, 2nd ed., New York: John Wiley & Sons, 1998.

4. T. H. Lee, *Planar Microwave Engineering: A Practical Guide to Theory, Measurement, and Circuits*, New York: Cambridge University Press, 2004.

5. R. J. P. Douville and D. S. James, Experimental study of symmetric microstrip bends and their compensation, *IEEE Trans. Microwave Theory Tech.*, Vol. 26, No. 3, 1978, pp. 175–182.

6. E. Levine and S. Shtrikman, Optimal designs of corporate-feed printed arrays adapted to a given aperture, The sixteenth conference of Electrical and Electronics Engineers in Israel, 1989, pp. 1–4.

7. E. J. Wilkinson, An *N*-way hybrid power divider, *IRE Trans. Microwave Theory and Tech.*, Vol. 8, No. 1, 1960, pp. 116–118.

8. W. H. Kummer, *Feeding and phase scanning*, in *Microwave Scanning Antennas*, 2, R. C. Hansen, ed., New York: Academic Press, 1964, pp. 2–102.

9. T. K. Anthony and A. I. Zaghloul, Designing a 32 element array at 76 GHz with a 33 dB taylor distribution in waveguide for a radar system, in *Antennas and Propagation Society International Symposium, 2009*, APSURSI '09. Charleston, SC: IEEE, 2009, pp. 1–4.

10. R. S. Elliott, *Antenna Theory and Design*, revised edition, Hoboken, NJ: John Wiley & Sons, 2003.

11. R. Hendrix, Aerospace system improvements enabled by modern phased array radar, Northrop Grumman Electronic Systems, October 2002, pp. 1–17.

12. R. E. Collin, *Antennas and Radiowave Propagation*, New York: McGraw-Hill, 1985.

13. T. A. Milligan, *Modern Antenna Design*, 2nd ed., *Hoboken*, NJ: Wiley-Interscience/ IEEE Press, 2005.

14. S. Silver, *Microwave Antenna Theory and Design*, 1st ed., New York: McGraw-Hill Book, 1949.

15. R. Elliott and L. Kurtz, The design of small slot arrays, *IEEE Trans. Antennas Propagat.*, Vol. 26, No. 2, 1978, pp. 214–219.

16. J. Frank and J. D. Richards, Phased array radar antennas, in *Radar Handbook*, M. I. Skolnick, ed., New York: McGraw-Hill, 2008, pp. 13.1–13.74.

17. J. Blass, Multidirectional antenna—A new approach to stacked beams. IRE International Convention Record, Vol. 8, Mar 1960, pp. 48–50.

18. J. L. Butler, Digital, matrix, and intermediate-frqeuency scanning, *Microwave Scanning Antennas*, R. C. Hansen, ed., New York: Academic Press, 1966, pp. 217–288.

19. J. Butler and R. Lowe, Beam-forming matrix simplifies design of electronically scanned antennas, *Electron. Design*, Vol. 9, April 12, 1961, pp. 170–173.

20. B. Pattan, The versatile Butler matrix, *Microwave J.*, Vol. 47, No. 11, 2004, pp. 126–130.

21. R. J. Mailloux, *Phased Array Antenna Handbook*, 2nd ed., Boston: Artech House, 2005.

22. A. Bhattacharyya, *Phased Array Antennas and Subsystems: Floquet Analysis, Synthesis, BFNs, and Active Array Systems*, Hoboken, NJ: Wiley-Interscience, 2006.

23. W. E. Kock, Metal-lens antennas, *Proc. IRE*, Vol. 34, No. 11, 1946, pp. 828–836.

24. J. Ruze, Wide-angle metal-plate optics, *Proc. IRE*, Vol. 38, No. 1, 1950, pp. 53–59.

25. J. Rao, Multifocal three-dimensional bootlace lenses, *IEEE Trans. Antennas Propagat.*, Vol. 30, No. 6, 1982, pp. 1050–1056.

26. M. Kales and R. Brown, Design considerations for two-dimensional symmetric bootlace lenses, *IEEE Trans. Antennas Propagat.*, Vol. 13, No. 4, 1965, pp. 521–528.

27. H. Gent, The bootlace aerial, *R. Radar Est. J.*, October 1957, pp. 47–57.

28. E. Brookner, Fundamentals of radar design, *Radar Technology*, E. Brookner, ed., pp. 3–66, Dedhan, MA: Artech House, 1979.

29. W. Rotman and R. Turner, Wide-angle microwave lens for line source applications, *IEEE Trans. Antennas Propagat.*, Vol. 11, No. 6, 1963, pp. 623–632.

30. D. M. Pozar, *Antenna Design Using Personal Computers*, Dedham, MA: Artech House, 1985.

31. Rotman Lens Designer, Remcom, 2009.

32. J. Dong and A. I. Zaghloul, A concept for a lens configuration for 360° scanning, *IEEE Antennas Wireless Propag. Lett.*, Vol. 8, pp. 985–988, 2009.

33. D. Berry, R. Malech, and W. Kennedy, The reflectarray antenna, *IEEE Trans. Antennas Propagat.*, Vol. 11, No. 6, 1963, pp. 645–651.

34. J. Huang and J. A. Encinar, *Reflectarray Antennas*, Hoboken, NJ: IEEE Press/ Wiley-Interscience, 2008.

35. N. E. Museum, RARF Antenna, 2009, display information.

36. R. M. Davis, C. C. Cha, S. G. Kamak et al., A scanning reflector using an off-axis space-fed phased-array feed, *IEEE Trans. Antennas Propagat.*, Vol. 39, No. 3, 1991. pp. 391–400.

37. R. Shore, Adaptive nulling in hybrid antennas, in *Antennas and Propagation Society International Symposium, 1993, AP-S. Digest*, Vol. 3, 1993, pp. 1342–1345.

38. N. Adatia, P. Balling, B. Claydon et al., A European contoured beam reflector antenna development. European Microwave Conference, 1980, pp. 95–99.

39. K. M. S. Hoo and R. B. Dybdal, Adaptive multiple beam antennas for communications satellites. IEEE Antennas and Propagation Society International Symposium, May 1990, pp. 1880–1883.

40. N. Amitay and H. Zucker, Compensation of spherical reflector aberrations by planar array feeds, *IEEE Trans. Antennas Propagat.*, Vol. 20, No. 1, 1972, pp. 49–56.

41. P. Lam, L. Shung-Wu, D. Chang et al., Directivity optimization of a reflector antenna with cluster feeds: A closed-form solution, *IEEE Trans. Antennas Propagat.*, Vol. 33, No. 11, 1985, pp. 1163–1174.

42. A. R. Cherrette, R. J. Acosta, P. T. Lam et al., Compensation of reflector antenna surface distortion using an array feed, *IEEE Trans. Antennas Propagat.*, Vol. 37, No. 8, 1989, pp. 966–978.

43. R. L. Haupt, Calibration of cylindrical reflector antennas with linear phased array feeds, *IEEE Trans. Antennas Propagat.*, Vol. 56, No. 2, 2008, pp. 593–596.

44. B. Tomasic, P. Detweiler, S. Schneider et al., Analysis and design of the hybrid, reflector-array antenna for highly directive, limited-scan application, *Antennas and Propagation Society International Symposium*, 2002, pp. 124–127.

45. R. Winston, Light collection within the framework of geometric optics, *J. Opt. Soc. Am.*, Vol. 60, 1970, pp. 245–247.

46. P. L. Detweiler and K. M. Pasala, A scattering matrix-based beamformer for wide-angle scanning in hybrid antennas, *IEEE Trans. Antennas Propagat.*, Vol. 52, No. 3, 2004, pp. 801–812.

47. O. M. Bucci, G. Franceschetti, G. Mazzarella et al., Intersection approach to array pattern synthesis, *IEE Proc. H, Microwaves Antennas Propagat.*, Vol. 137, No. 6, 1990, pp. 349–357.

48. J. J. Maciel, J. F. Slocum, J. K. Smith et al., MEMS electronically steerable antennas for fire control radars, *IEEE Aerosp. Electron. Syst. Mag.*, Vol. 22, No. 11, 2007, pp. 17–20.

49. W. E. Hord, Microwave and millimeter-wave ferrite phase shifters, http:// www.magsmx.com/pdf/Micro_MMW_PS.pdf, July 30, 2009.

50. D. Kuylenstierna, A. Vorobiev, P. Linner et al., Ultrawide-band tunable true-time delay lines using ferroelectric varactors, *IEEE Trans. Microwave Theory Tech.*, Vol. 53, No. 6, 2005, pp. 2164–2170.

51. D. C. Collier, Ferroelectric phase shifters for phased array radar applications, Proceedings of the Eighth IEEE International Symposium on Applications of Ferroelectrics, 1992, pp. 199–201.

52. D. N. McQuiddy, Jr., R. L. Gassner, P. Hull et al., Transmit/receive module technology for X-band active array radar, *Proc. IEEE*, Vol. 79, No. 3, 1991, pp. 308–341.

53. M. Borkowski, Solid-state transmitters, in *Radar Handbook*, M. I. Skolnick, ed., New York: McGraw-Hill, 2008, pp. 11.1–11.36.

54. D. N. McQuiddy, J. W. Wassel, J. B. LaGrange et al., Monolithic microwave integrated circuits: An historical perspective, *IEEE Trans. Microwave Theory Tech.*, Vol. 32, No. 9, 1984, pp. 997–1008.

55. F-22 Raptor Avionics August 4, 2009; http://www.globalsecurity.org/military/systems/aircraft/f-22-avionics.htm.

56. E. Lier and R. Melcher, A modular and lightweight multibeam active phased receiving array for satellite applications: Design and ground testing, *IEEE Antennas Propagat. Mag.*, Vol. 51, No. 1, 2009, pp. 80–90.

57. E. Brookner, Now: Phased-array radars: Past, astounding breakthroughs and future trends (January 2008), *Microwave J.*, Vol. 51, No. 1, 2008, pp. 31–48.

58. H. Steyskal and J. S. Herd, Mutual coupling compensation in small array antennas, *IEEE Trans. Antennas Propagat.*, Vol. 38, No. 12, 1990, pp. 1971–1975.

59. H. Steyskal, Digital beamforming at Rome laboratory, *Microwave J.*, Vol. 39, No. 2, 1996, pp. 100–124.

60. S. Drabowitch, *Modern Antennas*, 2nd ed., Dordrecht: Springer, 2005.

61. L. Pettersson, M. Danestig, and U. Sjostrom, An experimental S-band digital beamforming antenna, *IEEE Aerosp. Electron. Syst. Mag.*, Vol. 12, No. 11, 1997, pp. 19–29.

62. E. Brookner, Phased arrays and radars—Past, present, and future, *Microwave J.*, Vol. 49, No. 1, January 2006, pp. 1–17.

63. R. J. Mailloux and H. L. Southall, The analogy between the Butler matrix and the neural-network direction-finding array, *IEEE Antennas Propagat. Mag.*, Vol. 39, No. 6, 1997, pp. 27–32.

64. H. L. Southall, J. A. Simmers, and T. H. O'Donnell, Direction finding in phased arrays with a neural network beamformer, *IEEE Trans. Antennas Propagat.*, Vol. 43, No. 12, 1995, pp. 1369–1374.

65. R. L. Haupt, H. L. Southall, and T. H. O'Donnell, Biological beamforming, in *Frontiers in Electromagnetics*, D. H. Werner and R. Mittra, eds., New York: IEEE Press, 2000, pp. 329–370.

66. R. Sorace, Phased array calibration, *IEEE Trans. Antennas Propagat.*, Vol. 49, No. 4, 2001, pp. 517–525.

67. A. J. Boonstra and A. J. van der Veen, Gain calibration methods for radio telescope arrays, *IEEE Trans. Signal Processing*, Vol. 51, No. 1, 2003, pp. 25–38.

68. T. Takahashi, Y. Konishi, S. Makino et al., Fast measurement technique for phased array calibration, *IEEE Trans. Antennas Propagat.*, Vol. 56, No. 7, 2008, pp. 1888–1899.

8 Smart Arrays

A smart antenna adapts its receive and/or transmit pattern characteristics in response to the signals in the environment. At least two different antennas comprise a smart antenna, usually in the form of an antenna array. The amplitude and phase of the received signals are weighted and summed in such a way as to meet some desired performance expectation. This chapter describes several different kinds of smart antennas. The first is a retrodirective array that retransmits a received signal back in the same direction. Another type of smart array finds the angular location of signals in the environment, so they are called direction finding (DF) arrays. The simple Adcock array was presented in Chapter 2. In this chapter, arrays that can automatically detect many signals are presented. Antennas that automatically reject interference while still receiving the desired signal are important in an environment with many signals. It may place a null in the direction of an interference source or steer the main beam in the direction of a desired signal. The sidelobe canceller was the first approach to placing a null in the sidelobe of the array pattern. This concept was generalized to the adaptive array that requires signal measurements at each element. A digital beamformer is ideal for this approach due to the complete control of the signals at the elements. It is also possible to adaptively place nulls using a conventional corporate-fed array by using an optimization algorithm to minimize the total power output. The main beam is not nulled, because only some of the elements or a few bits in the amplitude/phase controls. Finally, the ultimate smart antenna has an adaptive transmit and adaptive receive array. MIMO (multiple input multiple output) communications systems have adaptive transmit and receive antennas. All of these approaches are described in this chapter.

8.1. RETRODIRECTIVE ARRAYS

Ideas for actual smart antennas did not originate until the 1950s when Van Atta invented retroirective and redirective arrays [1]. The Van Atta array receives and reradiates an incident wave in a predetermined direction with respect to the incident angle. Usually, this type of array adjusts the phase and

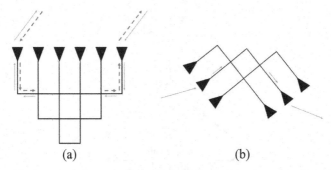

Figure 8.1. Diagram of a Van Atta arrays. **(a)** Retrodirective array. **(b)** Redirective array.

may also amplify the received signals such that it retransmits in the direction of the incident field. This array operates similar to a dihedral or trihedral corner reflector in which the signal reflects back to the transmitter. Less common is the redirective array that retransmits the received signal in a direction different from the received direction.

Figure 8.1a is a diagram of a Van Atta retrodirective array. Symmetric elements about the center of the array are connected by a transmission line. All transmission lines connecting the elements are of identical length, so the relative phase of the signals is maintained. Upon transmit, the received signals are now on opposite sides of the array, so the transmitted wavefront returns from direction that it came. Figure 8.1a follows the paths of signals received and transmitted by the end elements. Figure 8.1b is a redirective Van Atta array where the received signal is retransmitted in a different direction.

The received signals at the elements of an isotropic Van Atta array are given by

$$\left[e^{-jkdu_i(N-1)/2} \quad e^{-jkdu_i(N-2)/2} \quad \cdots \quad e^{jkdu_i(N-2)/2} \quad e^{jkdu_i(N-1)/2} \right] \tag{8.1}$$

where u_i is the angle of the incident field. Since the transmission lines are all the same length (l), a constant phase is added to the signals. By the time they reach the element on the other side of the array (assuming no loss), the signals are given by

$$\left[e^{jk[\ell+du_i(N-1)/2]} \quad e^{jk[\ell+du_i(N-2)/2]} \quad \cdots \quad e^{-jk[\ell+du_i(N-2)/2]} \quad e^{-jk[\ell+du_i(N-1)/2]} \right] \tag{8.2}$$

The phase shift between two adjacent elements in (8.2) is given by

$$\delta = -kdu_i \tag{8.3}$$

which implies that the transmit beam points in the same direction as the incident plane wave: u_i.

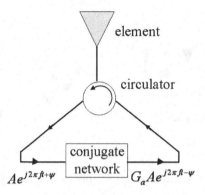

Figure 8.2. Active element in a retrodirective array.

Since the Van Atta array requires that symmetric elements in the array be connected by equal length transmission lines, this type of array will not work with nonplanar arrays or nonplanar wavefronts. An alternative is to have each element receive and then modify and retransmit the received signal in order to accommodate curved apertures, near-field sources, and broadband signals. An active retrodirective array has circulators and amplifiers in the transmission paths between elements as shown in Figure 8.2. This type of retrodirective array has a radar cross section given by [2]

$$\sigma = \frac{N^2 G_e^2 G_a \lambda^2}{4\pi} \qquad (8.4)$$

where N is the number of elements in the array, G_e is element gain, and G_a is amplifier gain minus losses of circulators. An active retrodirective array is useful as a transponder or as an enhanced radar target (Figure 8.2) [3]. This beacon automatically replies in the direction of the interrogation signal without prior knowledge of the interrogator location.

One approach to an active retrodirective array uses phase conjugation at the elements [4]. When a local oscillator (LO) heterodynes its signal with the RF signal, the following signal results [5]:

$$V_{IF} = V_{RF} \cos(2\pi f_{RF} t + \delta) \times V_{LO} \cos(2\pi f_{LO} t)$$
$$= \frac{1}{2} V_{RF} V_{LO} [\cos(2\pi (f_{LO} - f_{RF})t - \delta) + \cos(2\pi (f_{LO} + f_{RF})t + \delta)] \quad (8.5)$$

If the LO frequency is double the RF frequency, then the lower sideband is the phase conjugation of the received signal. The high sideband signal is easily filtered out, so it is not retransmitted. This approach becomes less practical as f_{RF} gets higher, because the LO source must operate at double f_{RF}, making the LO source very expensive [6].

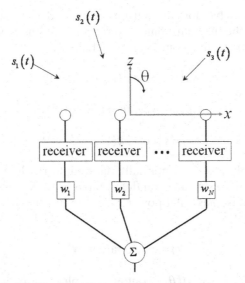

Figure 8.3. Diagram of an array with a receiver at each element.

Retrodirective arrays have applications in radio-frequency identification (RFID) [7], wireless communications, and RF transponders. All these applications require a signal be returned to the source. The signal may be amplified and the frequency or polarization may be slightly altered to distinguish it from the original signal. It is possible to have a two-dimensional retrodirective array that responds to both azimuth and elevation angles [8, 9].

8.2. ARRAY SIGNALS AND NOISE

Element signals are down-converted and sent to the computer (Figure 8.3). A computer forms an array steering vector to control the pattern. The time-dependent array output is

$$F(t) = w^\dagger X(t, \theta) \tag{8.6}$$

where

$$X(t, \theta_m) = A(\theta_m) s(t) + \mathbb{N}(t) \tag{8.7}$$

and w is the element weight vector, $s(t)$ is the signal vector, $A(\theta_m)$ is the array steering vector, and $N(t)$ is the noise vector. Neither the element weights nor the array steering vectors are time-dependent. The signal vector is the amplitude of the plane wave incident on the array at θ_m:

$$s(t) = [s_1(t) \quad s_2(t) \quad \cdots \quad s_M(t)]^T \tag{8.8}$$

The signals have a bandpass spectrum modulated by a suppressed carrier frequency due to the time harmonic field representations. Some possible signal representations to use in a computer model are

$$s_m(t) = s_m \text{sign}(\text{randn}(1, K)) \tag{8.9}$$

$$s_m(t) = s_m(\text{rand}(1, K) - 0.5) \tag{8.10}$$

$$s_m(t) = s_m \cos(2\pi(1:K)/K)e^{j\text{rand}} \tag{8.11}$$

The function randn returns a normally distributed random matrix, and rand returns a uniformly distributed random matrix with N rows and K columns. Gaussian white noise has a flat power spectrum having an amplitude σ_{noise}^2. It is calculated using

$$\mathbf{N}(t) = \sqrt{\sigma_{\text{noise}}}\,\text{randn}(N, K) \tag{8.12}$$

The array steering vector, $A(\theta_m)$, contains the phase of plane wave m at each of the elements. It varies with incident angle but not time. The steering vectors for each of the M incident signals are placed in the columns of a matrix, A.

$$A(\theta) = \begin{bmatrix} e^{jkx_1\cos\theta_1} & e^{jkx_1\cos\theta_2} & \cdots & e^{jkx_1\cos\theta_M} \\ e^{jkx_2\cos\theta_1} & e^{jkx_2\cos\theta_2} & \cdots & e^{jkx_2\cos\theta_M} \\ \vdots & \vdots & \vdots & \vdots \\ e^{jkx_N\cos\theta_1} & e^{jkx_N\cos\theta_2} & \cdots & e^{jkx_N\cos\theta_M} \end{bmatrix} \tag{8.13}$$

The output power is proportional to the array output squared.

$$P = \frac{1}{2}FF^{\dagger} = \frac{1}{2}E\left\{\left|A^{\dagger}(\theta)X(t,\theta)\right|^2\right\} = \frac{1}{2}A^{\dagger}R_T A \tag{8.14}$$

where R_T is the covariance matrix, $\dagger$ is the complex conjugate transpose of the vector, and $E\{\}$ is the expected value. In this chapter, we assume that all the signals are zero mean, stationary processes, so the covariance matrix is also the correlation matrix.

An estimate of the correlation matrix built from the signal and noise samples at K time intervals is given by

$$
\begin{aligned}
R_T &= E\left\{X(\kappa)X^{\dagger}(\kappa)\right\} \\
&= E\left\{X_s(\kappa)X_s^{\dagger}(\kappa)\right\} + E\left\{\mathbf{N}(\kappa)X_s^{\dagger}(\kappa)\right\} + E\left\{X_s(\kappa)\mathbf{N}^{\dagger}(\kappa)\right\} + E\left\{\mathbf{N}(\kappa)\mathbf{N}^{\dagger}(\kappa)\right\} \\
&\approx \frac{1}{K}\sum_{\kappa=1}^{K}X_s(\kappa)X_s^{\dagger}(\kappa) + \frac{1}{K}\sum_{\kappa=1}^{K}\mathbf{N}(\kappa)X_s^{\dagger}(\kappa) + \frac{1}{K}\sum_{\kappa=1}^{K}X_s(\kappa)\mathbf{N}^{\dagger}(\kappa) \\
&\quad + \frac{1}{K}\sum_{\kappa=1}^{K}\mathbf{N}(\kappa)\mathbf{N}^{\dagger}(\kappa) \\
&= R_s + R_{\text{noise}-s} + R_{s-\text{noise}} + R_{\mathbf{N}}
\end{aligned}
\tag{8.15}
$$

When the signal and noise are uncorrelated and enough samples are taken, then

$$R_T = R_s + R_{noise} \tag{8.16}$$

Since the noise is uncorrelated from element to element, the noise correlation matrix off-diagonal elements are zero when K is large enough, and the diagonal elements are the noise variance.

$$R_{noise} = \begin{bmatrix} \sigma_{noise}^2 & 0 & 0 \\ 0 & \ddots & 0 \\ 0 & 0 & \sigma_{noise}^2 \end{bmatrix} \tag{8.17}$$

Eigenvectors and eigenvalues of the correlation matrix correspond to the received signals and noise. An eigen decomposition of the correlation matrix is written as

$$R_{noise} = V_\lambda \begin{bmatrix} \lambda_1 & 0 & \cdots & 0 \\ 0 & \lambda_2 & \ddots & \vdots \\ \vdots & \ddots & \ddots & 0 \\ 0 & \cdots & 0 & \lambda_N \end{bmatrix} V_\lambda^{-1} = \sum_{m=1}^{M} \lambda_m V_\lambda[:,m] V_\lambda[:,m] + \sigma_{noise}^2 I_N \tag{8.18}$$

where V_λ is a matrix whose columns are the eigenvectors, λ_n is an eigenvalue associated with the eigenvector in column n, and I_N is an $N \times N$ identity matrix. The largest eigenvalues correspond to the signals, while the rest are associated with the noise. The signal eigenvectors span the signal subspace, while the noise eigenvectors span the noise subspace. All the noise eigenvalues equal the noise variance, σ_{noise}^2. They are not a function of angle. The signal eigenvalues, however, are a function of signal direction and signal amplitude. M uncorrelated signals incident on an N element array $(N > M)$ produces M signal eigenvalues and $N - M$ noise eigenvalues. Element patterns and element spacing affect the eigenvalues of the correlation matrix [10].

The signal eigenvectors associated with the correlation matrix are the array weights that have beams in the directions of the signals. The noise eigenvectors associated with the correlation matrix are the array weights that have nulls in the directions of the signals. Eigenvalues are related to the signal and noise powers. Array factors associated with eigenvector weights are called eigenbeams [11]. For a uniform linear array along the x axis, eigenbeam m is calculated from eigenvector m by

$$F_m(\theta) = \sum_{n=1}^{N} V_\lambda[n,m] e^{jk(n-1)d\sin\theta} \tag{8.19}$$

The correlation matrix is the crux of most adaptive array schemes, so it is worth further investigation before presenting any adaptive antenna algorithms. A few examples will help.

Example. Compare the eigenvectors and eigenvalues of the correlation matrix of a 4-element uniform array with half-wavelength spacing under the following scenarios:

   a. 1-V/m plane wave incident at 20°.
   b. 1-V/m plane wave incident at –50° and a 1-V/m plane wave incident at 30°.
   c. 1-V/m plane wave incident at –50° and a 2-V/m plane wave incident at 30°.

Assume $\sigma_{noise} = 0.01$.

   a. The eigenvectors and eigenvalues associated with 1-V/m plane wave incident at 20° are displayed in Table 8.1. The first three eigenvalues equal σ_{noise} while the last one, corresponding to the signal, is much higher. The amplitude of V_4 is uniform while the phase has a progression equal to the phase needed to steer the eigenbeam to 20°.

$$\theta_s = \sin^{-1}\left(\frac{1.0762}{kd}\right) = 20°$$

Figure 8.4 is a plot of the four eigenbeams. The signal eigenbeam is a uniform array factor steered to 20°. The noise eigenbeams all have nulls at 20°.

   b. The eigenvectors and eigenvalues associated with 1-V/m plane wave incident at –50° and a 1-V/m plane wave incident at 30° are displayed in Table 8.2. The first two eigenvalues are equal to σ_{noise} while the second two, corresponding to the signals, are much higher. The eigenbeams associated with V_3 and V_4 have peaks at –50° and 30° because

TABLE 8.1. Eigen Analysis of the Correlation Matrix for a 4-Element Uniform Array with a 1-V/m Plane Wave Incident at 20°

Eigenvectors							
V_1		V_2		V_3		V_4	
Amplitude	Phase	Amplitude	Phase	Amplitude	Phase	Amplitude	Phase
0.2364	2.2504	0.4856	–6.0233	0.6777	–0.8105	0.4990	3.2150
0.5232	–0.7366	0.5877	–3.5063	0.3601	–0.5136	0.5012	2.1462
0.4502	–2.8798	0.4914	–1.5152	0.5521	0.3777	0.5010	1.0762
0.6838	0	0.4211	0	0.3259	0	0.4988	0
Eigenvalues							
λ_1		λ_2		λ_3		λ_4	
0.0099		0.0103		0.0109		1.3418	

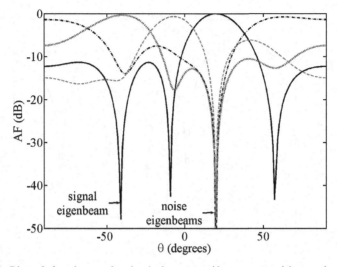

Figure 8.4. Plot of eigenbeams for the 4-element uniform array with one signal present at $\theta = 20°$.

TABLE 8.2. Eigen Analysis of the Correlation Matrix for a 4-Element Uniform Array with a 1-V/m Plane Wave Incident at −50° and a 1-V/m Plane Wave Incident at 30°

Eigenvectors							
V_1		V_2		V_3		V_4	
Amplitude	Phase	Amplitude	Phase	Amplitude	Phase	Amplitude	Phase
0.6169	5.0193	0.1334	1.7328	0.5586	0.3545	0.5382	−4.0009
0.3420	2.3395	0.6950	2.2633	0.5167	0.2259	0.3647	−1.3021
0.3426	2.6843	0.6940	−0.4834	0.5898	0.8114	0.2304	−1.5727
0.6206	0	0.1323	0	0.2703	0	0.7241	0
Eigenvalues							
λ_1		λ_2		λ_3		λ_4	
0.0101		0.0108		0.2628		2.9598	

$$\theta_s = \sin^{-1}\left(\frac{0.8114}{kd}\right) = -50°$$

$$\theta_s = \sin^{-1}\left(\frac{-1.5727}{kd}\right) = 30°$$

Figure 8.5 is a plot of the four eigenbeams. The noise eigenbeams all have nulls at −50° and 30°.

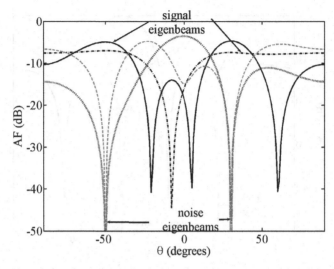

Figure 8.5. Plot of eigenbeams for the 4-element uniform array with 1-V/m plane wave incident at −50° and a 1-V/m plane wave incident at 30°.

TABLE 8.3. Eigen Analysis of the Correlation Matrix for a 4-Element Uniform Array with a 1-V/m Plane Wave Incident at −50° and a 2-V/m Plane Wave Incident at 30°

			Eigenvectors				
V_1		V_2		V_3		V_4	
Amplitude	Phase	Amplitude	Phase	Amplitude	Phase	Amplitude	Phase
0.5900	4.8935	0.2263	−0.0702	0.5579	0.3535	0.5380	−3.9996
0.3877	2.8131	0.6707	1.4889	0.5171	0.2247	0.3639	−1.3072
0.3880	2.0843	0.6700	−1.5496	0.5900	0.8093	0.2290	−1.5733
0.5925	0	0.2236	0	0.2706	0	0.7251	0

		Eigenvalues	
λ_1	λ_2	λ_3	λ_4
0.0101	0.0108	0.4083	7.5103

c. The eigenvectors and eigenvalues associated with 1-V/m plane wave incident at −50° and a 2-V/m plane wave incident at 30° are displayed in Table 8.3. As in Table 8.2, the first two eigenvalues are equal to σ_{noise} while the second two, corresponding to the signals, are much higher. The eigenbeams associated with V_3 and V_4 have peaks at −50° and 30°. This time, the peaks in the patterns are not of equal height: One beam has a higher peak at −50° and the other beam has a higher peak at 30°. Figure 8.6 is a plot of the four eigenbeams. The noise eigenbeams all have nulls at −50° and 30°.

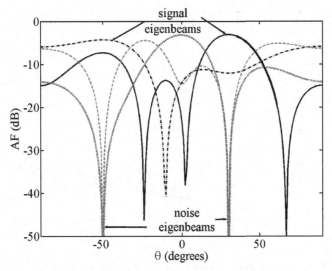

Figure 8.6. Plot of eigenbeams for the 4-element uniform array with 1-V/m plane wave incident at −50° and a 2-V/m plane wave incident at 30°.

Example. The eigenvalues of the correlation matrix relate to the strength of the signals, the noise, and the separation between the signals. A 4-element uniform array with half-wavelength spacing has up to two signals incident upon it. Plot the relationship between the eigenvalues and the standard deviation of the noise, the eigenvalues and the separation distance between two signals of equal strength, and the eigenvalues and the ratio of the strength of two signals.

When two signals of strength 1-V/m modeled by (8.9) are incident at 0° and 40° with $\sigma_{noise} = 0.01$, the correlation matrix is

$$R_T = \begin{bmatrix} 1.80 & 0.55 + j0.82 & 0.33 - j0.67 & 1.80 - j0.20 \\ 0.55 - j0.82 & 2.09 & 0.67 + j1.06 & 0.36 - j0.70 \\ 0.33 + j0.67 & 0.67 - j1.06 & 2.13 & 0.53 + j0.84 \\ 1.80 + j0.20 & 0.36 + j0.70 & 0.53 - j0.84 & 1.88 \end{bmatrix}$$

and has the eigenvalues $[0.0078\ 0.0102\ 3.3786\ 4.5022]^T$.

The two signals correspond to the two largest eigenvalues. Increasing σ_{noise} from 0 to 1.98 when only one signal is present at $\theta_1 = 0°$ causes the eigenvalues to increase as shown in Figure 8.7. The eigenvalues are sorted from the largest, λ_1 (signal), to the smallest, λ_4. As σ_{noise} increases, so do all the eigenvalues. Variations in the eigenvalues increase as σ_{noise} increases.

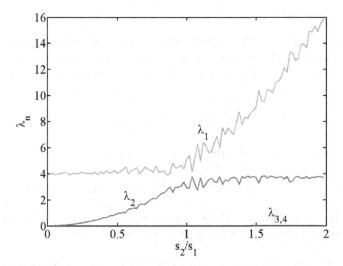

Figure 8.9. Plot of eigenvalues versus signal strength ratio for two equally weighted signals are incident on the 4-element array.

incident at $\theta_1 = 0°$ and $0 \le s_2 \le 1.98s_1$ incident at $\theta_2 = 40°$. Figure 8.9 shows the eigenvalues graphed as a function of s_2/s_1. When $s_2 = 0$, then $\lambda_1 = 4$ and $\lambda_2 = 0$. Increasing s_2 until it equals s_1 results in λ_1 staying approximately constant but λ_2 increases. When $s_2 > s_1$ then λ_2 stays approximately constant while λ_1 increases. The largest eigenvalue always corresponds to the most powerful signal.

The same principles for forming the correlation matrix for linear arrays apply to arbitrary arrays as well. The array steering vector becomes a function of θ and ϕ rather than a single angle.

8.3. DIRECTION OF ARRIVAL ESTIMATION

Direction finding was already introduced in Chapter 2. This chapter presents several signal processing algorithms for automatically determining the locations of the signals incident upon an array [12]. All the algorithms presented here assume that a calibrated receiver is at each element in the array. A calibrated receiver ensures that the relative signal amplitudes and phases are maintained between the elements. Otherwise, the correlation matrix would not be correct, so the eigenvectors and eigenvalues would be wrong.

Since an array's ability to resolve signals depends upon the beamwidth of the array, other high-resolution algorithms have been developed, so that even small arrays can resolve closely spaced signals. They resolve $N − 1$ signals by using narrow nulls in place of the wide main beam. These are also sensitive to noise, however.

8.3.1. Periodogram

One way to determine the signals present in the vicinity of an array is to scan the beam over the region of interest and plot the output power as a function of angle. The relative array output power as a function of angle can be found by

$$P(\theta) = A^\dagger(\theta) R_T A(\theta) \tag{8.20}$$

where the uniform array steering vector is given by

$$A(\theta) = e^{jkx_n \cos\theta}, \theta_{min} \le \theta \le \theta_{max} \tag{8.21}$$

A plot of the output power versus angle is known as a periodogram. Resolving closely spaced signals is limited by the array beamwidth.

Example. Demonstrate the effect of separation angle between two sources using an 8-element uniform array with $\lambda/2$ spacing.

When $\theta_1 = -30°$ and $\theta_2 = 30°$, two distinct peaks occur in the periodogram (Figure 8.10). Moving θ_1 to 10° still maintains two distinct peaks, but the dip between them has become shallow. When θ_1 is at 20°, then only one peak occurs halfway between the two signals. Thus, if the signals are close together, then they appear as one signal from a direction between the two signals.

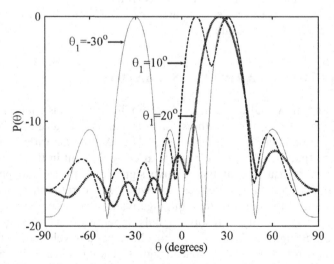

Figure 8.10. Plot of the periodogram of the 8-element uniform array when $\theta_1 = -30°$, 10°, 20°, and $\theta_2 = 30°$.

8.3.2. Capon's Minimum Variance

The periodogram basically uses the main beam of the array to determine signal locations. Since the main beam is wide, especially for small arrays, the ability to separate multiple signals or accurately locate a signal is not very good. Using nulls to locate signals is much more desirable, because the nulls have a narrow angular extent. Capon's method is the maximum likelihood estimate of the power arriving from a desired direction while all the other sources are considered interference [13]. Thus, the goal is to minimize the output power while forcing the desired signal to remain constant. The signal-to-interference ratio is maximized by the array weights

$$w = \frac{R_T^{-1} A}{A^\dagger R_T^{-1} A} \tag{8.22}$$

The resulting spectrum is given by

$$P(\theta) = \frac{1}{A^\dagger(\theta) R_T^{-1} A(\theta)} \tag{8.23}$$

Capon's method does not work well when the signals are correlated.

Example. An 8-element uniform array with $\lambda/2$ spacing has three signals incident upon it: $s_1(-60°) = 1$, $s_2(0°) = 2$, and $s_3(10°) = 4$. Find the Capon spectrum.

Figure 8.11 shows the result of Capon's method superimposed on the periodogram. Capon's method can better distinguish between two closely spaced sources and the spectrum between the source peaks is very low. The periodogram cannot separate the signals at $0°$ and $10°$.

8.3.3. MUSIC Algorithm

MUSIC is an acronym for *MU*ltiple *SI*gnal *C*lassification [14]. MUSIC assumes the noise is uncorrelated and the signals are either uncorrelated or mildly correlated. When the array calibration is perfect and the signals uncorrelated, then the MUSIC algorithm can accurately estimate the number of signals, the angle of arrivals, and the signal strengths [15]. The MUSIC spectrum is given by

$$P(\theta) = \frac{A^\dagger(\theta) A(\theta)}{A^\dagger(\theta) V_\lambda V_\lambda^\dagger A(\theta)} \tag{8.24}$$

where V_λ is a matrix whose columns contain the eigenvectors of the noise subspace. The eigenvectors of the noise subspace correspond to the $N - N_s$ smallest eigenvalues of the correlation matrix.

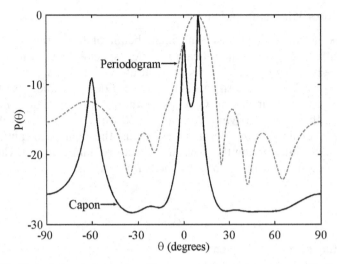

Figure 8.11. Plot of the 4-element uniform array Capon spectrum with three signals present.

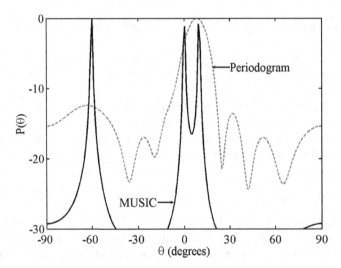

Figure 8.12. Plot of the 8-element uniform array MUSIC spectrum with three signals present.

Example. An 8-element uniform array with $\lambda/2$ spacing has three signals incident upon it: $s_1(-60°) = 1$, $s_2(0°) = 2$, and $s_3(10°) = 4$. Find the MUSIC spectrum.

Figure 8.12 shows the MUSIC spectrum superimposed on the periodogram. MUSIC easily distinguishes between two closely spaced sources and the spectrum between the source peaks is lower than in the Capon spectrum.

The MUSIC algorithm is not very robust, so many improvements have been proposed. One popular modification, called root-MUSIC, accurately locates the direction of arrival by finding the roots of the array polynomial [16]. The denominator of (8.24) can be written as

$$A^\dagger(\theta)V_\lambda V_\lambda^\dagger A(\theta) = \sum_{n=M+1}^{M-1} c_n z^n \qquad (8.25)$$

where

$$z = e^{jknd\sin\theta}$$

$$c_n = \sum_{r-c=n} V_\lambda V_\lambda^\dagger = \text{sum of } n\text{th diagonal of matrix } V_\lambda V_\lambda^\dagger$$

The summation in (8.25) is over the diagonals of $V_\lambda V_\lambda^\dagger$. Coefficient n is the sum of the elements in diagonal n. Roots of the polynomial, z_m, close to the unit circle correspond to the poles of the MUSIC spectrum. Solving for the angle of the phase of the roots of the polynomial in (8.25) produces

$$\theta_m = \sin^{-1}\left(\frac{\arg(z_m)}{kd}\right) \qquad (8.26)$$

Roots close to the unit circle correspond to actual signals while the rest are spurious. The $V_\lambda V_\lambda^\dagger$ has $2N-1$ diagonals, so the polynomial has $2N-2$ roots.

Example. An 8-element uniform array with $\lambda/2$ spacing has three signals incident upon it: $s_1(-60°) = 1$, $s_2(0°) = 2$, and $s_3(10°) = 4$. Find the location of the signals using the root MUSIC algorithm.

The magnitude and angle of the roots are calculated using $|z_m|$ and (8.26) to get the values in Table 8.4. The three signal directions are accurately identified as the roots that lie on the unit circle ($|z_m| \approx 1$). A unit circle plot of all 14 roots is shown in Figure 8.13.

TABLE 8.4. The Roots Found Using Root MUSIC[a]

m	$|z_m|$	θ_m (degrees)	Signal?	m	$|z_m|$	θ_m (degrees)	Signal?
1	2.05	−34.21	No	8	0.99	0.08	Yes
2	1.82	−22.50	No	9	1.01	−60.10	Yes
3	1.64	33.12	No	10	0.99	−60.10	Yes
4	1.66	53.59	No	11	0.61	33.12	No
5	1.00	10.0	Yes	12	0.60	53.59	No
6	1.00	10.0	Yes	13	0.55	−22.50	No
7	1.01	0.08	Yes	14	0.49	−34.21	No

[a]The ones close to the unit circle represent the correct signal directions.

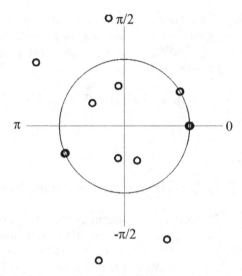

Figure 8.13. Unit circle representation of all the roots found using root MUSIC.

8.3.4. Maximum Entropy Method

The maximum entropy method (MEM) is also called the all-poles model or the autoregressive model. MEM is based on a rational function model of the spectrum having all poles and no zeros [17]. Hence, it can accurately reproduce sharp resonances in the spectrum. The MEM spectrum is given by [18]

$$P(\theta) = \frac{1}{A^\dagger(\theta) R_T^{-1}[:,n] R_T^{\dagger-1}[:,n] A(\theta)} \qquad (8.27)$$

where n is the nth column of the inverse correlation matrix. The selection of n gives slightly different results.

Example. An 8-element uniform array with $\lambda/2$ spacing has three signals incident upon it: $s_1(-60°) = 1$, $s_2(0°) = 2$, and $s_3(10°) = 4$. Find the MEM spectrum.

Figure 8.14 shows the MEM spectrum superimposed on the periodogram. MEM has even sharper spectral lines than the MUSIC spectrum.

8.3.5. Pisarenko Harmonic Decomposition

The smallest eigenvector (λ_N) minimizes the mean squared error of the array output with the constraint that the norm of the weight vector equals one. The Pisarenko harmonic decomposition (PHD) spectrum is [19]

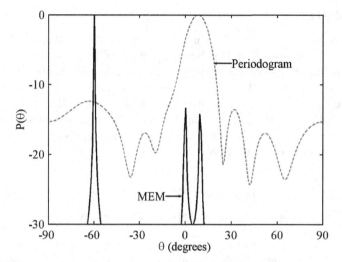

Figure 8.14. Plot of the 4-element uniform array MEM spectrum with three signals present.

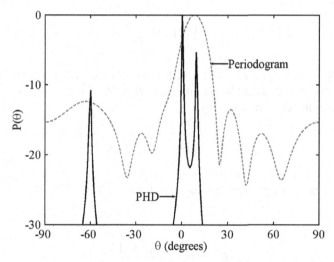

Figure 8.15. Plot of the 4-element uniform array PHD spectrum with three signals present.

$$P(\theta) = \frac{1}{\left| A^{\dagger}(\theta) e_1 \right|} \tag{8.28}$$

Example. An 8-element uniform array with $\lambda/2$ spacing has three signals incident upon it: $s_1(-60°) = 1$, $s_2(0°) = 2$, and $s_3(10°) = 4$. Find the PHD spectrum.

Figure 8.15 shows the PHD spectrum superimposed on the periodogram. PHD has very sharp spectral lines like MEM.

8.3.6. ESPRIT

ESPRIT is an acronym for *E*stimation of *S*ignal *P*arameters via *R*otational *I*nvariance *T*echniques [20]. It is based upon breaking an *N*-element uniform linear array into two overlapping subarrays with $N-1$ elements. One subarray starts at the left end of the array, and the other starts at the right end of the array. The $N-2$ shared elements in the middle are called matched pairs. ESPRIT makes use of the phase displacement between the two subarrays to calculate the angle of arrivals for the signals. The signals at the elements in the two subarrays are written as

$$
\begin{aligned}
X_1(\kappa) &= A_1 s(\kappa) + N(\kappa) \\
X_2(\kappa) &= A_2 s(\kappa) + N(\kappa) = A_1 \Phi s(\kappa) + N(\kappa)
\end{aligned}
\tag{8.29}
$$

where the diagonal matrix, Φ has elements $\Phi_{m,m} = e^{jkd\sin\theta_m}$. The eigenvectors of the full array are in matrix V_λ. A new $N \times M$ matrix with only the signal eigenvectors is defined by

$$
V_s = V_\lambda [1:N, 1:M]
\tag{8.30}
$$

The first $N-1$ rows are associated with the first subarray, while the last $N-1$ rows are associated with the second subarray. A matrix Ψ relates the two subarray eigenvector matrixes by

$$
V_s[1:N-1, 1:M] = \Psi V_s[2:N, 1:M]
\tag{8.31}
$$

Solving for Ψ and equating its eigenvalues to Φ results in the signal angle estimates of

$$
\theta_m = \sin^{-1}\left(\frac{\arg\left(\lambda_m^\Psi\right)}{kd} \right)
\tag{8.32}
$$

where λ_m^Ψ = eigenvalues of Ψ.

Example. An 8-element uniform array with $\lambda/2$ spacing has three signals incident upon it: $s_1(-60°) = 1$, $s_2(0°) = 2$, and $s_3(10°) = 4$. Estimate the incident angles using ESPRIT.

First calculate the correlation matrix, then find the eigenvectors associated with the three signals.

$$V_3 = \begin{bmatrix} -0.2981 - 0.0000i & -0.5109 - 0.0000i & -0.2513 + 0.0000i \\ -0.2967 + 0.1666i & -0.3576 + 0.0594i & 0.3588 - 0.2779i \\ -0.2097 + 0.3127i & -0.2840 + 0.0643i & -0.1276 + 0.2624i \\ -0.0565 + 0.3852i & -0.1554 - 0.0297i & 0.0633 - 0.3765i \\ 0.1100 + 0.3767i & -0.1231 - 0.0931i & 0.1494 + 0.3602i \\ 0.2517 + 0.2828i & -0.1546 - 0.2559i & -0.1812 - 0.2271i \\ 0.3166 + 0.1246i & -0.1997 - 0.2998i & 0.3966 + 0.2041i \\ 0.2946 - 0.0457i & -0.3451 - 0.3684i & -0.2493 + 0.0485i \end{bmatrix} \begin{matrix} \text{Subarray 1} \\ \\ \\ \\ \text{Subarray 2} \end{matrix}$$

Next, compute Ψ to get

$$\begin{bmatrix} 0.87 - j0.45 & -0.31 - j0.061 & 0.096 + j0.022 \\ 0.07 - j0.067 & 0.95 - j0.063 & -0.35 + j0.32 \\ -0.0085 + j0.0008 & -0.13 - j0.025 & -0.88 + j0.40 \end{bmatrix}$$

The eigenvalues of ψ are found and substituted into (8.32) to find an estimate of the angle of arrival.

$$\theta_m = [-59.91° \quad 0.06° \quad 10.0°]$$

8.3.7. Estimating and Finding Sources

Many of the direction finding algorithms need to know the exact number of sources incident upon the array. One way is to just pick the eigenvalues that exceed some threshold set above the noise. The number of eigenvalues that exceed the threshold equals the number of sources. One algorithm for estimating M is as follows [21]:

1. Estimate the correlation matrix from K time samples.
2. Find and sort the eigenvalues ($\lambda_1 > \lambda_2 > \cdots > \lambda_N$).
3. Find M that minimizes [21]

$$K(N-M)\ln\left[\frac{\sum_{n=M+1}^{N} \lambda_n}{(N-M)\left(\prod_{n=M+1}^{N} \lambda_n\right)^{\frac{1}{N-M}}}\right] + f(M,N) \qquad (8.33)$$

where

$$f(M,N) = \begin{cases} M(2N-M) & \text{AIC} \\ 0.5M(2N-M)\ln N & \text{MDL} \end{cases}$$

AIC = Akaike's information criterion

MDL = minimum description length

8.4. ADAPTIVE NULLING

As the wireless users crowd the frequency spectrum, interference becomes more common. When the main beam gain times the desired signal is less than the sidelobe gain times the interference signal, then the interference overwhelms the desired signal. An adaptive antenna adjusts its antenna pattern to steer the main beam in the direction of the desired signal while placing nulls in the direction of the interference. The original adaptive antenna, called a sidelobe canceler, was invented by Howells and Applebaum [22]. It was mainly intended for radar systems. A similar adaptive algorithm was independently developed by Widrow [23]. This least mean square (LMS) algorithm became the canonical adaptive algorithm primarily intended for communications systems. Many signal processing algorithms have their roots in the LMS algorithm. They all require a calibrated receiver at each element in the array. Another approach that relies on random search algorithms works with a corporate fed array with one receiver [24]. These approaches minimize the output power of the array. If only a few of the elements or the least significant bits of the weights are used to perform the nulling, then nulls cannot be placed in the main beam but can be placed in the sidelobes [25]. The advent of global optimization algorithms have made this approach more attractive [26].

This section is divided into three parts. First, the radar approaches called sidelobe blanking and sidelobe canceling are presented. These techniques do not require a pilot signal or replica of the desired receive signal to work. The second set of algorithms rely upon the signal correlation matrix. They are similar to the direction finding algorithms, except one of the signal steering vectors corresponds to the desired signal, while the others correspond to the interference. Some examples that demonstrate the operation of these types of algorithms are presented. Finally, the random search type algorithms are presented, because of their attractive minimal hardware requirements.

8.4.1. Sidelobe Blanking and Canceling

A sidelobe blanker eliminates unwanted interference entering the sidelobes of a radar array [27] (Figure 8.16). It consists of a high-gain radar antenna and a lower-gain auxiliary antenna. The gain of the auxiliary antenna must exceed the gain of the highest sidelobe of the high-gain antenna as shown in Figure 8.17. A software algorithm tests three hypotheses in the detection decision [28]:

1. A target in the main beam results in a large signal in the main channel but causes a small signal in the auxiliary channel, because the main beam

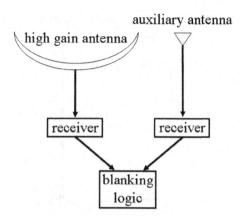

Figure 8.16. Sidelobe blanker.

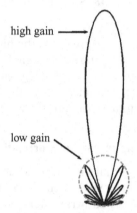

Figure 8.17. The main antenna has a high gain, while the auxiliary antenna has a gain slightly greater than the maximum sidelobe level of the main antenna.

gain is much higher than the auxiliary antenna gain. The sidelobe blanking processing allows this signal to pass.

2. If the signal received by the auxiliary antenna is larger than the signal received by the main antenna, then the sidelobe blanking processing suppresses that signal.

3. No signal is present

Figure 8.18 is a diagram of a sidelobe canceler that consists of a high-gain antenna pointing at the desired signal and one or more low-gain auxiliary antennas [29]. The gain of the small antennas is approximately the same as the gain of the peak sidelobes of the high-gain antenna. Appropriately weighting the low-gain antenna signal and subtracting it from the high-gain antenna

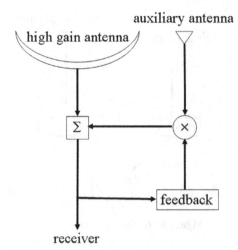

Figure 8.18. Sidelobe canceler.

signal results in canceling the interference in the high-gain antenna. One low-gain antenna is needed for each interfering signal [30]. The low-gain antenna minimally perturbs the main beam of the high-gain antenna. The sidelobe blanker is a decision-making system, whereas the sidelobe canceler actually mixes signals in order to eliminate the interference.

8.4.2. Adaptive Nulling Using the Signal Correlation Matrix

The sidelobe canceler assumes that the weighted signals from N small auxiliary antennas cancel the interference signals entering the sidelobes of a high-gain antenna. If the large antenna is an array, then the array elements could also serve as the auxiliary antennas. This configuration is called an adaptive array [30]. As shown for direction finding, the correlation matrix is useful for determining the directions of signals incident on the array.

8.4.2.1. Optimum Element Weights. Making use of the correlation matrix for adaptive arrays is slightly different than for direction finding, because not all the received signals are treated equally. In adaptive nulling applications, one signal is desired while all the other signals cause interference and should be rejected. If the desired signal arrives at the elements from the angle (θ_s, ϕ_s), then the difference in phase between the signals at the N elements is given by the array steering vector

$$A_s = \begin{bmatrix} e^{jk(x_1 u_s + y_1 v_s + z_1 w_s)} \\ e^{jk(x_2 u_s + y_2 v_s + z_2 w_s)} \\ \vdots \\ e^{jk(x_N u_s + y_N v_s + z_N w_s)} \end{bmatrix} \tag{8.34}$$

M undesired signals strike the array at angles (θ_m, ϕ_m). The phase difference between interference signal m at all the elements is given by the array steering vector

$$A_m = \begin{bmatrix} e^{jk(x_1 u_m + y_1 v_m + z_1 w_m)} \\ e^{jk(x_2 u_m + y_2 v_m + z_2 w_m)} \\ \vdots \\ e^{jk(x_N u_m + y_N v_m + z_N w_m)} \end{bmatrix}, \qquad m = 1, 2, \ldots, M \qquad (8.35)$$

In the time domain, the signals present at the elements are given by

$$\begin{aligned} s(t) &= \text{desired signal} \\ i_m(t) &= \text{signal from interference source } m \\ N(t) &= \text{white Gaussian noise} \end{aligned} \qquad (8.36)$$

Discrete time samples of these signals, $s(\kappa)$, $i_m(\kappa)$, and $N(\kappa)$, add together to form a discrete sample of the total signal at each element.

$$\begin{aligned} X(\kappa) &= A_s s(\kappa) + \sum_{m=1}^{M} A_m i_m(\kappa) + N(\kappa) \\ &= X_s(\kappa) + X_i(\kappa) + N(\kappa) \end{aligned} \qquad (8.37)$$

Multiplying these signals by their corresponding element weights and adding them together results in the array output.

$$F = w^\dagger X(\kappa) \qquad (8.38)$$

In terms of power the output is

$$P = FF^* = E\left\{ \left| w^\dagger X(\kappa) \right|^2 \right\} = w^\dagger R_T w \qquad (8.39)$$

where

$$\begin{aligned} R_T &= E\{X(\kappa) X^\dagger(\kappa)\} \\ &= E\{X_s(\kappa) X_s^\dagger(\kappa)\} + E\{X_i(\kappa) X_i^\dagger(\kappa)\} + E\{N(\kappa) N^\dagger(\kappa)\} \\ &= R_s + R_i + R_N \end{aligned} \qquad (8.40)$$

A good figure of merit for evaluating how well an array receives the desired signal while rejecting the interference and noise is the signal-to-noise ratio [31]:

$$\text{SNR} = \frac{w^\dagger R_s w}{w^\dagger (R_i + R_n) w} \qquad (8.41)$$

The signal received by the array differs from the desired reference or pilot signal, $d(\kappa)$, by

$$\varepsilon(\kappa) = d(\kappa) - w^\dagger(\kappa) X(\kappa) \tag{8.42}$$

The mean square error of (8.42) is

$$E\{\varepsilon^2(\kappa)\} = E\{d^2(\kappa)\} + w^\dagger(\kappa) R_T(\kappa) w(\kappa) - 2w^\dagger(\kappa) q(\kappa) \tag{8.43}$$

where the signal correlation vector is defined to be

$$q(\kappa) = E\{d(\kappa)\mathbf{s}(\kappa)\} \tag{8.44}$$

Taking the gradient of the mean square error with respect to the weights results in

$$\nabla_w E\{\varepsilon^2(\kappa)\} = 2R_T(\kappa) w(\kappa) - 2q(\kappa) \tag{8.45}$$

The optimum weights make the gradient zero, so

$$2R_T(\kappa) w_{\text{opt}}(\kappa) - 2q(\kappa) = 0 \tag{8.46}$$

Solving for the optimum weights yields the Wiener–Hopf solution [32]:

$$w_{\text{opt}}(\kappa) = R_T^{-1}(\kappa) q(\kappa) \tag{8.47}$$

Finding the solution requires two very important pieces of information: the signal at each element to form the correlation matrix and the desired signal for the signal correlation vector.

8.4.2.2. Least Mean Square Algorithm. Although the Wiener–Hopf solution is not a practical approach to adaptive antennas, it forms the mathematical basis of the adaptive least mean square (LMS) algorithm [23]. The LMS algorithm uses the method of steepest descent to find the minimum of $\varepsilon^2(\kappa)$. In the method of steepest descent, the new weight vector is found by adding a step size times the negative of the gradient of $\varepsilon^2(\kappa)$ to the current weight vector. The gradient is with respect to the weights.

$$w(\kappa+1) = w(\kappa) + \mu\left[-\nabla_w E\{|d(\kappa) - w^\dagger(\kappa) X(\kappa)|^2\}\right] \tag{8.48}$$

Taking the gradient yields

$$w(\kappa+1) = w(\kappa) + \mu s(\kappa)[d(\kappa) - w^\dagger s(\kappa)] \tag{8.49}$$

where μ is the step size that absorbs the -2 that results from taking the gradient in (8.48). Note that the expected values conveniently disappear due to the difficulty in implementing the expected value operator in real time. Instead, the instantaneous values replace the expected values. LMS convergence speed is proportional to the size of μ. If μ is too small, then convergence is slow, while if μ is too large, then the algorithm will overshoot the optimum weights. The algorithm is stable when [32]

$$0 \le \mu \le \frac{2}{\lambda_{\max}} \tag{8.50}$$

where $\lambda_{\max}$ is the maximum eigenvalue of the correlation matrix. Its convergence speed slows as the ratio of the maximum to minimum eigenvalues increases, because (8.43) has a long, narrow valley that slows the gradient's progress toward the minimum [33].

Selecting an adequate reference or pilot signal is important to the success of the LMS algorithm. If the pilot signal is the signal itself, then the output reproduces the signal in the optimal mean square error sense and eliminates the noise. Of course, the actual signal would not be used as the pilot signal, because if the actual signal is known, then there is no reason to have the adaptive system. The LMS algorithm is typically used in communications systems where the desired signal is present but not in radar systems where the desired signal is not present. Some considerations regarding the reference signal include the following [24]:

1. It needs to be highly correlated with the desired signal and uncorrelated with the interference signals.
2. It should have similar directional and spectral characteristics as those of the desired signal.

Example. An 8-element uniform array with $\lambda/2$ spacing has the desired signal incident at $0°$ and two interference signals incident at $-21°$ and $61°$. Use the LMS algorithm to place nulls in the antenna pattern. Assume $\sigma_n = 0.01$. The signal is represented by (8.11) and the interference by (8.9).

After $K = 500$ iterations, the antenna pattern appears in Figure 8.19. It has a directivity of 8.7 dB, which is lower than the 9.0 dB of the 8-element uniform array. Figure 8.20 shows the LMS signal superimposed on the real signal as a function of iteration. Figure 8.21 and Figure 8.22 are the LMS weights. They converge in about 150 iterations.

8.4.2.3. Sample Matrix Inversion Algorithm.
A single time sample of the correlation matrix does not accurately represent the correlation matrix to use in (8.47). The sample matrix, $\hat{R}_T$, is the time average estimate of the correlation matrix [34]. This estimate comes from an average of K samples of the received element signals

$$\hat{R}_T = \frac{1}{K} \sum_{\kappa=1}^{K} x(\kappa)x^\dagger(\kappa) \tag{8.51}$$

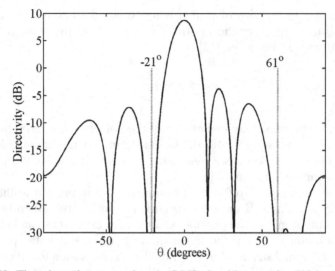

Figure 8.19. The adapted pattern after the LMS algorithm ran for 500 iterations. The signal is at 0° and the interference is at −21° and 61°.

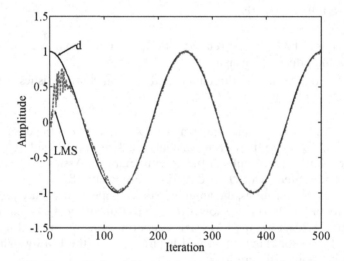

Figure 8.20. The LMS signal (dashed line) and the actual signal (solid line).

and the correlation vector is

$$\hat{\mathbf{q}}(\kappa) = \frac{1}{K} \sum_{\kappa=1}^{K} d^{\dagger}(\kappa) \mathbf{s}(\kappa) \qquad (8.52)$$

At the kth time sample, the weights are given by

$$W(\kappa) = \hat{R}_T^{-1}(\kappa) \hat{q}(\kappa) \qquad (8.53)$$

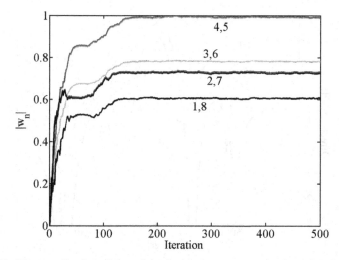

Figure 8.21. The amplitude weights of the 8-element array versus iteration for the LMS algorithm.

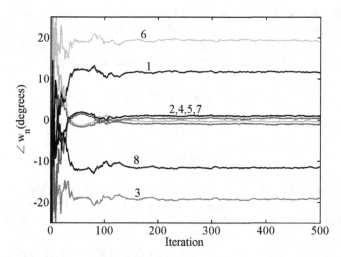

Figure 8.22. The phase weights of the 8-element array versus iteration for the LMS algorithm.

Example. An 8-element uniform array with $\lambda/2$ spacing has the desired signal incident at $0°$ and two interference signals incident at $-21°$ and $61°$. Use the SMI algorithm to place nulls in the antenna pattern. Assume $\sigma_n = 0.01$. The signal is represented by (8.11) and the interference by (8.9).

The SMI pattern improves as the number of samples increases. Figure 8.23 shows the adapted pattern for $K = 10, 25,$ and 50 samples. The array directivity increases from 5.9 dB ($K = 10$) to 8.0 dB ($K = 25$) to 8.5 dB ($K = 50$).

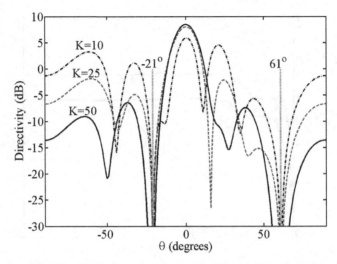

Figure 8.23. The SMI adapted pattern after $K = 10, 25,$ and 50 samples.

8.4.2.4. Recursive Least Squares Algorithm.
The recursive least squares (RLS) algorithm recursively updates the correlation matrix such that more recent time samples receive a higher weighting than past samples [31]. A straightforward implementation of the algorithm is written as

$$R_T(\kappa) = x(\kappa)x^\dagger(\kappa) + \alpha R_T(\kappa-1) \qquad (8.54)$$

and the correlation vector is

$$\mathbf{q}(\kappa) = d^\dagger(\kappa)\mathbf{s}(\kappa) + \alpha q(\kappa-1) \qquad (8.55)$$

where the forgetting factor, α, is limited by $0 \le \alpha \le 1$. Higher values of α give more weight to previous values than do lower values of α. An even better relationship calculates an update for the inverse of the correlation matrix [31]

$$R_T^{-1}(\kappa+1) = \alpha^{-1}R_T^{-1}(\kappa) - \frac{R_T^{-1}(\kappa)X(\kappa+1)X^\dagger(\kappa+1)R_T^{-1}(\kappa)}{\alpha^2\left[1 + X^\dagger(\kappa+1)R_T^{-1}(\kappa)X(\kappa+1)/\alpha\right]} \qquad (8.56)$$

with the weights given by

$$w(\kappa+1) = w(\kappa) + R_T^{-1}(\kappa+1)X(\kappa+1)\left[s(\kappa+1) - X^\dagger(\kappa+1)w(\kappa)\right] \qquad (8.57)$$

Example. An 8-element uniform array with $\lambda/2$ spacing has the desired signal incident at $0°$ and two interference signals incident at $-21°$ and $61°$. Use the RLS algorithm to place nulls in the antenna pattern. Assume $\sigma_n = 0.01$. The signal is represented by (8.11) and the interference by (8.9).

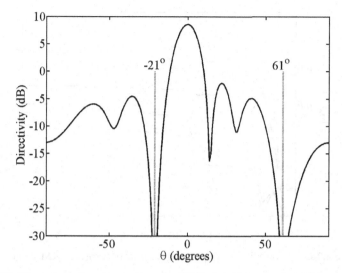

Figure 8.24. The adapted pattern after the RLS algorithm ran for 25 iterations. The signal is at 0° and the interference is at −21° and 61°.

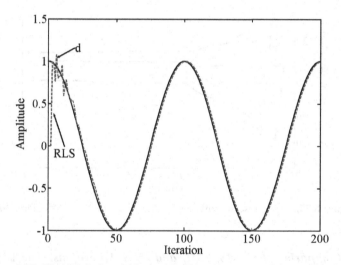

Figure 8.25. The RLS signal (dashed line) and the actual signal (solid line).

After $K = 25$ iterations and $\alpha = 0.9$, the antenna pattern appears in Figure 8.24 with a directivity of 8.6 dB. Figure 8.25 shows the RMS signal superimposed on the real signal as a function of iteration. Figure 8.26 and Figure 8.27 are the RMS weights. They converge in about 15 iterations.

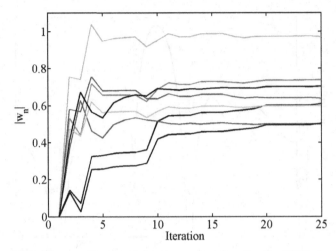

Figure 8.26. The amplitude weights versus iteration for the RMS algorithm.

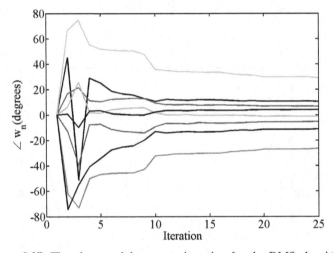

Figure 8.27. The phase weights versus iteration for the RMS algorithm.

8.4.2.5. Comparing the LMS, SMI, and RLS Algorithms. The LMS algorithm slowly converges when the ratio of the maximum to minimum eigenvalue of the correlation matrix is large. A large eigenvalue ratio implies that there are narrow valleys in the cost surface [33], so the method of steepest descent converges slowly. Since the amplitude of the eigenvalues of the correlation matrix are proportional to the interference signal powers, the LMS algorithm converges fast when the interference powers are similar and slow when they are not. Figure 8.28 shows the adapted pattern after 1000 iterations with a null placed at 21° but not at −21°. The 0-dB desired signal was incident

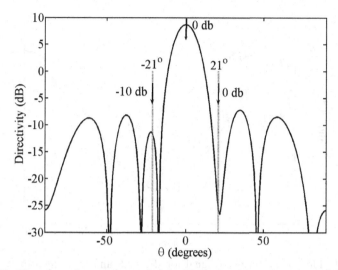

Figure 8.28. The adapted array factor for the LMS algorithm after $K = 1000$ iterations when a 0-dB signal is incident at $0°$, and a -10-dB interference is incident at $-21°$ and a 0-dB interference is incident at $21°$.

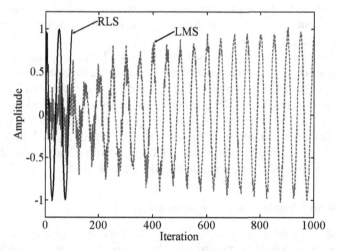

Figure 8.29. Plot of the RLS signal (100 iterations) versus the LMS signal as a function of iteration when a 0-dB signal is incident at $0°$ and a -12-dB interference is incident at $-21°$ and $61°$.

on the 8-element array at $0°$, while a -10-dB interference was incident at $21°$ and a 0 dB interference was incident at $-21°$.

Figure 8.29 is a plot of the received signal as a function of iteration for the RLS and LMS algorithms when the 0-dB desired signal was incident on the 8-element array at $0°$ while 12-dB interference signals are incident at $-21°$ and

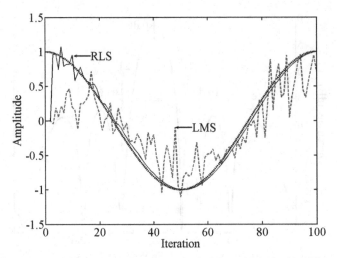

Figure 8.30. Plot of the received signal for the RLS and LMS algorithms when the noise is $\sigma_{noise} = 0.1$.

61°. Note that the RLS algorithm converges within a few iterations, while the LMS takes many iterations. The forgetting factor plays an important role in speeding convergence.

The algorithms are also sensitive to noise. Increasing the noise from $\sigma_{noise} = 0.01$ to $\sigma_{noise} = 0.1$ significantly reduces the effectiveness of the adaptive algorithms. Figure 8.30 is a plot of the received signal recovered by the LMS and RLS algorithms. The LMS is much more sensitive to noise than the RLS algorithm. Figure 8.31 shows the adapted patterns after 100 iterations. The RLS algorithm places deep nulls at −21° and 61°, while the SMI algorithm has a null at 61° but a small sidelobe at −21°. On the other hand, the LMS algorithm has a null at −21° but a sidelobe still remains at 61°.

8.4.3. Adaptive Nulling via Power Minimization

Another class of adaptive nulling algorithms adjusts the array weights until the total output power is minimized. These algorithms are cheap to implement, because they use the existing array architecture without expensive additions, such as digital beam forming. Since a digital beamformer is not needed, array calibration is simpler too. Their drawbacks include slow convergence and possibly high pattern distortions. This approach only works if the desired signal is not present or if the gain of the cancellation beams in the adaptive algorithm is small compared to the main beam gain. Sidelobe cancelers are an example of limiting the nulling to the sidelobes, because the gain of the auxiliary antennas are too small to have a major impact on the main beam, but large enough to cancel sidelobes.

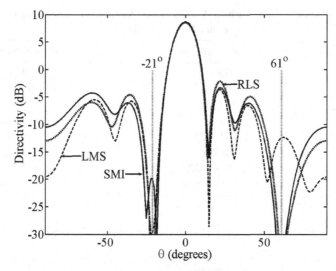

Figure 8.31. Adapted array factors for the SMI, RLS, and LMS algorithms after K = 50, 200, and 1000 iterations when a 0-dB signal is incident at 0° and a −12-dB interference is incident at −21° and 61°.

Making only a few of the array elements adaptive prevents the destruction of the main beam but allows nulls to be placed in the sidelobes [35]. Enough elements are selected to place a null in the highest sidelobe without significantly distorting the main beam. A second approach forms an approximate numerical gradient and uses a steepest descent algorithm to find the minimum output power [36]. As long as the weight changes are small and the sidelobes are low, little main beam distortion occurs while placing the nulls. This approach has been implemented experimentally but is slow and can get trapped in a local minimum. As a result, the best weight settings to achieve appropriate nulls are usually not found. A final approach limits the array weight settings. Large reductions in the amplitude weights are required in order to reduce the main beam. Consequently, if only small amplitude and phase perturbations are allowed, then a null cannot be placed in the main beam but can be placed in the sidelobes. Lower sidelobes require smaller perturbations to the weights in order to place the null. Using only a few least significant bits of the digital phase shifter and attenuator prevents the algorithm from placing nulls in the main beam. The amplitude and phase associated with each bit of a digital weight (up to 8 bits) is shown in Table 8.5. The lower bits are quite capable of disrupting the main beam. For instance, giving half the elements a 180° phase shift (bit 1) will place a null in the peak of the main beam.

Example. A 20-element, 20-dB, $\bar{n}$ = 3 Taylor linear array with elements spaced half a wavelength apart has 6-bit amplitude and phase weights. If the

TABLE 8.5. Amplitude and Phase Values of the Nulling Bits when There Are a Total of 6 bits for the Amplitude and Phase Weights

Weight Bit	Amplitude weight	Phase weight (degrees)
1	0.5	180
2	0.25	90
3	0.125	45
4	0.0625	22.5
5	0.03125	11.25
6	0.015625	5.625
7	0.0078125	2.8125
8	0.00390625	1.40625

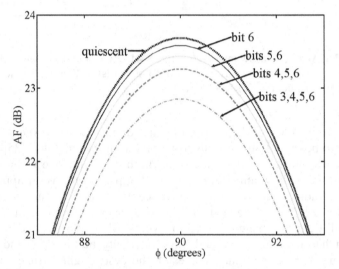

Figure 8.32. Maximum main beam reduction possible when 1 to 4 least significant bits out of 6 total bits in an amplitude weight are used to null a signal at $\phi = 90°$. 0 bits is the quiescent pattern.

only source enters the main beam, then the adaptive algorithm tries to reduce the main beam in order to reduce the total output power. Show how using 1 through 4 least significant bits (bits 3 to 6) alters the main beam.

Figure 8.32 (amplitude weights) and Figure 8.33 (phase weights) show the main beam reduction when the following bits from Table 8.5 are used:

1. bit 6
2. bits 5,6
3. bits 4,5,6
4. bits 3,4,5,6

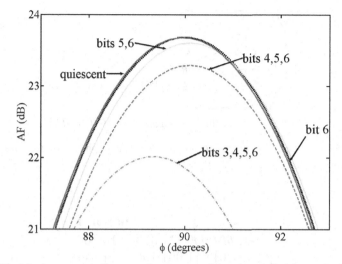

Figure 8.33. Maximum main beam reduction possible when 1 to 4 least significant bits out of 6 total bits in a phase weight are used to null a signal at $\phi = 90°$. 0 bits is the quiescent pattern.

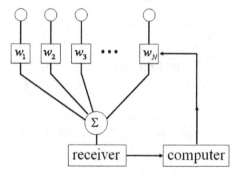

Figure 8.34. Diagram of an adaptive antenna that minimizes the total output power.

A maximum reduction of 1 dB is possible using four least significant bits of amplitude. Using one through three least significant bits results in very little perturbation to the main beam. Unlike amplitude-only nulling, phase-only nulling causes beam squint. The phase had more effect on the main beam than did amplitude. This example demonstrates that adaptive nulling with the least significant bits would not result in significant degradation to the main beam.

A diagram of the adaptive array appears in Figure 8.34. The array has a corporate feed with variable weights at each element. The phase shifters in the weights are available for beam steering and calibration as well as nulling. The amplitude weights are used for low-sidelobe tapers and calibration. This

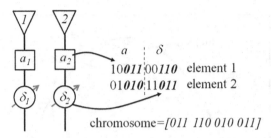

Figure 8.35. The least significant bits of the amplitude phase weights are put in a chromosome.

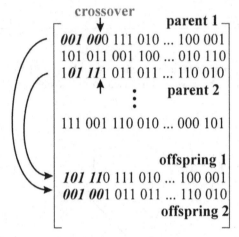

Figure 8.36. Two parents are selected from the mating pool. Two offspring are created using single-point crossover and placed into the population matrix to replace discarded chromosomes.

adaptive array configuration is much simpler and cheaper than the digital beamforming array required by other adaptive algorithms.

8.4.3.1. Amplitude and Phase Adaptive Nulling. To demonstrate the concept of adaptive nulling through power minimization, consider an array with 5-bit amplitude and phase weights at each element. To prevent main beam nulling and severe pattern distortion, the genetic algorithm that performs the adaptive nulling controls only the 3 least significant bits of each weight. A vector called a chromosome stores the adaptive bits as shown in Figure 8.35. All the chromosomes under consideration make up the N_{pop} rows of the population matrix. In this case, there are 3 bits/weight, 2 weights per element, and N_{adap} adaptive elements in the array, so the population matrix is $N_{pop} \times 6N_{adap}$. Each chromosome in the population is then sent from the computer to the antenna to adjust the weights, and the output power is measured

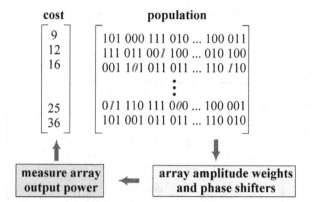

Figure 8.37. Random bits in the population are mutated (italicized bits). The chromosomes are sent to the array one at a time and the total output power measured. Each chromosome then has an associated output power.

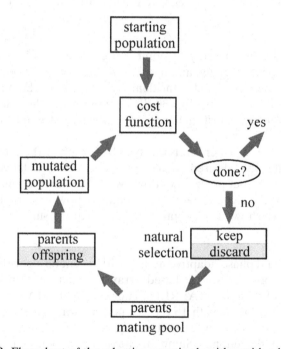

Figure 8.38. Flow chart of the adaptive genetic algorithm with a linear array.

and stored as the cost associated with the chromosome (Figure 8.37). Low-cost chromosomes become parents and mate to form offspring as shown in Figure 8.36 (single point crossover used here). Mutations occur inside the population (italicized digits in Figure 8.37). This process continually adjusts the antenna pattern by placing nulls in the sidelobes while having minimal impact on the main beam as shown in Figure 8.38.

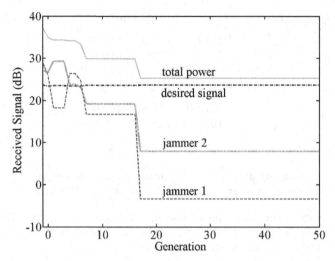

Figure 8.39. Signal levels as a function of generation for the phase-only algorithm.

Example. A 20-element array of point sources spaced 0.5λ apart has 6-bit amplitude and phase weights and a 20-dB, $\bar{n} = 3$, low-sidelobe Taylor amplitude taper. The desired signal is incident on the peak of the main beam and is normalized to one or 0 dB. Two 30-dB jammers enter the sidelobes at 111° and 117°. The genetic algorithm has a population size of 8 and a mutation rate of 10%.

Figure 8.39 are the power levels received by the array. The total power level decreases while the desired signal power remains relatively constant. Sometimes, one jammer power goes down while the other jammer power goes up. The ratio of the signal power to the jammer power is graphed in Figure 8.40. Figure 8.41 shows the adapted antenna pattern superimposed on the quiescent pattern.

Amplitude- and phase-adaptive nulling with a genetic algorithm was experimentally demonstrated on a phased array antenna developed by the Air Force Research Laboratory (AFRL) at Hanscom AFB, MA [37]. The antenna has 128 vertical columns with 16 dipoles per column equally spaced around a cylinder that is 104 cm in diameter (Figure 8.42). Figure 8.43 is a cross-sectional view of the antenna. Summing the outputs from the 16 dipoles forms a fixed elevation main beam pointing 3° above horizontal. Only eight columns of elements are active at a time. Consecutive eight-elements form a 22.5° arc (1/16 of the cylinder), with the elements spaced 0.42λ apart at 5 GHz. Each element has an 8-bit phase shifter and 8-bit attenuator. The phase shifters have a least significant bit equal to 0.0078125π radians. The attenuators have an 80-dB range with the least significant bit equal to. 3125 dB. The antenna has a quiescent pattern resulting from a 25-dB, $\bar{n} = 3$ Taylor amplitude taper.

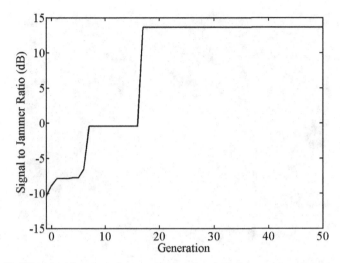

Figure 8.40. The signal-to-interference ratio for the phase-only adaptive algorithm with two jammers at 111° and 117°.

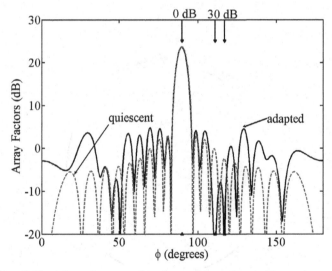

Figure 8.41. Adapted pattern for phase-only nulling with two jammers.

Phase shifters compensate for the curvature of the array and unequal path lengths through the feed network.

A 5-GHz continuous-wave source served as the interference. Only the four least significant bits of the phase shifters and attenuators were used. The genetic algorithm had a population size of 16 chromosomes and used single-

Figure 8.42. Experimental adaptive cylindrical array (R. L. Haupt, Adaptive nulling with a cylindrical array, AFRL-SN-RS-TR-1999-36, March 1999).

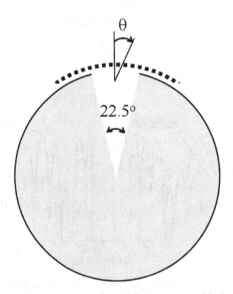

Figure 8.43. The cylindrical array has 128 elements, with 8 active at a time.

point crossover. Only one bit in the population was mutated every generation, resulting in a mutation rate of 0.1%. Nulling tended to be very fast with the algorithm placing a null down to the noise floor of the receiver in less than 30 power measurements. Two cases of placing a single null are presented here. The first example has the interference entering the sidelobe at 28°, and the

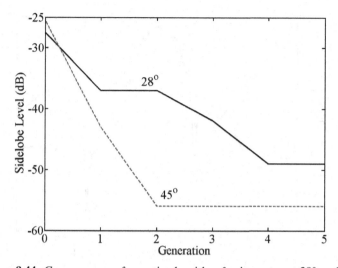

Figure 8.44. Convergence of genetic algorithm for jammers at 28° and 45°.

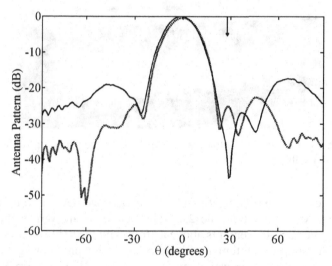

Figure 8.45. Null placed in the far-field pattern at 28°.

second example has the interference at 45°. Figure 8.44 plots the sidelobe level at 28° and 45° as a function of generation. The resulting far-field pattern measurements are shown in Figures 8.45 and 8.46 superimposed on the quiescent pattern. These examples demonstrate that the genetic algorithm quickly places nulls in the sidelobes in the directions of the interfering signals by minimizing the total power output.

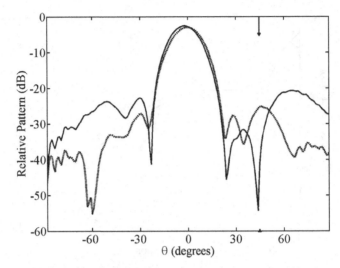

Figure 8.46. Null placed in far-field pattern at 45°.

Figure 8.47. Photograph of 8-element adaptive array (courtesy of Andrea Massa, University of Trento).

The particle swarm optimization algorithm has also been used for adaptive nulling [38]. As an example, the 2.4-GHz 8-element array of eight equally spaced ($d = \lambda/2$) dipole elements above a ground plane is shown in Figure 8.47 served as an adaptive antenna with amplitude and phase weights at each element. A passive RF power combiner with seven microstrip Wilkinson power combiners sent the output power to a spectrum analyzer in order to estimate the SINR (signal-to-interference-plus-noise ratio). The particle swarm algorithm adjusted the weights through a vector modulator in order to maximize the SINR. Figure 8.48 shows plots of the computed and measured adapted antenna patterns.

8.4.3.2. Phase-Only Adaptive Nulling. A phased array may or may not have variable amplitude weights but always has phase shifters for beam steering and calibration. Since the phase weights already exist, why not use the phase

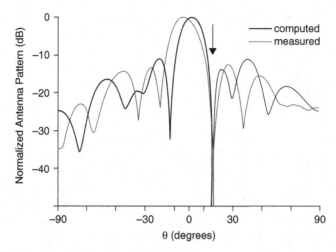

Figure 8.48. Computed and measured adapted antenna patterns (courtesy of Andrea Massa, University of Trento).

shifters as adaptive weights? The theory behind phase-only nulling first appeared in reference 39. The authors present a beam-space algorithm derived for a low-sidelobe array and assume that the phase shifts are small. When the direction of arrival for all the interfering sources is known, then cancellation beams in the directions of the sources are subtracted from the original pattern. Adaptation consists of matching the peak of the cancellation beam with the culprit sidelobe and subtracting [40].

The new phase settings that minimize the output power can be found by using the method of steepest descent [25].

$$\delta_n(\kappa+1) = \delta_n(\kappa) + \mu \frac{P(\kappa) - P(\kappa-1)}{\Delta(\kappa)} \tag{8.58}$$

where $P(\kappa)$ is the array output power at time step κ, $\delta_n(\kappa)$ is the phase shift at element n, $\Delta(\kappa)$ is the small phase increment, and

$$\mu = \frac{\Delta^2}{\sqrt{\sum_{n=1}^{N} [P(\kappa) - P(\kappa-1)]^2}}$$

This algorithm was tested for phase-only simultaneous nulling of the sum and difference patterns of an 80-element linear array of H-plane sectoral horns [25]. The sum channel had a 30-dB Taylor taper, and the difference channel had a 30-dB Bayliss taper. Both channels shared the 8-bit beam steering phase shifters. No source was present in the main beam, but one CW source was aimed at the sidelobes in the quiescent sum and difference patterns at 23°. If

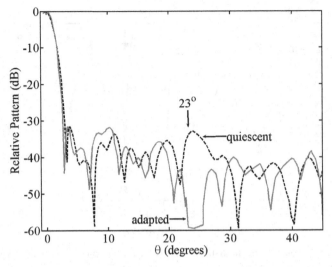

Figure 8.49. Sum patterns due to adaptive nulling in the sum pattern only.

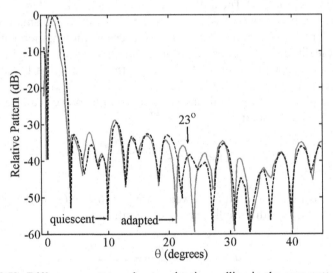

Figure 8.50. Difference patterns due to adaptive nulling in the sum pattern only.

the algorithm is only used to minimize the sum channel output, then the resulting sum pattern appears in Figure 8.49 with the difference pattern in Figure 8.50. A null appears in the sum pattern but not the difference pattern. Minimizing the output from both channels results in the patterns shown in Figures 8.51 and 8.52. This time, nulls are placed in both patterns.

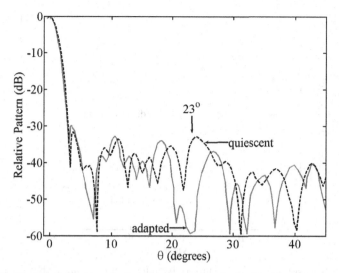

Figure 8.51. Sum patterns due to simultaneous adaptive nulling in the sum and difference patterns.

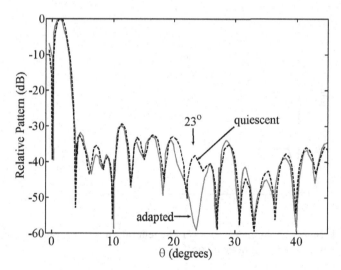

Figure 8.52. Difference patterns due to simultaneous adaptive nulling in the sum and difference patterns.

The gradient method is slow, because the phase at each element is serially toggled for a power measurement every iteration. Also, the steepest descent algorithm can get stuck in a local minimum. A genetic algorithm was first proposed for phase-only adaptive nulling in reference 41.

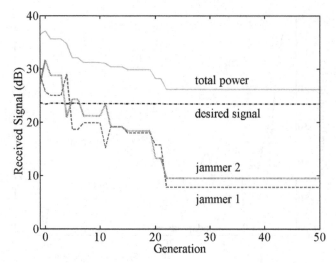

Figure 8.53. Convergence of the phase-only adaptive nulling algorithm using a genetic algorithm.

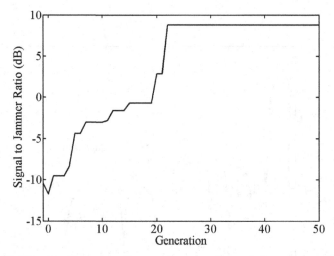

Figure 8.54. Signal-to-jammer ratio versus generation of the phase-only adaptive nulling algorithm using a genetic algorithm.

Example. A 20-element array of point sources spaced 0.5λ apart has 6-bit phase shifters and a 20-dB, $\bar{n} = 3$ low-sidelobe Taylor amplitude taper. The desired signal is incident on the peak of the main beam and is normalized to one or 0 dB. Two 30-dB jammers enter the sidelobes at 111° and 117°. The genetic algorithm has a population size of 8 and a mutation rate of 10%.

The algorithm successfully placed nulls at both angles. The convergence is shown in Figure 8.53 while the increasing signal to jammer ratio appears in Figure 8.54. Figure 8.55 shows the adapted pattern compared to the quiescent pattern.

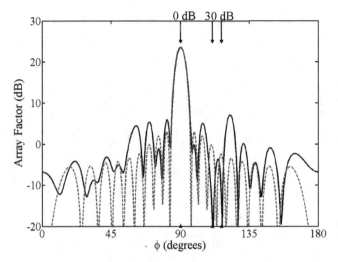

Figure 8.55. Adapted array pattern for phase-only adaptive nulling algorithm using a genetic algorithm.

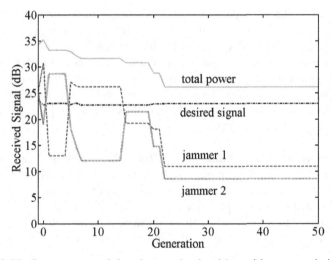

Figure 8.56. Convergence of the phase-only algorithm with symmetric jammers.

Moving to the case of two 30-dB interference sources at 50° and 130° confronts the adaptive algorithm with the problem of symmetric interference sources. The genetic algorithm could only null one of the interference sources with three least significant phase bits, so a minimum of four had to be used. Adding a fourth bit resulted in nice convergence as shown in Figure 8.56. As noted previously, four phase bits results in noticeable main lobe degradation as shown in Figure 8.57.

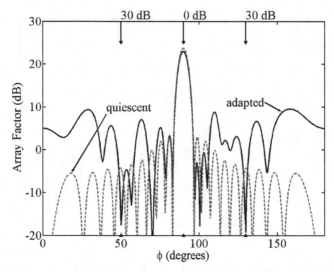

Figure 8.57. Adapted pattern for phase-only nulling with symmetric jammers.

Phase-only nulling has the advantage of simple implementation. The trade-off is that more bits must be used to null interference signals that are at symmetric locations about the main beam. The additional nulling bits result in higher distortions in the main beam and sidelobes. Small phase shifts produce symmetric cancellation beams that are 180° out of phase. When they are added to the quiescent pattern to produce a null at one location, the symmetric sidelobe increases. Allowing larger phase shifts [42] or adding amplitude control overcomes this problem.

8.4.3.3. Amplitude-Only Adaptive Nulling. Although not very common, an array can have variable amplitude weights at the elements without phase shifters. Vu suggested moving conjugate zeros on the unit circle in the direction of interfering sources [43]. This approach is not adaptive and depends upon the ability of finding the locations of the interfering sources. All the zeros that are not used to place nulls can then be used to control the rest of the array factor.

The simulated array consists of eight vertically polarized dipoles spaced 0.075λ with a variable amplitude weight at the two elements on each end of the array. It is a narrow band system operating at 2 GHz. Two signals are incident upon the array. A 0-dB desired signal appears at $\phi = 90°$ and an undesired signal appears at $\phi = 68°$. The GA found the amplitude settings for the four dipoles as appears in Table 8.6. The three-dimensional quiescent and adapted patterns are shown in Figure 8.58. Table 8.7 indicates that the gain decreases by 1.2 dB while the sidelobe in the direction of the interference decreases by 19.2 dB. As a result, the signal-to-noise ratio (SNR) increases

TABLE 8.6. The Quiescent and Adapted Amplitude Weights for the 8-Element Array

Element:	1	2	3	4	5	6	7	8
Quiescent:	1	1	1	1	1	1	1	1
Adapted:	0.02	0.27	1	1	1	1	0.81	0.13

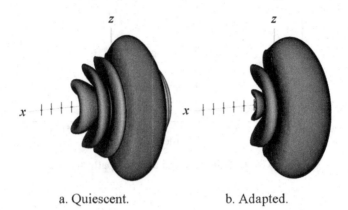

a. Quiescent.   b. Adapted.

Figure 8.58. (a) Quiescent and (b) adapted patterns for the 8-element amplitude-only dipole array.

TABLE 8.7. Array Pattern Statistics for the Quiescent and Adapted Arrays

	Quiescent	Adapted
Gain (dB)	12.3	11.1
SLL at 68° (dB)	−0.6	−19.8
SNR (dB)	−4.2	31.9

from −4.2 dB to 31.9 dB. Pattern cuts for the quiescent and adapted arrays are shown in Figure 8.59.

A 2.2-GHz, 8-element array was made from monopole/switch elements as shown in Figure 8.60. The elements have a variable switch controlled by an IR LED [44]. The corporate feed consists of 8 low-loss, phase-stable coaxial cables and a broadband 8-to-1 power combiner. Only the two edge elements on either side of the array are used by a genetic algorithm to place nulls. The measured S_{11} of the array is below −10 dB from 2.11 to 2.53 GHz for a bandwidth of 18.4%. Figures 8.61 and 8.62 are the measured amplitude and phase of S_{12} for the adaptive elements in the array (the numbers on the plots correspond to the element numbers in Figure 8.60) as a function of LED current.

Figure 8.63 shows the adapted pattern superimposed on the quiescent pattern when only a 10-dBm signal at −35° is present. The main beam loses

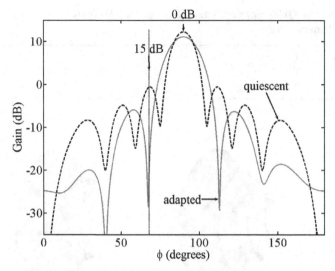

Figure 8.59. Quiescent and adapted pattern cuts.

Figure 8.60. Experimental linear array of broadband monopole elements with variable IR switches for amplitude control.

3.9 dB while the null is 42 dB below the sidelobe at −35°. A second example shows the results of placing two signals of 15 dBm at −19° and −35° (Figure 8.64). No signal is incident upon the main beam. In this case, the main beam is reduced by 3.6 dB. The sidelobe at −19° goes down 13 dB, and the sidelobe at −35° is reduced by 9 dB.

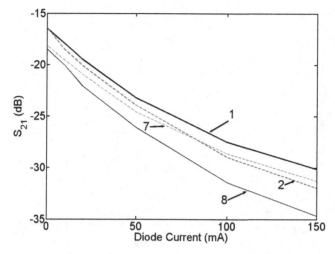

Figure 8.61. Measured S_{21} of the elements as a function of diode current.

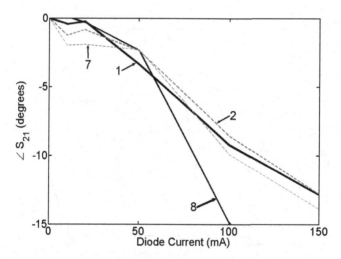

Figure 8.62. Measured phase of S_{12} of the elements as a function of diode current.

8.5. MULTIPLE-INPUT MULTIPLE-OUTPUT (MIMO) SYSTEM

A typical communications system has one antenna for transmit and one antenna for receive, which is known as a single-input single-output system. Another version has a single antenna on transmit and an array on receive. This version is known as single input and multiple output. The converse of an array on transmit with a single antenna on receive is a multiple input single output system. These types of systems have driven the need for antenna arrays for many years.

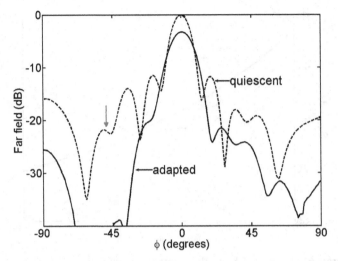

Figure 8.63. Adapted and quiescent far-field patterns when a signal is incident at −35°.

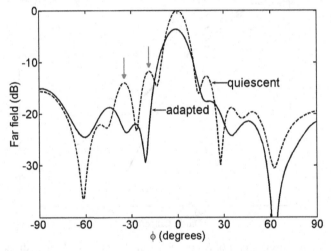

Figure 8.64. Adapted and quiescent far-field patterns when a signal is incident at −19° and −35°.

A multiple-input multiple-output (MIMO) system uses an array for transmit as well as another array for receive (Figure 8.65). As a result, both antennas can be smart or adaptive to increase the amount of data transferred over the communications channel. MIMO was developed to counteract the problems associated with multipath in wireless communications. A MIMO system increases its capacity in rich multipath environments by exploiting the spatial properties of the multipath channel, thereby offering an additional dimension

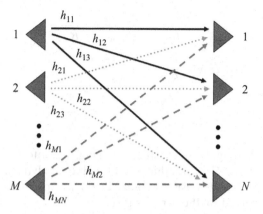

Figure 8.65. Signal paths between transmit and receive antennas in a MIMO system.

that enhances communication performance. A beam is synthesized to transmit data to a single user while placing nulls in the directions of the other users. An array provides antenna or spatial diversity, because more than one element transmits/receives the same signal from different locations. Separating the antenna elements causes the signals from M transmitting antennas to take different paths to the N receiving antennas. The received signals are a function of the transmitted signals, the channel paths, and the noise [45].

$$r = Hs + \mathbb{N} \qquad (8.59)$$

where $\mathbb{N}$ is the noise vector and H is the channel matrix given by

$$H = \begin{bmatrix} h_{11} & h_{12} & \cdots & h_{1N} \\ h_{21} & h_{22} & & \vdots \\ & & \ddots & \\ h_{M1} & \cdots & & h_{MN} \end{bmatrix} \qquad (8.60)$$

and the channel matrix elements, h_{mn}, are the transfer functions describing the channel between transmit antenna m and receive antenna n. In order to recover the transmitted data, s, an accurate estimate of H is needed. The transmitted data are then calculated from the received data by inverting the channel matrix

$$s = H^{-1}r \qquad (8.61)$$

If the medium were free space, then these channel matrix elements would just be the free-space Green's function and H is written as

$$H = \begin{bmatrix} \dfrac{e^{-jkR_{11}}}{R_{11}} & \dfrac{e^{-jkR_{mn}}}{R_{mn}} & \dfrac{e^{-jkR_{mn}}}{R_{mn}} \\[2ex] \dfrac{e^{-jkR_{21}}}{R_{21}} & \dfrac{e^{-jkR_{mn}}}{R_{mn}} & \\[2ex] \dfrac{e^{-jkR_{mn}}}{R_{MN}} & & \dfrac{e^{-jkR_{mn}}}{R_{mn}} \end{bmatrix} \tag{8.62}$$

Multipath, noise, fading, Doppler shift, coupling, and interference all contribute to variations in the H matrix that are difficult to analytically or numerically predict. Consequently, H is usually found through experimentation.

The power received by the array is given by

$$P = r^{\dagger} r = s^{\dagger} H H^{\dagger} s \tag{8.63}$$

The $M \times M$ matrix $HH^{\dagger}$ can be decomposed as

$$HH^{\dagger} = V_{\lambda} \begin{bmatrix} \lambda_1 & 0 & 0 \\ 0 & \ddots & 0 \\ 0 & 0 & \lambda_M \end{bmatrix} V_{\lambda}^{\dagger} \tag{8.64}$$

The singular value decomposition of H is given by [46]

$$H = U_{SVD} D V_{SVD}^{\dagger} \tag{8.65}$$

where

$$D = \begin{bmatrix} \sqrt{\lambda_1} & 0 & 0 \\ 0 & \ddots & 0 \\ 0 & 0 & \sqrt{\lambda_M} \end{bmatrix}$$

$$\sqrt{\lambda_m} = \text{singular values}$$

$$U_{SVD}, V_{SVD} = \text{singular vectors}$$

The singular values are just the square root of the eigenvalues in (8.64). Substituting (8.65) into (8.59) results in

$$r = U_{SVD} D V_{SVD}^{\dagger} s + \mathbb{N} \tag{8.66}$$

This equation can be written as

$$r' = Ds' + \mathbb{N}' \tag{8.67}$$

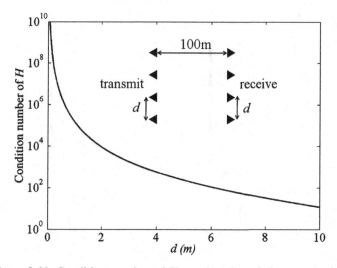

Figure 8.66. Condition number of H as a function of element spacing.

where

$$r' = U_{SVD}^{-1}r$$

$$s' = V_{SVD}^{\dagger}s$$

$$\mathbb{N}' = U_{SVD}^{-1}\mathbb{N}$$

There are g parallel independent radio subchannels between the transmit and receive antennas, where g is the rank of H. The rank of a matrix is the number of nonzero singular values.

Example. A MIMO system has a 3-element array of isotropic point sources spaced d apart on transmit and receive. The system operates at 2.4 GHz and the arrays are 100 m apart and face each other. Show how the condition number of H changes as the element spacing increases.

Figure 8.66 shows how the condition number of H decreases as the element spacing for both the transmit and receive arrays increase. Increasing the element spacing in only the transmit or receive array also decreases the condition number but not as fast. The lower the condition number, the more accurate the inversion of H is.

8.6. RECONFIGURABLE ARRAYS

A reconfigurable array changes its performance characteristics by using switches, such as MEMS or PIN diodes, to connect elements to adjacent

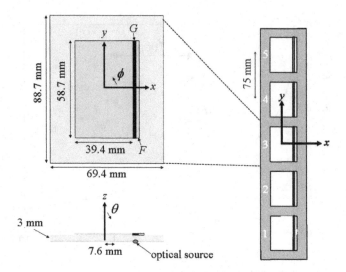

Figure 8.67. Diagram of the adaptive reconfigurable patch array.

structures. The 5-element array in Figure 8.67 has elements that are rectangular patches made from a PEC that is 58.7 × 39.4 mm. The substrate is a slab of optically transparent fused quartz with $\varepsilon_r = 3.78$ backed by a PEC groundplane. The substrate is 88.7 × 69.4 mm and is 3 mm thick. The patch has a thin strip of silicon (G) 58.7 × 2 mm with $\varepsilon_r = 11.7$. The right edge of the patch is a thin strip of PEC (F) 58.7 × 4.2 mm. A laser or LED beneath the groundplane illuminates the silicon through small holes in the groundplane or by making the groundplane from a transparent conductor. The silicon conductivity is a function of the light intensity. A graph of the amplitude of the return loss is shown in Figure 8.68 for the following conductivities: 0, 2, 5, 10, 20, 50, 100, 200, and 1000 S/m. At 2 GHz, there is a distinct resonance when the silicon has no conductivity. As the conductivity increases, the resonance is at 1.78 GHz. The amount of power delivered to the patch at 2 GHz reduces as the conductivity increases, so the photoconductive silicon acts as an amplitude control to that element.

The element spacing is 75 mm or 0.5λ. If the silicon insets all have a conductivity of zero, then the array is uniform with a far-field pattern shown in Figure 8.69. This quiescent pattern has a gain of 12.81 dB and a relative peak sidelobe level of 13.84 dB. The element patterns of the uniform array are shown in Figure 8.70. The average gain of these patterns at boresight is 6.14 dB.

Increasing the conductivity of the silicon in a patch decreases the product of the patch gain times the power delivered to the patch. Carefully tapering the illumination of the LEDs creates an amplitude taper. An array pattern with equal sidelobes results when the conductivity has values of [16 5 0 5 16] S/m (Figure 8.69). The antenna pattern has a gain of 10.4 dB with a peak rela-

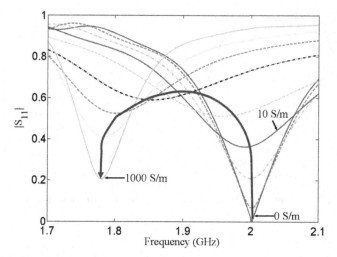

Figure 8.68. Plots of the magnitude of s_{11} for silicon conductivities of 0, 2, 5, 10, 20, 50, 100, 200, and 1000 S/m.

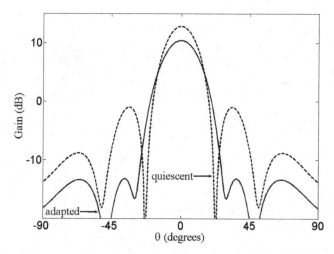

Figure 8.69. The quiescent pattern is the dashed line and has all the conductivities set to 0. The adapted pattern is the solid line and has the silicon conductivities set to [16 5 0 5 16] S/m.

tive sidelobe level of 23.6 dB. The upper-left element patterns correspond to the uniform array. Applying the illumination taper produces the element patterns shown in Figure 8.70.

The antenna in Figure 8.71 has a Z-shaped active microstrip element in the center of 18 identical, equally spaced parasitic metallic elements. Its operating frequency is 2.45 GHz. The parasitic elements have PIN diodes (activated

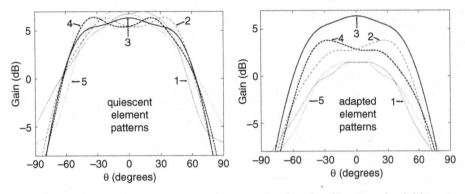

Figure 8.70. The element patterns of the array that has the silicon conductivities set to [16 5 0 5 16] S/m.

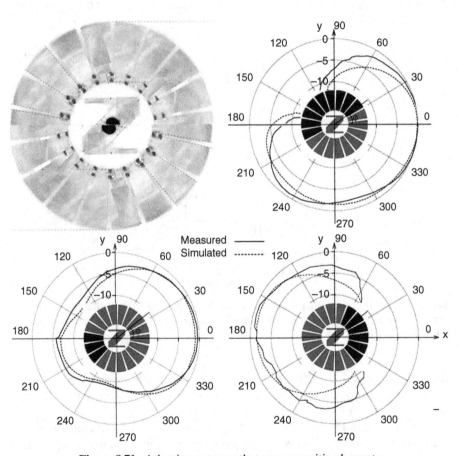

Figure 8.71. Adaptive antenna that uses parasitic elements.

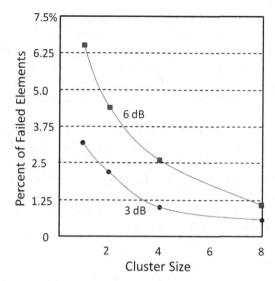

Figure 8.72. Plot of the percent element failures required to raise the average sidelobe level of an 8000 element circular array with a 40-dB Taylor amplitude taper and triangular element spacing with $d = 0.5\lambda$ by 3 or 6 dB for several cluster sizes.

elements in black and deactivated elements in gray color) that connect adjacent elements. Connecting several adjacent parasitic elements steers the main lobe in azimuth. The beamwidth depends upon the number of parasitic elements connected together. The dark parasitic elements in Figure 8.71 correspond to the ones that are connected by the PIN diodes. Plots of the computed and measured far-field patterns are also shown in Figure 8.71.

As noted in Chapter 2, element failures reduce gain and increase sidelobe levels. At some point, the failures cause the system to shut down. To develop an acceptable maintenance schedule and to reduce the probability of a catastrophic array failure, the mean time between failure of the array should be maximized by minimizing the component failure rate and selecting an appropriate array architecture.

The cost of an array consists of the production cost (design, purchase and/ or fabrication of parts, assembly, and testing of the antenna) and the life-cycle cost (cost over an antenna's operational lifetime to replace or repair all failed components in the antenna). The life-cycle cost is highly dependent upon the component MTBF. Usually, the array has a very high passive component MTBF, so its contribution to the life-cycle cost is ignored here. On the other hand, the MTBF of active components, such as T/R modules and power supplies, drive the life-cycle cost of an antenna.

Performance degradation is defined in terms of increased peak and/or average sidelobes from associated with the component failures. If one T/R module feeds multiple elements, then the T/R module failure results in a cluster of element failing. The larger the cluster, the more devastating the failure is. Figure 8.72 is a plot of the percent element failures required to raise the average sidelobe level by 3 or 6 dB for several cluster sizes for an 8000-element circular array with a 40-dB Taylor amplitude taper and triangular element spacing with $d = 0.5\lambda$ [47]. Using components with high MTBF, reducing the number of elements, and adding redundancy increases the array MTBF.

In order to reduce the effects of element failures, the element weights can be recalculated to bring the sidelobe levels down. Techniques for finding these new element weights are described in reference 48 for corporate arrays and for digital beamforming arrays [49]. Recently, more sophisticated approaches using a genetic algorithm [50] and an iterative method using the inverse Fourier transform [51]. In all cases, the location of the failed elements must be known to calculate the weight corrections.

REFERENCES

1. L. C. V. Atta, *Electromagnetic Reflector*, 2,908,002, U.S.P. Office, 1959.
2. S. Drabowitch, *Modern Antennas*, 2nd ed., Dordrecht: Springer, 2005.
3. M. Skolnik and D. King, Self-phasing array antennas, *IEEE Trans. Antennas and Propagat.*, Vol. 12, No. 2, 1964, pp. 142–149.
4. C. Pon, Retrodirective array using the heterodyne technique, *IEEE Trans. Antennas and Propagat.*, Vol. 12, No. 2, 1964, pp. 176–180.
5. R. Y. Miyamoto and T. Itoh, Retrodirective arrays for wireless communications, *IEEE Microwave Mag.*, Vol. 3, No. 1, 2002, pp. 71–79.
6. S. Lim and T. Itoh, A 60 GHz retrodirective array system with efficient power management for wireless multimedia sensor server applications, *IET Microwaves Antennas Propagat.*, Vol. 2, No. 6, 2008, pp. 615–625.
7. T. Y. Y. L. Chiu, W. S. Chang, Q. Xue, C. H. Chan, Retrodirective array for RFID and microwave tracking beacon applications, *Microwave Opt. Technol. Lett.*, Vol. 48, No. 2, 2006, pp. 409–411.
8. B. Nair and V. F. Fusco, Two-dimensional planar passive retrodirective array, *Electron. Lett.*, Vol. 39, No. 10, 2003, pp. 768–769.
9. M. K. Watanabe, R. N. Pang, B. O. Takase et al., A 2-D phase-detecting/heterodyne-scanning retrodirective array, *IEEE Trans. Microwave Theory Tech.*, Vol. 55, No. 12, 2007, pp. 2856–2864.
10. R. Compton, Jr., On eigenvalues, SINR, and element patterns in adaptive arrays, *IEEE Trans. Antennas Propagat.*, Vol. 32, No. 6, 1984, pp. 643–647.
11. A. W. Rudge, *The Handbook of Antenna Design*, 2nd ed., London: P. Peregrinus on behalf of the Institution of Electrical Engineers, 1986.
12. S. Chandran, *Advances in Direction-of-Arrival Estimation*, Boston: Artech House, 2006.
13. J. Capon, High-resolution frequency-wavenumber spectrum analysis, *Proc. IEEE*, Vol. 57, No. 8, 1969, pp. 1408–1418.

14. R. Schmidt, Multiple emitter location and signal parameter estimation, *IEEE Trans. Antennas and Propagat.*, Vol. 34, No. 3, 1986, pp. 276–280.

15. R. Schmidt and R. Franks, Multiple source DF signal processing: An experimental system, *IEEE Trans. Antennas Propagat.*, Vol. 34, No. 3, 1986, pp. 281–290.

16. A. Barabell, Improving the resolution performance of eigenstructure-based direction-finding algorithms, IEEE International Conference on Acoustics, Speech, and Signal Processing, 1983, pp. 336–339.

17. J. P. Burg, The relationship between maximum entropy spectra and maximum likelihood spectra, *Geophysics*, Vol. 37, No. 2, 1972, pp. 375–376.

18. R. T. Lacoss, Data adaptive spectral analysis methods, *Geophysics*, Vol. 36, No. 4, 1971, pp. 661–675.

19. V. F. Pisarenko, The retrieval of harmonics from a covariance function, *Geophys. J. Int.*, Vol. 33, No. 3, 1973, pp. 347–366.

20. A. Paulraj, R. Roy, and T. Kailath, A subspace rotation approach to signal parameter estimation, *Proc. IEEE*, Vol. 74, No. 7, 1986, pp. 1044–1046.

21. L. C. Godara, *Smart Antennas*, Boca Raton, FL, CRC Press, 2004.

22. P. Howells, Explorations in fixed and adaptive resolution at GE and SURC, *IEEE Trans. Antennas Propagat.*, Vol. 24, No. 5, 1976, pp. 575–584.

23. B. Widrow, P. E. Mantey, and L. J. Griffiths et al., Adaptive antenna systems, *Proc. IEEE*, Vol. 55, No. 12, 1967, pp. 2143–2159.

24. R. A. Monzingo, T. W. Miller, and Knovel (Firm), *Introduction to Adaptive Arrays*, Scitech, 2004.

25. R. L. Haupt, Adaptive nulling in monopulse antennas, *IEEE Trans. Antennas Propagat.*, Vol. 36, No. 2, 1988, pp. 202–208.

26. R. L. Haupt and D. H. Werner, *Genetic Algorithms in Electromagnetics*, Hoboken, NJ: IEEE Press/Wiley-Interscience, 2007.

27. M. I. Skolnik, *Radar Handbook*, New York: McGraw-Hill, 2007.

28. H. M. Finn, R. S. Johnson, and P. Z. Peebles, Fluctuating target detection in clutter using sidelobe blanking logic, *Aerospace and Electronic Systems, IEEE Transactions on*, Vol. AES-7, No. 1, 1971, pp. 147–159.

29. A. Farina, Single sidelobe canceller: theory and evaluation, *IEEE Trans. Aerosp. Electron. Syst.*, Vol. AES-13, No. 6, 1977, pp. 690–699.

30. S. Applebaum, Adaptive arrays, *IEEE Trans. Antennas and Propagat.*, Vol. 24, No. 5, 1976, pp. 585–598.

31. F. B. Gross, *Smart Antennas for Wireless Communications: With MATLAB*, New York: McGraw-Hill, 2005.

32. R. T. Compton, *Adaptive Antennas: Concepts and Performance*, Philadelphia: Prentice-Hall, 1987.

33. W. H. Press and Numerical Recipes Software (Firm), *Numerical recipes in FORTRAN*, Cambridge University Press, 1994.

34. I. Gupta, SMI adaptive antenna arrays for weak interfering signals, *IEEE Trans. Antennas and Propagat.*, Vol. 34, No. 10, 1986, pp. 1237–1242.

35. D. Morgan, Partially adaptive array techniques, *IEEE Trans. Antennas and Propagat.*, Vol. 26, No. 6, 1978, pp. 823–833.

36. R. Haupt, Adaptive nulling in monopulse antennas, *IEEE Trans. Antennas and Propagat.*, Vol. 36, No. 2, 1988, pp. 202–208.

37. R. L. Haupt and H. Southall, Experimental adaptive cylindrical array, *Microwave Journal*, 1999, pp. 291–296.

38. M. Benedetti, R. Azaro, and A. Massa, Experimental validation of fully-adaptive smart antenna prototype, *Electron. Lett.*, Vol. 44, No. 11, 2008, pp. 661–662.

39. C. Baird and G. Rassweiler, Adaptive sidelobe nulling using digitally controlled phase-shifters, *IEEE Trans. Antennas Propagat.*, Vol. 24, No. 5, 1976, pp. 638–649.

40. H. Steyskal, Simple method for pattern nulling by phase perturbation, *IEEE Trans. Antennas and Propagat.*, Vol. 31, No. 1, 1983, pp. 163–166.

41. R. L. Haupt, Phase-only adaptive nulling with a genetic algorithm, *IEEE Trans. Antennas Propagat.*, Vol. 45, No. 6, 1997, pp. 1009–1015.

42. R. Shore, Nulling a symmetric pattern location with phase-only weight control, *IEEE Trans. Antennas and Propagat.*, Vol. 32, No. 5, 1984, pp. 530–533.

43. T. B. Vu, Method of null steering without using phase shifters, *IEE Proc. Microwaves Opt. Antennas, H*, Vol. 131, No. 4, 1984, pp. 242–245.

44. J. R. Flemish, H. W. Kwan, R. L. Haupt et al., A new silicon-based photoconductive microwave switch, *Microwave Optical. Technol. Lett.*, IEEE Aerospace Conference, Vol. 51, No. 1, 2009, pp. 248–252.

45. S. M. Alamouti, A simple transmit diversity technique for wireless communications, *IEEE J. Selected Areas Commun.*, Vol. 16, No. 8, 1998, pp. 1451–1458.

46. M. A. Jensen and J. W. Wallace, A review of antennas and propagation for MIMO wireless communications, *IEEE Trans. Antennas Propagat.*, Vol. 52, No. 11, 2004, pp. 2810–2824.

47. A. K. Agrawal and E. L. Holzman, Active phased array design for high reliability, *IEEE Trans. Aerosp. Electron. Syst.*, Vol. 35, No. 4, 1999, pp. 1204–1211.

48. T. J. Peters, A conjugate gradient-based algorithm to minimize the sidelobe level of planar arrays with element failures, *IEEE Trans. Antennas and Propagat.*, Vol. 39, No. 10, 1991, pp. 1497–1504.

49. R. J. Mailloux, Phased array error correction scheme, *Electron. Lett.*, Vol. 29, No. 7, 1993, pp. 573–574.

50. S. Seong Ho, S. Y. Eom, S. I. Jeon et al., Automatic phase correction of phased array antennas by a genetic algorithm, *IEEE Trans. Antennas Propagat.*, Vol. 56, No. 8, 2008, pp. 2751–2754.

51. W. P. N. Keizer, Element Failure Correction for a large monopulse phased array antenna with active amplitude weighting, *IEEE Trans. Antennas and Propagat.*, Vol. 55, No. 8, 2007, pp. 2211–2218.

INDEX

Antenna Arrays: A Computational Approach, by Randy L. Haupt
Copyright © 2010 John Wiley & Sons, Inc.

Printed in the United States
By Bookmasters